Moon Bound

Choosing and Preparing NASA's Lunar Astronauts

Other Springer-Praxis books of related interest by Colin Burgess

NASA's Scientist-Astronauts
with David J. Shayler
2006
ISBN 978-0-387-21897-7

Animals in Space: From Research Rockets to the Space Shuttle
with Chris Dubbs
2007
ISBN 978-0-387-36053-9

The First Soviet Cosmonaut Team: Their Lives, Legacy, and Historical Impact
with Rex Hall, M.B.E.
2009
ISBN 978-0-387-84823-5

Selecting the Mercury Seven: The Search for America's First Astronauts
2011
ISBN 978-1-4419-8404-3

Colin Burgess

Moon Bound

Choosing and Preparing NASA's Lunar Astronauts

Published in association with
Praxis Publishing
Chichester, UK

Colin Burgess
12 Madison Place
Bonnet Bay
New South Wales 2226
Australia

SPRINGER–PRAXIS BOOKS IN SPACE EXPLORATION

ISBN 978-1-4614-3854-0 ISBN 978-1-4614-3855-7 (eBook)
DOI 10.1007/978-1-4614-3855-7
Springer New York Heidelberg Dordrecht London

Library of Congress Control Number: 2013932138

Cover design: Jim Wilkie
Project copy editor: David M. Harland
Typesetting: BookEns, Royston, Herts., UK

Printed on acid-free paper

Springer is part of Springer Science+Business Media (www.springer.com)

Contents

This book is respectfully dedicated to the five NASA astronauts from Groups 2 and 3 who lost their lives before they had their chance to realize their dreams of one day gazing down on our blue planet from orbit, or flying to the Moon.

Charles Arthur Bassett II, USAF
Roger Bruce Chaffee, USN
Theodore Cordy Freeman, USAF
Elliot McKay See, Jr.
Clifton Curtis Williams, Jr., USMC

And to the memory of Neil A. Armstrong,
the first human to set foot on the Moon
(5 August 1930–25 August 2012)

Acknowledgements

The genesis of this book goes back several years to a casual conversation I was enjoying in a London pub with good friend and prolific space flight author David Shayler, in which we were discussing the make-up of the finalist group for NASA's Mercury astronauts. At the time there were still five names missing from David's list. These names later came my way by the kindness of retired USAF Lt. Col. Walter B. ("Sully") Sullivan, Jr. He not only supplied those missing names from nearly five decades back, but proved to be a valued friend and invaluable helper as I put together my book on that story.

When David found out I was investigating this new book on NASA's second and third astronaut groups he once again kindly opened his extensive files, sending me the names of the finalists for those groups, which he had happily unearthed while conducting a random search for other material. Although the Internet and instant communication have proved a boon for writers and other investigators, David is one of those people who strongly believe in good old "digging in the dust", as he so delightfully calls it. During a visit to Houston several years earlier, he was working in the National Archives and Records Administration in Fort Worth, Texas, researching NASA's Gemini program through various documents supplied to Rice University by the Johnson Space Center (JSC). Within the hundreds of General Subject files he was flipping through in one of many archive boxes, in this case No. 382, he came across a 15 December 1966 letter from the School of Aerospace Medicine, Brooks Air Force Base in Texas. It was a detailed costing for aeromedical evaluations held in 1962 and 1963 for prospective Gemini astronauts, showing the average examination cost per candidate to have been $788. More importantly, however, there was an attachment to the letter giving the names and dates of all candidates' medical evaluations between 1960 and 1966. From these lists, David was able to supply me with the names, ranks and examination dates for all 32 Group 2 candidates and all 34 Group 3 candidates. As always, I am truly indebted to David for kindly allowing me access to his records and for his encouragement and support in writing this book.

Mentioned above is "Sully" Sullivan. Back in 1959, then a lieutenant, he was the appointed USAF liaison officer for the 32 Mercury finalists who were ordered to the

Wright Aerospace Development Center in Dayton, Ohio. In addition to working with these men on their day-to-day schedules and other administrative work, he developed long term friendships with many of them. This proved a blessing when – early on – I was attempting to convince the unsuccessful finalists to assist me in putting their biographies in a book. Soon after the book *Selecting the Mercury Seven: The Search for America's First Astronauts* was released by Springer-Praxis, I asked Sully, my own 'liaison officer', if he would help me with this follow-up volume, to which he readily agreed. As we located each of the non-selected test pilots from both groups – or their surviving family members – Sully would make the initial contact and introduce me, which smoothed the way for my later contact with them. Over recent years we have become firm friends, and it is a great pleasure to acknowledge in this book (as in the previous one) the impressive, meticulous, and resolute work he has done for me, and ultimately you, the reader. It is little wonder to me that he was selected to assist as the candidates' liaison officer back in 1959.

Many thanks also go to fellow space historian Michael Cassutt, for once again stepping up to the plate and contributing the Foreword to this book – a task which he achieved with such a wonderfully incisive style for the first book. Being a biographer of astronauts and cosmonauts himself – when he is not busy writing his own books or television scripts – I know how much Michael enjoyed reading the manuscripts for both books. I treasure his friendship and opinions.

My principal thanks must, of course, go to the surviving members of both NASA finalist groups who not only responded to my initial enquiries, but went on to provide information and photographs with the utmost enthusiasm for what I was doing. They made it an interesting and pleasurable experience, and I am immensely grateful and appreciative. Profound thanks therefore go to: *Col. Reginald R. Davis (USAF); Col. Roy S. Dickey (USAF); Thomas E. Edmonds; John M. Fritz; Capt. William J. Geiger (USMC); Orville C. Johnson; RADM George M. Furlong, Jr. (USN); Lt. Col. Samuel M. Guild, Jr. (USAF); VADM William E. Ramsey (USN); RADM Robert H. Shumaker (USN); RADM L. Robert Smith (USNR); Lt. Col. Robert E. Solliday (USMC); Col. Alfred H. Uhalt, Jr. (USAF); Lt. Col. Robert J. Vanden-Heuvel (USAF); M/Gen. Kenneth W. Weir (USMC); and Cmdr. Richard L. Wright (USN).* Special posthumous thanks also go to Capt. John R. C. Mitchell, a Group 2 finalist who assisted me with *Selecting the Mercury Seven*, but sadly passed away shortly before it was published.

The years have taken their toll, and not all of the finalists are with us today, so I turned instead to family members – their widows, sons, daughters, nieces, nephews, and those who knew them best in either their service or private lives. Sincere thanks to all of you for helping me to write about those magnificent men who are no longer in our midst. They were: *Maj. Michael J. Adams (USAF)* – Michelle Evans; *Capt. Roland E. Aslund (USN)* – Diana Aslund and Joan Cudeback; *Capt. Tommy I. Bell, Jr.* – Carolyn Bell Phillips; *Capt. Carl Birdwell, Jr. (USN)* – Bob Birdwell; *Capt. John K. Cochran (USMC)* – Ken Cochran and Kathleen Cochran Clayton; *Donald G. Ebbert* – Greg Ebbert; *Capt. David L. Glunt, Jr. (USN)* – Ann Glunt; *Cmdr. William P. Kelly, Jr. (USN)* – Barbara Kelly; *Capt. Marvin G. McCanna, Jr. (USN)* – Trey and Mary McCanna; *Capt. John R. C. Mitchell (USN)* – Katherine Nickel;

Capt. Alexander K. Rupp (USAF) – Karen Rupp Deming and Bill McWilliams; *Capt. John D. Yamnicky (USN)* – Jann Yamnicky, Jennifer Yamnicky, Lorraine Yamnicky Dixon, Mark Yamnicky, Judy and Lee Bausch, L/Gen. George D. Miller (USAF, Ret.), Garnett Bailey, Craig Rutter, Dolores Sebastian, Carmine Sebastian, Dick Liljestrand, Elizabeth Carroll Foster, Dennis Plautz, Joe Sutliff and Harry Errington.

Invaluable help in researching and compiling biographical information also came from family members of other non-selected finalists, and I am immensely grateful to Terry Vanden-Heuvel Jones, Jim Fritz and Terry Kirkpatrick Loewen.

Some years back I co-authored a book with Kate Doolan for the University of Nebraska Press called *Fallen Astronauts: Heroes Who Died Reaching for the Moon*, in which we gave comprehensive biographies of several of the astronauts who were selected in NASA's Group 2 and 3, so I would like to thank those family members once again for their generosity, hospitality and memories. My extended thanks go to: *Maj. Charles A. Bassett II (USAF)* – Jeannie Bassett-Robinson, Karen (Bassett) Stevenson, Peter Bassett and Bill Bassett; *Lt. Cmdr. Roger B. Chaffee (USN)* – Martha Chaffee and Sheryl Chaffee Marshall; *Capt. Theodore C. Freeman (USAF)* – Faith Freeman Herschap, Anna Mae Freeman Thompson and Perry McGinnis; *Maj. Edward G. Givens, Jr. (USAF)* – Morgan and Cathrine Doyle, and Ed Givens III; *Lt. Col. Virgil I. Grissom (USAF)* – Betty Grissom and Scott Grissom; *Elliot M. See, Jr.* – Marilyn See, Sally See Kneuven, Sally See Llewellyn and the late Neil Armstrong; *Lt. Col. Edward H. White II (USAF)* – Jeanne Whatley, Bonnie Baer and Ed White III; *Maj. Clifton C. Williams, Jr. (USAF)* – Beth Williams, Gertrude Williams, Catherine Williams and Jane Dee Williams.

Individually, and collectively, I acknowledge all those who assisted so readily in the compilation of this book, supplying the gems of information and photographs so vital in turning a few scattered pages into the book that you now have before you. In no particular order, my sincere appreciation goes to Jerry Zacharias, RADM John R. ("Smoke") Wilson, Jr., Tracy Kornfeld, David Shugarts, Kate Doolan, Morris J. Herbert (West Point Association of Graduates), Clayton Adams, Neal Thompson, Joachim Becker at SPACEFACTS, Jere Allen, Neil Corbett (the "Tartan Terror"), David Mazurek, Gary Verver, Al Hallonquist, Will ("Tiny") Tomsen, Steve Townes, Walter Price, Dr. Robert Voas, Nadine Wisely (Blue Angels Association), Colin Babb (Naval Aviation News), Jennifer Bryan (U.S. Naval Academy, Nimitz Library archivist), the entirely wonderful Brigitte Tamashiro at the U.S. Air Force Test Pilot School, and Paula Smith, Jan Schell and Becki Hoffman at the Society of Experimental Test Pilots (SETP).

Several pioneering Gemini and Apollo astronauts readily lent their support to this book through personal and telephone interviews, as well as providing much helpful information at Novaspace's Spacefest IV in Tucson, Arizona in 2012. Firstly, many thanks go to Bill Anders. At Spacefest, Alan Bean, Gene Cernan, Walt Cunningham, Ed Mitchell, Dick Gordon, Dave Scott and Al Worden kindly assisted with answers to my questions and offered stories of their own. For allowing me to attend the event, my gratitude is extended to the amazing Spacefest organizers Kim and Sally Poor.

In January 2012 contact was established with a principal player in the selection of the Group 2 and 3 astronauts, panel member Warren J. North, and I was eager for his insights into the process. It was therefore with great sadness that I discovered he had passed away just weeks later. His input would have been extremely helpful, and he is saluted in this book.

As always, long-time friend and co-author on other books, Francis French, looked through the manuscript and not only reported errors and typos but also made certain recommendations that improved the text.

My final thanks go to those who were responsible for the publication of this book, and their ongoing support of the author. To Maury Solomon, Editor of Physics and Astronomy at Springer Books (New York) for her friendship and belief in the merits of this book, and her assistant Megan Ernst for sorting out the many difficulties that arose with patience and a reassuring professionalism; to Clive Horwood and Romy Blott of Praxis in the U.K.; to my astute copyeditor David M. Harland; and to Jim Wilkie for his superb cover artwork. As previously, this great team has transformed my stories into a superbly crafted book.

Illustrations

Front cover: Group 3 astronaut C.C. Williams in backup training for the Gemini X mission.

Back cover (left to right): Neil Armstrong with the Lunar Landing Training Vehicle; Elliot See during a water egress exercise; Gene Cernan, Roger Chaffee and Charlie Bassett on a geological field trip.

Chapter 11

Chapter 12

Prologue

At 2:00 p.m. on the afternoon of 9 April 1959, seven apprehensive test pilots – now chosen as Mercury astronauts – were introduced to the press amid unexpected hype and adulation at NASA's temporary headquarters in the Dolley Madison House, Washington, D.C. From outset to announcement, the selection of America's first cadre of astronauts had been conducted in strict secrecy.

A year earlier, in April 1958, and in the light of recent achievements in space by the Soviet Union, U.S. President Dwight D. Eisenhower sent to Congress a bill calling for the immediate establishment of a civilian aeronautics and space agency. The bill was presented and supported, resulting in the passage of the Space Act of 1958, which in turn led to the creation of NASA, the National Aeronautics and Space Administration, on 1 October that year.

One of the first tasks of the new space agency was to implement the selection and training of a small group of outstanding pilots willing to fly into space aboard a capsule that was being designed for the American manned space program, known as Project Mercury. The job of defining and then undertaking a program to select these potential space pilots fell to NASA's Space Task Group, then located at the agency's Langley Research Center in Hampton, Virginia. Accordingly, an astronaut selection committee was assembled.

This eclectic committee consisted of Charles J. Donlan, who was Assistant Director for Project Mercury and headed the candidate screening committee; Warren J. North, formerly a test pilot and engineer with the National Advisory Committee for Aeronautics (NACA) and now NASA's Chief of Space Flight Programs; and flight surgeons, Dr. Stanley C. White, MD, Maj., U.S. Air Force, and Dr. William S. Augerson, MD, Capt., U.S. Army. Additionally, there were two psychologists, Dr. Allen O. Gamble of the National Science Foundation and the Manpower Evaluation Development Office at NASA Headquarters, and Dr. Robert B. Voas, USN, and two psychiatrists, Dr. George E. Ruff, MD, Capt., USAF, at that time Chief of the Stress and Fatigue Section of the Aero Medical Laboratory (AML) at Wright-Patterson AFB, and Dr. Edwin Z. Levy, MD, Capt., USAF. Thus all branches of the military had an active involvement in the selection process. These eight men set in motion an initial screening of military records and later carried out interviews and testing of the selected candidates. In setting out their parameters the question became a matter of precisely who, and with what qualifications, ought they to seek.

One major problem for the committee was satisfactorily resolved in December 1958 when President Eisenhower decreed that the nation's first astronauts had to be drawn from the ranks of military test pilots. The advantages were obvious; test pilots were already familiar with the rigors of military life, they were available at very short notice; and their full flight and medical records were readily accessible.

Initially, the committee was contemplating a selection pool of around 150 pilots, from which a nominal group of 36 finalists would be chosen to undertake physical, psychological and stress testing. It was originally planned that twelve would then be selected to undertake a nine-month training and qualification program, at the end of which the top six candidates would be selected as the nation's first astronauts. As Dr. White from the committee explained, as they began to pound out the exact criteria, they required individuals who were not only in top physical condition but had also demonstrated the capability to remain calm and work through tough and dangerous assignments. And stamina was an important factor, because the men had to have a good response to stressful situations and be able to withstand it over a period of time.

Specific limitations were then defined; the candidates had to possess a university degree; be a graduate of a test pilot school; be in superb condition both mentally and physically; have around 1,500 hours in high performance jets; be no taller than 5 foot 11 inches (as governed by the dimensions within a Mercury capsule); and be under forty years of age. Initially, the age limit had been set at 35 years, but the rigorous qualifications caused it to be raised to 39.

In the first week of January 1959 a meeting was convened at NASA Headquarters in Washington, at which it was decided to use the Lovelace Clinic in Albuquerque, New Mexico for comprehensive medical testing of the chosen candidates. The clinic was a non-government facility and the results of the examinations would become the property of NASA – not the military. It was felt this would offer reassurance to the pilot candidates that any poor results, which could potentially jeopardize their ability to continue in that service, would not go on their service records. It was also agreed that the ensuing stress and related tests would be at the Aero Medical Laboratory of the Wright Air Development Center (WADC) in Dayton, Ohio.

With these decisions in hand, the selection committee arranged with the Pentagon to retrieve and review the personal records of those who had graduated from test pilot schools in the previous ten years, examining them for basic requirements and a minimum number of flight hours. At the end of this process, they had the names of 508 potential candidates. Next it was necessary to cross-check these records against medical files in order to substantially narrow the field. Eventually, in what became known as Phase One of the operation, the names and records of 110 men were set aside as meeting the minimum qualifications: 58 Air Force pilots, 47 Navy officers, and five from the Marine Corps. Each of the 110 candidates was ranked in terms of his overall qualifications. Several factors were taken into account, such as total flying time, total testing experience, ratings of senior instructors at the test pilot schools – even the age and number of their children.

The committee's final task in this phase of the operation was to place the reviews in ranking order – best through least qualified – then split the files into three working groups of around 35 men, with the most promising in the first group. Charles Donlan then notified NASA of the results.

Literally within days of the initial screening, invitations were sent out to the top 35 candidates, requesting their presence at a briefing session and interviews in the Pentagon on Monday, 2 February, for what would begin Phase Two of the selection process. The orders were issued by the Chief of Naval Operations or the Air Force Chief of Staff, as appropriate. The candidates were instructed to leave their uniforms at home, and not to discuss their top-secret orders or the nature of the briefing with anyone. The second group would be called a week later, and the third group a week after that.

After informative briefings by service heads and NASA representatives on Project Mercury and the opportunity to apply to become astronauts, each candidate was told he could decline without prejudice to his military career. Those that opted to proceed would spend the rest of the week undergoing interviews and preliminary suitability tests. It was soon realized that to screen all 110 candidates would put an unnecessary strain on the resources of the selection team. After the second round of briefings and interviews, a total of 69 men had been processed. Of that number, 16 had declined, 6 were found to be too tall, and another 15 had been eliminated by one or more of the tests. According to Dr. Allen Gamble, he and Bob Voas found that they had 32 well-qualified candidates who had passed every test with flying colors. With a nominal 12 astronaut positions on offer, and a surprisingly high volunteer rate from the first two groups, it was decided not to summon the remaining group of 41 candidates, as they had not ranked quite as high on their records.

After batteries of tests had been carried out at the Lovelace Clinic and the Wright Air Development Center – as described in full in the author's earlier book, *Selecting the Mercury Seven: The Search for America's First Astronauts* – all of the medical, physical, psychological and stress test results were given to the selection committee. With the number of positions on offer reduced to 6, but with 7 firm candidates, the committee faced the near-impossible task of finding a reason to exclude one man. To remedy this dilemma, Dr. Robert R. Gilruth, head of the Space Task Group, elected to accept them all, and these were the men proudly presented to the assembled press in Washington, and through them the world, in April 1959.

Three years later, on 18 April 1962, NASA announced that it would be selecting a second cadre of astronauts following the tremendous successes and acceleration of the manned program. More pilots were now needed as Mercury transitioned into the two-man Gemini program.

This time, there was some policy reorientation. Mercury astronauts Alan Shepard and Donald ("Deke") Slayton were appointed to the selection panel, which also contained Warren J. North, a member of the Mercury selection panel. Slayton had recently been named coordinator of astronaut activities (i.e. chief astronaut) after his disqualification from flight assignment owing to a minor heart irregularity. As he so rightly pointed out in his later memoir, the panel could probably have simply gone back to the group of finalists from the Lovelace and Wright-Patterson exams in 1959 and hired another group right there from the 25 who did not make the final cut, but in the end it was decided not to do this.

According to Slayton, the panel devised a set of criteria for the second astronaut group that would enable the selections to be made with far less fuss. First of all, the invitation was opened to include civilians with experience as a jet test pilot, and to those with scientific as well as engineering backgrounds. The physical requirements would also be revised, as the planned Gemini and Apollo spacecraft were intended to be slightly larger than the Mercury craft. It was therefore decided to raise the height limit by an inch to 6 feet. Additionally, as these new programs would extend beyond the planned 3 years of Project Mercury, the age limit was reduced from 40 to 35.

When the deadline of 1 June 1962 rolled around there were 252 applications on Slayton's desk. Another one arrived a little late, but the panel wisely decided that the applicant was too well qualified to be refused for tardiness. He was a well-respected civilian X-15 pilot named Neil Armstrong.

The experiences of the Mercury Seven had demonstrated what was required of the nation's astronauts. Given the incredible appeal and outstanding challenge of the job, once again the nation's finest test pilots lined up hoping to become one of NASA's renowned "star voyagers". This is their story.

Foreword

On Monday afternoon, 17 September 1962, Tom Stafford, a captain in the U.S. Air Force, was celebrating his 32nd birthday in unusual circumstances. He was sitting on stage in an auditorium at the University of Houston, Texas, being introduced as one of a group of nine new astronauts selected by the National Aeronautics and Space Administration.

Alongside him were three other Air Force test pilots, Major Frank Borman and Captains Jim McDivitt and Ed White, as well as three naval aviators, Lieutenant Commander Jim Lovell and Lieutenants Pete Conrad and John Young. There were also two civilian test pilots, Neil Armstrong from NASA and Elliot See from General Electric.

Stafford knew some of these men; Borman, McDivitt and White had been his students at the Air Force Test Pilot School within the past two years. He had met Conrad and Armstrong, too, prior to arriving in Houston the day before.

But personal relationships were not on his mind that day. What Stafford thought as he looked to his left and right was: "One of us is going to be the first man to walk on the Moon."

It was an insight that no human could have had prior to that September day – or since.

That group of nine men, all test pilots between the ages of 31 and 36, had been deliberately selected by NASA to serve as the primary pilots for the Apollo program.

They hadn't been selected just for their flying skills, though that was an important factor. They were selected for their intelligence, for their ability to serve as project engineers for the command module and lunar module of the Apollo spacecraft that would hopefully take them to the Moon before the decade was out, and then return them safely to Earth.

NASA already had seven astronauts in the Mercury program. But those men were approaching the end of their original tours of duty; the space agency expected some or even most of them to return to their military careers ... certainly it did not plan for them to remain in the program for another seven years.

It was this new group – the Nine – that was tasked with developing and flying Apollo.

Within a year, the Nine would be joined by the Fourteen, a mixed group of test pilots, operational pilots and research pilots whose role would be to support the Nine in development work and serve as additional crew members.

However, it was this Nine – Armstrong, Borman, Conrad, Lovell, McDivitt, See, Stafford, White and Young – who would be the superstars of the Race to the Moon, experiencing both its high points (spacewalks, rendezvous, lunar orbit, lunar landing) and its low points (accidental death).

Colin Burgess' *Moon Bound* explores their story, and those of the Fourteen, in a new and exciting fashion. He also gives us a new perspective on the Nine and the Fourteen by presenting the stories of the men who, for one reason or another, did not make the cut – the men who were, in Tom Wolfe's cruel-but-accurate phrase, "left behind". Some of these pilots went on to highly successful careers in the military, becoming generals and admirals. Others died in combat or aircraft-related accidents. Some simply continued their careers and eventually made the transition to a well-earned retirement ... and likely wondered, "What if ...?"

Chapter Six, 'The Boy From Barren Run', tells the fascinating and tragic tale of naval aviator John Yamnicky. His story alone is worth whatever you paid for this book.

The strength of *Moon Bound* is no surprise to readers of the history of human space flight, because Colin has established himself as one of our best writers on the subject. In addition to the valuable overviews of Mercury, Vostok, Gemini, Soyuz and Apollo (*Into That Silent Sea* and *In the Shadow of the Moon*, both co-authored with Francis French), he has written about Australia's astronauts, NASA's scientist-astronauts, Russia's cosmonauts, Teacher-in-Space Christa McAuliffe, and – in my personal favorite, *Fallen Astronauts* – those men who were selected but didn't live to see the lunar landing.

I must also mention his *Selecting the Mercury Seven: The Search for America's First Astronauts*, which is a vital precursor to *Moon Bound*.

His other work, notably on the triumphs and tragedies of the Australian military in World War II, and his professional knowledge of the world of aviation, give him a unique perspective on the lives and careers of these men.

Colin has also been dogged and energetic in pursuing new information, not just on the non-selected men, few of whom have ever been profiled, but also on the selection process, medical tests and training of the Nine and the Fourteen, and those who came after them.

Open the pages. Prepare for launch. Take the *Moon Bound* voyage.

Michael Cassutt
August 2012
Los Angeles

Part One

1

Announcements and volunteers

Three years after the selection of the seven Mercury astronauts in the early months of 1959 there was a significant acceleration in NASA's manned space flight program. It was fueled by massive additional funding following President John F. Kennedy's bold – even audacious – commitment to the nation, directing NASA to place men on the Moon by the end of the decade. This, in turn, had been sparked by a provocative and unsettling number of headline-grabbing space "firsts" in the Soviet Union's burgeoning space program. These factors, coupled with an expected increase in the frequency of U.S. space missions, created the need for five to ten more astronauts to commence training for the complex two-man flights that would build on the success of Project Mercury.

A CALL FOR MORE ASTRONAUTS

The organization known as the Space Task Group (STG) was a working assemblage of NASA engineers – primarily from the Langley Research Center in Hampton, Virginia – who had been tasked in late 1958 with superintending America's manned space flight programs. In 1960, Dr. Robert R. Gilruth, Director of the STG, had advocated a follow-up, single-pilot program to Project Mercury, which at that time was within months of launching the first American into space. Designated Mercury Mark II, Gilruth anticipated an extension of the work completed by the original Mercury program that would build upon the agency's understanding and techniques of human space flight, while also maintaining an American presence in space. NASA had its seven Mercury astronauts, selected the previous year, so those plans could conceivably continue. However, despite some earlier setbacks, this was a time when the United States fully expected to become the predominant nation in the exploration of space, with an emphasis on manned missions. Those illusions were shattered overnight on 12 April 1961 when the Soviet Union launched Yuri Gagarin on a single-orbit circuit of the globe, barely three weeks ahead of the eagerly anticipated suborbital mission of astronaut Alan Shepard.

On 25 May 1961, less than three weeks after Shepard's successful ballistic space

Dr. Robert R. Gilruth, Director of NASA's Space Task Group and later the Manned Spacecraft Center. (Photo: NASA)

shot, President Kennedy pledged the United States to an historic undertaking to land American astronauts on the Moon by the end of that decade. A titanic and massively expensive manned space race to the Moon had now begun. In January 1962, eight months after President Kennedy's stirring and inspiring challenge, Mercury Mark II vanished into history as NASA's next bold vision evolved into what became known as Project Gemini, a two-man spacecraft that would serve as a crucial, interim step between the relatively simple Project Mercury and the extremely demanding lunar landing program of Project Apollo.

Everyone, it seemed, wanted to get in on the action of what President Kennedy had prophesized as "this new ocean" of human space activity. Congressman Olin Teague and his supporters began calling for an "astronaut academy" similar to those established for the American military, but the administrators of NASA decided this was an unnecessary waste of time and resources when they were now in the position to conduct interviews and complete selection procedures in-house.

THE RECRUITMENT PROCESS BEGINS

In the last week of March 1962, NASA Administrator James E. Webb announced the selection process for a new group of astronauts to augment the seven-strong Mercury astronaut team would begin "within the next few weeks". Anyone possessing the right qualifications and experience, including an extensive number of hours logged in high-performance jet aircraft – preferably in experimental test work – was eligible to apply, and this time the process would be open to civilians as well as members of the military.

However, in anticipation of the question, Webb expressed doubt that any women would qualify. "I do not think that we will be anxious to put a woman or any other person of particular race or creed into orbit just for the purpose of putting them there."[1] This declaration came in the wake of unofficial astronaut tests undertaken by the accomplished female pilot Geraldyn ("Jerrie") Cobb at the Lovelace Clinic in New Mexico, the private facility where the Mercury astronaut candidates had been subjected to extensive and intrusive medical examinations and tests. Cobb scored impressive results in the same tests, and shortly thereafter another twelve selected women pilots had joined her in passing the Stage I physical examinations. This resulted in media-driven pressure on NASA for the civilian space agency to include women in future astronaut recruitment drives. But these ambitions would soon fade after a special subcommittee of the House Committee on Science and Astronautics decreed that female aviators were ineligible, mainly because they were not permitted to become military jet test pilots – one of the specific requirements for NASA's astronaut recruitment.

Webb also hinted that the prerequisite age limit might be lowered from 40 to 35, as the newer astronauts were expected to be involved in space training and flights for many more years than those selected to join the Mercury team in 1959. Additionally, he suggested that the slightly more voluminous Gemini spacecraft would enable the 5 foot 11 inch height restriction to be eased a little.

The Mercury astronaut selection process had been carried out in complete secrecy. Only those invited to attend preliminary service-related and NASA briefings were eligible, should they later decide to volunteer. However, things were vastly different for the second group. This time applications were openly encouraged, with the minimum qualification standards being published and disseminated to aircraft manufacturers, government agencies, military services, the Society of Experimental Test Pilots, and even the news media.

It was fully anticipated that a large number of suitable applicants would react to the invitation, submitting their names and credentials to NASA. As Warren J. North from the Manned Spacecraft Center's Flight Crew Support Division later surmised, "The advanced engineering aspects and the challenging nature of manned space flight has appealed to this country's engineering pilots. Consequently, the crew selection has been a process of screening highly experienced pilot volunteers, men who have demonstrated over a period as long as 15 years their ability to operate and analyze high-performance aircraft – machines which, like spacecraft, consist of complex propulsion, electrical, mechanical and hydraulic systems."[2]

SETTING PARAMETERS

On 18 April 1962, NASA formally announced that it would accept applications from both military and civilian test pilots for the second group of trainee astronauts. If successful, and following an initial training period, they were expected to participate in support operations for Project Mercury and then join the Mercury astronauts in piloting the Gemini missions – the first of which was expected at that time to fly in 1964. They might subsequently be required to act as commanders on early Apollo flights.

The applicants had to submit to the Manned Spacecraft Center (MSC) in Houston a number of forms describing in detail their academic credentials, flight and work experience. The five basic candidate standards they had to address were listed as:

- Be experienced jet test pilots, preferably currently engaged in flying high-performance jet aircraft.
- Have attained experimental test pilot status through military service, the aircraft industry or NASA, or have graduated from a military test pilot school.
- Have earned a degree in physical or biological sciences or in engineering.
- Be a citizen of the United States, under 35 years of age at the time of selection, and six feet or less in height.
- Be recommended by their parent organizations.

The announcement also stated that the training program of the new astronauts would include working in cooperation with design and development engineers, simulator flying, centrifuge training, additional scientific training, and flights in high-performance aircraft.

The selection timetable stated that applications from any individuals meeting the qualifications would be accepted by the Director of the Manned Spacecraft Center until the 1 June 1962 deadline. The following month, those pilots making it through to the next stage would be interviewed and assigned written examinations on their engineering and scientific knowledge. At the conclusion of that screening phase, selected applicants would be given a thorough examination by a group of medical specialists. The final selection would be made and announced in the fall.[3]

AN ASTRONAUT SELECTING ASTRONAUTS

Donald ("Deke") Slayton was one of the original Mercury astronauts selected in April 1959. However, fate had many twists in store for the former Air Force test pilot after he began full-time training for the Mercury-Atlas 7 (MA-7) mission aboard a spacecraft he had already named *Delta 7*. His worst fears, and those of any military pilot, were confirmed when the flight surgeons detected a slight anomaly with his physical condition. Tests revealed Slayton to be suffering from a mild heart condition labeled as an idiopathic atrial fibrillation. He had always been able to overcome this known problem with a little exercise, but NASA was understandably skittish about sending someone into orbit who was deemed to be less than 100 per cent fit, and he was subsequently removed from the flight. Slayton, totally devastated, was not only replaced on his orbital mission by Scott Carpenter but also barred from operating any high-performance aircraft without an accompanying co-pilot.

On 16 March 1962 Slayton endured a miserable press conference in which he had to answer reporters' questions concerning why he had been forced to withdraw from his Mercury mission. Having banished Slayton from active astronaut status for an indefinite period, NASA had to find him a suitable job. As it turned out, they needed a senior manager to run the astronaut office and Slayton was temporarily handed the job of Coordinator of Astronaut Activities. Ironically, one of his first tasks was to help to select the new group of astronauts.

According to Slayton, he developed a set of criteria for the second astronaut group that would allow the selection program to proceed with minimum fuss. There would be changes made from the Mercury group, he said, later explaining that "the Gemini and Apollo spacecraft would be slightly bigger than Mercury, so we raised the height limit to six feet even – or we'd have lost Tom Stafford. We allowed candidates to have a degree in the biological sciences."[4]

As the Apollo program was planned to extend over many years and a vast number of flights, the age limit was also lowered from 40 to 35. "I figured we'd need about ten new astronauts to make up the Gemini and first Apollo crews," Slayton went on, "knowing full well that another selection would be needed in a year or two."[5] His prediction of further astronaut recruitment the following year proved to be correct.

"To be honest," Slayton also declared, "I could probably have just gone back to that group of finalists who went through the Lovelace and Wright-Patterson exams in 1959 and hired another group right there. I'm glad we didn't; that second group of astronauts is probably the best all-around group ever put together."[6]

Given his new position, Slayton's responsibility was to chair a panel convened for the purpose of interviewing the candidates, together with fellow Mercury astronaut Alan Shepard and Warren North, who had served on the Mercury selection panel. Walter C. Williams, the Associate Director of the Manned Spacecraft Center, would sit in as an observer on some panel sessions. Formerly with NACA and Associate Director of the Space Task Group, Williams was responsible for many facets of Project Mercury, including factory check-outs, pre-launch preparations, launch and in-flight activities, recovery operations and post-flight analysis as well as future flight planning.

While a great deal is known about Slayton and Shepard, the two astronauts on the selection panel, the same cannot be said for the third active member. It is worth digressing to examine the qualifications of Warren North to join the panel and assist in choosing the nation's next astronauts.

At the time of the selection process for the Group 2 astronauts, Warren James North was Chief of the Flight Crew Operations Division at the Manned Spacecraft Center, which had responsibility for astronaut training and crew integration into the design and operation of the spacecraft and launch vehicles. Married to the former Leah Pendleton, a schoolteacher, and the father of three children, he had enjoyed an exciting career prior to his tenure with NASA.

A native of Winchester, Illinois, North had his college education interrupted by the war, and he spent the years 1943-45 as a pilot in the air corps. After receiving his

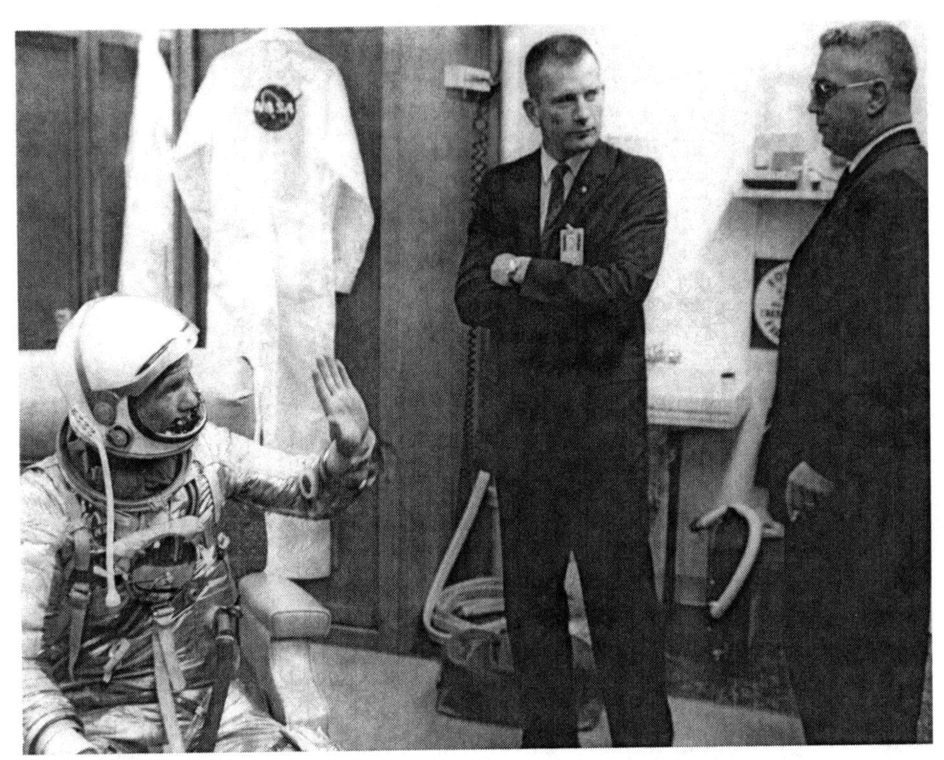

Deke Slayton (middle) and MSC Associate Director Walter Williams in a pre-launch discussion with Mercury astronaut John Glenn. (Photo: NASA)

bachelor's in aeronautical engineering from Purdue University, Indiana, he began his career with NACA that same year by joining the Lewis Flight Propulsion Laboratory in Cleveland, Ohio. He spent his early years with NACA in the dual capacity of an engineer and engineering test pilot, before transferring in 1953 to the Aerodynamics Noise Branch to conduct research on turbojet noise reduction.

The following year, North received his master of science degree in aeronautical engineering from the Case Institute of Technology in Cleveland. Then, in 1955, he was awarded an Institute of Aeronautical Sciences (IAS) Flight Test Engineering Fellowship to Princeton University, New Jersey, where he furthered his academic career in aeronautical engineering, particularly in flight testing.

After receiving his master of arts degree from Princeton he returned to the Lewis Laboratory and was appointed Assistant Chief of the Aerodynamics Noise Branch. Two years later he was promoted to head the stability group of the Missile Design Panel, and served in this position until his appointment to NASA on 1 October 1958 in charge of "manned satellites" in the Office of Space Flight Programs. It was here that he began assisting in the coordination of Project Mercury, and this in turn led to his appointment to Flight Crew Operations.

North would become increasingly involved in the selection and training of the Project Mercury astronauts. During this period, in addition to being on the selection

During a tour of MSC in October 1964, Warren North (right) shows off the new center's facilities to Drs. Wernher von Braun and Joachim Kuettner from the Marshall Space Flight Center. (Photo: NASA)

panel, he went through many of the same tests as the Mercury finalists, joining them in sessions in the centrifuge, the heat chamber, and in zero-G flying. Little wonder that he became known around the office as an "unofficial astronaut".

In addition to his work in selecting and training astronauts, North was also deeply involved in planning for Project Gemini, supervising the preparation of plans and making many significant contributions to the design and development of the Gemini spacecraft.[7]

OUTLINING THE TRAINING PROGRAM

For those pilots ultimately selected into the astronaut corps, an intensive training program was scheduled to commence in mid-October 1962. According to Associate Director of the Manned Spacecraft Center, Walter Williams:

"The early phases of this training program will familiarize them with the Mercury spacecraft, launch vehicle and operational techniques. They will then receive spacecraft and launch vehicle briefings on Gemini and Apollo. As they become more familiar with Gemini and Apollo, they will be assigned, together with the current Mercury pilots, to help establish design and operational concepts.

"Concurrent with the project-oriented aspects of the program, the men will attend basic science lectures one or two days per week. Because of their previous academic and occupational experience, most of the courses will be of the refresher type. The basic program will place special emphasis on space navigation, computer theory, flight mechanics, astronomy, physics of the upper atmosphere and space, bioastronautics, advanced propulsion systems, aerodynamics, guidance and control, space communications, global meteorology and selenology.

"During later phases of the training program, the pilots will work with static and dynamic simulators to establish detailed flight operational procedures.

"NASA has established a special aircraft operations group in Houston to provide proficiency flying for the pilots. T-33 and F-102 aircraft are being assigned.

"Although the early phases of this training program were tailored primarily for the new pilots, the Mercury pilots will be integrated with the new group immediately, and all will train together insofar as is practical."[8]

CANDIDATES COME FORWARD

Unlike the U.S. Navy and U.S. Marines, who submitted the names of all of their test pilots meeting NASA's requirements, the Air Force conducted its own evaluation of qualified pilots to select those they determined should be submitted.

One of the selected candidates was Capt. Robert Smith, whose service dossier had been boosted by a personal note of endorsement from aviation pioneer and war hero Gen. James H. ("Jimmy") Doolittle. As Smith recalls, after the Air Force's in-house selection process, the service submitted eleven names to the space agency, which he mentioned as being the total number of astronauts needed by NASA. This may have seemed somewhat arrogant to NASA, because the Navy submitted a larger group of initial candidates, all of whom satisfied the requirements. NASA also had its own candidates, including a highly experienced test pilot named Neil Armstrong. "Likely, politics demanded some of each [service], except for the Marines who had only two applicants," Smith stated.[9] His recollection seems a little erroneous in this as well, as four of the Group 2 finalists came from the Marine Corps. Additionally, NASA had not settled on eleven as the definitive number of astronauts required; rather, they had publicly stated they were seeking between five and ten to join the Mercury astronauts in training.

The eleven candidate names submitted to NASA by the Air Force were:

Maj. Frank F. Borman II
Capt. Michael Collins
Capt. Roy S. Dickey
Capt. Joe H. Engle
Maj. Neil P. Garland
Capt. James A. McDivitt
Capt. Francis G. Neubeck
Capt. Robert W. Smith
Capt. Thomas P. Stafford

Capt. Alfred H. Uhalt, Jr.
Capt. Edward H. White II

As Smith further recalled in his memoirs, "Our final preparation was what we dubbed 'Charm School' in which a PhD in semantics or some such thing and three others trained us for the NASA interviewing and the final board assessment. We had to dress in civvies for the occasion of our 'finals', a simulated interview, and were critiqued on our demeanor, our responses, and even our attire. We were sent into an ultimate evaluation, up to and including a group interview with General Curtis LeMay, Chief of Staff, in his office. One thing I took from that meeting was the fear that the Chief held over senior officers, and there were many tales. One of our gang asked the three-star general, DCS/Personnel who was leading us to the Chief's office whether we should salute individually or line up and do it as a group. His cop-out was to do whatever felt best under the circumstances."[10]

Another of the Air Force candidates, Capt. Tom Stafford, also recalls the 'charm school' experience, with advice on how to dress, speak, and conduct themselves in dealing with NASA. "What I remember most about that briefing," he declared, "is that many of us kept slipping off to another room to watch a television showing Scott Carpenter making the second Mercury orbital flight. It was May 24, 1962."[11]

Eventually the eleven USAF candidates were fully briefed and ready. Their names were submitted to NASA and most made it through to the next phase of the operation, with the exception of Joe Engle and Neil Garland, who were rejected by NASA for the later evaluation and selection process.

The eleven selected USAF astronaut candidates standing behind their seated, unnamed Air Force briefing team. From left: Ed White, Tom Stafford, Mike Collins, Greg Neubeck, Bob Smith, Neil Garland, Frank Borman, Jim McDivitt, Joe Engle, Roy Dickey and Al Uhalt. (Photo: USAF)

A LATE INCLUSION

Technically, civilian X-15 test pilot Neil Armstrong should not have become an astronaut in NASA's second cadre. Obviously preoccupied with other matters – including the recent death of his daughter, Karen Anne – he actually submitted his application about a week after the NASA deadline of Friday, 1 June. Fortunately he had a friend and ally in the right place at the right time.

Dick Day had worked with Armstrong at Edwards AFB as a flight simulation expert. In February 1962 he transferred to MSC, becoming Assistant Director of the Flight Crew Operations Division. In this capacity he worked with the selection panel for the second group of astronauts, and one of his clerical tasks was to prepare and present all the applications to the panel.

"I really don't know why Neil delayed his application," Day told Armstrong's biographer James Hansen in 2003, "but he did, and all the applications came to me, since I was the head of flight crew training. But he had done so many things so well at Edwards. He was so far and away the best qualified, more than any other, certainly as compared to the first group of astronauts. We wanted him in."[12]

Only Day and Walt Williams – who had also strongly urged Armstrong to apply – knew what happened next. It was a simple matter of slipping the late application into the stack of candidate résumé folders prior to the selection panel's first meeting. The rest, as they say, is history.

MARINE CORPS APPLICANT

In October 1957, U.S. Marine Corps Capt. Ken Weir was stunned to read in the *Washington Post* that the Soviets had launched a satellite which was currently racing around the world and making unhindered passes over the United States of America. "Then they had a capsule with a dog circling the Earth. The United States put some German scientists to work and soon had a monkey riding around in a spacecraft. That was the extent of my exposure to outer space activities."

A year later the same newspaper brought Weir some information on the selection of the seven original NASA astronauts, which he frankly admits he knew nothing about beforehand. At the same time, he was busy instructing more than a hundred 2nd lieutenants in the Marine Rifle Platoon in a Helicopter Assault course at the Marine Corps Officer Basic School. One of the men happened to enquire whether he wanted to be an astronaut. "My knee-jerk answer was, 'Of course, very much so, but I was passed over.' Unfortunately, many of the 2nd lieutenants were left with the impression that I had somehow been good enough as an infantry tactics and pack animal instructor with wings to be considered along with numerous candidates for the U.S. space program, but just barely missed the cut."

In 1961, Weir had just completed his course at the U.S. Navy Test Pilot School at Patuxent River, Maryland when the Naval Air Test Center (NATC) flight surgeon gave him a routine electrocardiogram examination during his annual physical. To Weir's surprise and consternation the surgeon informed him he was immediately

Capt. Ken Weir, USMC, taken circa 1957 while attached to Marine Attack Squadron VMA-251. (Photo: USMC)

grounded because he had found what were referred to as inverted T-waves, which might be the result of different cardiac and non-cardiac conditions. Even today, fifty years on, cardiologists sometimes find inverted T-waves in people whose tests are completely normal, and they simply do not know why they occur.

"He didn't have any idea what that meant, being new to the science," Weir griped, "but he made my life miserable anyway, and then he told me I also had high blood pressure. Two weeks later, after a massive dose of stress tests and electrocardiograms at the National Naval Medical Center in Bethesda, Maryland, I was cleared for full flight test pilot operations."

Exotic flight testing of that era focused on extremely high-performance airplanes including the X-15 and the upcoming Dyna-Soar (X-20) programs. Without doubt, the ultimate goal of most test pilots of the time was to be involved with jet- and rocket-propelled airplanes that operated at high Mach numbers and extremely high altitudes, capable of pulling lots of Gs, and delivering devastating weapons. Weir was anxious to be in that number.

"The USAF had just started up their Aerospace Research Pilot School (ARPS) at the Air Force Flight Test Center, Edwards Air Force Base, California, and somehow the Navy had obtained one slot in the second class. A group of NATC Patuxent River Navy and Marine test pilots were sent to the Naval Aviation Medical facility at Pensacola, Florida for testing and evaluation. I was one of a dozen or so individuals that were subjected to the enhanced physical exams and psychological testing. The group included Carl ("Tex") Birdwell and Bill Ramsey, as well as future NASA astronauts John Young and Al Bean.

"A couple of times I was very vocal in protesting the ridiculous examinations until I noticed that the flight surgeons or technicians were suddenly taking copious notes every time I offered an objection. John Young was grounded on the spot for some obscure vision deficiency. Nobody, it seems, cared about my inverted T-waves or my hypertension. We all returned to Patuxent River. John Young told them to stuff the Aerospace Research Pilot School – all he wanted was his flight status back." Weir then revealed a ribald episode concerning one of his colleagues.

"Tex Birdwell, who had been the outstanding student in his Test Pilot School and was my boss at the Flying Qualities and Performance Branch of the Flight Test Division, was selected to be the Navy's representative in the second class at ARPS. He was departing for Edwards the day after attending the annual Marine Corps Mess Night at the Piney Point Officers Club, where he had participated in considerable celebration with far too much wine and spirits. He proceeded to center-punch his TR-3 sports car into a guard rail on one of the bridges crossing a creek next to the golf course. The headlights wrapped around a pillar and the engine was pushed all the way back between the seats.

"When he regained consciousness on the hospital operating table in his mess dress, covered with mud, seaweed, grease, debris and lots of blood, Tex asked the doctor what he was doing. The doctor said he was trying to get all the glass out of his face. Tex promptly tried to help with his greasy hands. The doctor gave up, saying Tex wasn't very good looking to begin with, finished sewing him up, then wrapped him mummy-like around his head and shoulders. Off Tex went, and that is the way the Navy's representative arrived at Edwards for astronaut training.

"When he reported into ARPS, the renowned Air Force Commandant, Col. Chuck Yeager, could not believe what kind of ancient Egyptian mummy the Navy had sent him to turn into a space pilot. Tex healed and did quite well as a student among some of the real stalwarts in the Air Force test pilot community, each of whom went on to very interesting and rewarding flight test careers pushing the envelopes of research airplane performance."

At the same time, while fully involved in the testing of the F4H-1F Phantom jet, one of the Navy test pilots asked Weir if he had applied for the astronaut program. A notice about the recruitment of a second group of astronauts had apparently been out for several days, but Weir did not know anything about it. Curious, he went to see the commanding officer (CO) of the Marine Air Detachment at Patuxent River to inquire about this opportunity. The CO said that if Weir was interested, he would add his name to those of the other Marines who were applying. Eventually he received the application forms, filled them in as best he could, located the requested records, and sent everything in via official channels.

"A few weeks later I received a large manila envelope with a letter informing me that I had been screened for further processing, a whole bunch of other forms for background investigations and other required records, and was ordered to proceed to Brooks Air Force Aero Medical Facility in San Antonio, Texas to take the astronaut physical. Several of my friends received the disappointing small white envelope letter informing them that they had not been selected for further screening."[13]

One of the fortunate ones to receive a similar order for Brooks was Navy Lt. Charles ("Pete") Conrad, Jr., then stationed at the U.S. Naval Test Center at Patuxent River as a test pilot, flight instructor, and performance engineer. Conrad had been a finalist in the Mercury selection process, but even though he had scored highly his often flippant and querulous manner during the testing had not impressed some of the examiners and he had missed out in the final cut.

One Thursday afternoon Conrad was enjoying a drink in the noisy officers' club at Pax River (as it was known) when two old Navy buddies walked in – astronauts Alan Shepard and Wally Schirra. They spotted Conrad and straightaway joined him at his table. After exchanging a few good-natured barbs, Shepard told his Navy friend that NASA was about to announce a search for a new lot of astronauts and they had been talking Pete up to everyone at the space agency. "They all know the Conrad legend now," Shepard said with a broad, toothy smile that vanished as he continued. "They also know you're a shit-hot pilot. You want to give it another shot?"

Conrad saw that Shepard was now being serious, so he gave him a straight, simple answer, "Yeah."[14]

Now, for the second time, Pete Conrad was back in contention to fly into space.

MEANWHILE, IN THE USSR

Three years earlier, amid great secrecy, a group of twenty young Air Force pilots had been selected as the Soviet Union's first cosmonaut team, receiving their space flight training at a hidden facility outside of Moscow. No one beyond a select few knew of them, or their identities, most of which would not be released to the public until they were actually in orbit on a mission.

By way of contrast, the U.S. space program was an open book. The names and other details of the seven Mercury astronauts had been revealed with great fanfare at a media conference in Washington, D.C. on 9 April 1959, and even though NASA later withheld the name of the first American selected to fly into space for a time, the identity of each astronaut, their anticipated launch date and mission objectives were announced well in advance. The agency had adopted and maintained this policy in consultation with the White House, and although it was a workable policy agreeable to an astronaut-adoring public, it would cause continuing frustrations on occasions when the Soviet Union took full advantage of this knowledge to launch cosmonauts on similar in-your-face missions just ahead of the United States, even surpassing the announced goals of the NASA flights where possible.

STRIKING COMPARISONS

However, early in the selection of astronauts and cosmonauts there were essential philosophical differences, and in many ways the American astronauts were superior pilots to their counterparts in the Soviet Union, and also better qualified. Unlike the cosmonauts, whose considerably less flight experience was mostly gained in MiG-15 and MiG-19 fighter jets, the men selected to be NASA astronauts were highly trained military test pilots with solid engineering educational backgrounds and hundreds of hours of flight experience in supersonic or advanced aircraft.

Furthermore, NASA's astronauts would train with the expectation of maintaining a degree of control over their spacecraft, and in that training phase would enjoy a considerable and influential degree of input into the design features, cockpit layout, and further development of their spacecraft with the manufacturers. By comparison, the first Soviet cosmonauts were virtual passengers in their Vostok spacecraft, which were controlled from the ground. Apart from monitoring instruments and performing a small number of innocuous onboard experiments, they had very little control over the vehicle or its systems.

Initially, Soviet cosmonauts were selected from jet-qualified pilots in the Soviet Air Force and Navy. Engineering skills were not mandatory, and although test pilot experience was deemed beneficial it was not a requirement. Finding qualified pilots who were in good health and small enough to fit into the spacecraft was therefore a much simpler task than the battery of intrusive medical, psychological, physiological and stress tests that were imposed on America's astronaut candidates.

COSMONAUTS LOST TO THE PROGRAM

While they were every bit as competitive as their American counterparts, a certain arrogance and lack of discipline in the first cosmonaut team – often associated with lax supervision and excessive alcohol consumption – would create massive problems for the Soviet space hierarchy, whose tolerance was short-fused. Out of the twenty men selected in the first Soviet cosmonaut team, four would be summarily dismissed for breaches of strict rules without flying into space. A physical problem discovered during centrifuge training unfortunately cost another cosmonaut trainee his place in the team, while yet another suffered a serious and disqualifying neck injury during a simple diving accident in shallow water away from the training center. Still another suffered a disqualifying stomach ulcer at a crucial time in his career. Lastly, and most horrifying of all, 24-year-old cosmonaut trainee Valentin Bondarenko died of extensive burns suffered when he was engulfed in flames during a training exercise inside an oxygen-enriched pressure chamber. He died several hours later in utmost agony.

Of the twenty pilots chosen to fly into space as part of the first Soviet cosmonaut team, only twelve would actually realize that ambition. By comparison, all seven of NASA's Mercury astronauts would eventually make it into space, with five of them making more than one flight.

OTHER SPACE PROGRAMS, OTHER SPACE PILOTS

Even as plans moved ahead for the selection of new cadres of astronauts by NASA, it should be noted that Project Mercury was not the first or only U.S. space program. Other military and civilian programs were either running concurrently or envisaged.

Project MISS

In March 1958, members of the U.S. Air Force's Air Research and Development Command (ARDC) proposed a priority initiative to launch a pilot into space well ahead of the Russians. The pilot would be sent into space, possibly into orbit, aboard a manufactured capsule mounted on top of a military intercontinental

Three of the proposed Project MISS space pilots at Edwards AFB, circa 1958. From left: Neil Armstrong, Capt. Iven Kincheloe, Jr., and Maj. Robert White. Armstrong and Kincheloe would later be selected for the X-15 program, but Kincheloe died in an F-104A accident on 26 July 1958 before the X-15 flew. His place in the program was taken by backup pilot Robert White. (Photo: AFTC/HO, Edwards AFB)

ballistic missile (ICBM), possibly a two-stage Thor rocket. From its original title of Project 7969, the plan became the Man in Space Soonest (MISS) program. As the project developed, the Air Force sought the cooperation of NACA, which had been working on similar concepts, and on 16 June 1958 a tentative agreement was signed to proceed using the Atlas rocket as the launch vehicle.

Moving right along, the Air Force put together and submitted a list of suitable service test pilots as potential launch subjects. Those proposed as Project MISS pilots were:

Neil A. Armstrong, 27, NACA
William B. Bridgeman, 42, Douglas Aircraft Company
A. Scott Crossfield, 36, North American Aviation
Iven C. Kincheloe, Jr., 29, USAF
Jack B. McKay, 34, NACA
Robert A. Rushworth, 33, USAF
Joseph A. Walker, 36, NACA
Alvin S. White, 38, North American Aviation
Robert M. White, 33, USAF

However, plans to have a pilot launched into space by June 1960 began to unravel when NACA became increasingly involved in the creation of a civilian space agency that would become NASA and on 5 November 1958 create the Space Task Group for Project Mercury. But even before this, on 1 August the Advanced Research Projects Agency (ARPA) canceled the allotted funding and Project MISS was abandoned.[15]

Aerospace Research Pilot School

In 1959, with a number of military and civilian space programs on the drawing board, crucial changes would be made to the curriculum at the U.S. Air Force Test Pilot School at Edwards AFB. The X-15 rocket plane was already undergoing flight performance testing at Edwards, and it was apparent that the USAF would soon have a need for a dedicated manned space flight operation. That year, Capt. Ed Givens and his civilian instructor William Schweickhard took the concept of a full aerospace course to the school's commandant, Maj. Richard Lathrop. After much deliberation he decided there was considerable merit in the idea, and asked his special assistant Maj. Thomas McElmurry to help to get the project up and running. The following year Maj. Frank Borman joined the team as well. He became a champion of the aerospace course and helped out where he could.

On 5 June 1961 the proposal became a reality with Class I comprising five student aerospace pilots, all graduates of the Test Pilot School: Maj. Frank Borman, Maj. Robert Buchanan, Capt. James McDivitt, Maj. Thomas McElmurry, and William Schweickhard. On 12 October that year the Experimental Flight Test Pilot School became the USAF Aerospace Research Pilot School (ARPS), and its role was to help Air Force pilots to gain the qualifications to become astronauts, or USAF astronaut-designees as they became known.

The second ARPS class was announced on 20 April 1962, and the nominated pilots began their studies two months later. This time there were eight participants: Captains Charles Bock, Albert Crews, Robert McIntosh, Robert Smith and William Twinting; and Majors Byron Knolle and Don Sorlie. Initially, USN Lt. Cmdr. Lloyd Hoover was the eighth member of Class II, but he withdrew and was replaced by Lt. Cmdr. Carl ("Tex") Birdwell.

Many of the graduates from the first two ARPS classes would go on to become involved in projects such as the X-15, X-20 and Lifting Body programs, while two members of the first class – Borman and McDivitt – would become NASA Group 2 astronauts.

On 22 October 1962, while NASA was concentrating on Project Gemini, the third ARPS class was selected, comprising ten Air Force officers: Captains Alfred Atwell, Charles Bassett II, Michael Collins, Joe Engle, Edward Givens, Gregory Neubeck, James Roman and Alfred Uhalt, along with Majors Tommy Benefield and Neil Garland. This class, which received specific X-20 training as part of the course, was later joined by an eleventh member, Capt. Ernst Volgenau. Three members of Class III would later become NASA astronauts – Bassett, Collins and Givens.

The fourth and final ARPS class began in May 1963, made up of fourteen USAF officers and a single representative each from the Navy and Marines. They were Air Force Captains Michael Adams, Tommy Bell, William Campbell, Edward Dwight, Frank Frazier, Theodore Freeman, James Irwin, Frank Liethen, Lachlan Macleay, James McIntyre, Robert Parsons, Alexander Rupp, David Scott, and Russell Scott, along with USN Lt. Walter Smith and USMC Major Kenneth Weir.

While a number of military astronaut-designees managed to make their way into NASA's civilian astronaut corps, the Air Force eventually shut down its X-20 Dyna-Soar and MOL programs. With them went the chances for any of the others – apart from those already engaged in the X-15 program – to make it into space.[16]

X-20 (Dyna-Soar)

An initiative of the U.S. Air Force, the delta-wing X-20 or Dyna-Soar (a contraction of Dynamic Soaring) was to have been developed as a manned vehicle capable of placing military astronauts into space to conduct either an independent mission or to rendezvous with a space station already in orbit. On completion of the flight's mostly covert objectives the retrorockets would fire to return the craft to Earth, concluding with a horizontal landing on a runway under the pilot's control.

Plans initially called for the X-20 to be placed into orbit by being launched atop a Titan I booster rocket, but as the weight of the spacecraft rapidly increased during its development, the more powerful Titan III became the preferred booster. Following two years of feasibility studies, on 9 November 1959 the Air Force selected Boeing Aircraft as the chief contractor for the winged boost-glide spacecraft.

Earlier, the need for highly skilled pilots resulted in ten active NASA and USAF pilots being selected to undergo secret physical tests in August 1959. On 1 April 1960, seven of these candidates were chosen to train for military missions that were expected to begin within years. They were:

Neil A. Armstrong, NASA
William H. Dana, NASA
Capt. Henry C. Gordon, USAF
Capt. William J. Knight, USAF
Capt. Russell L. Rogers, USAF
Milton O. Thompson, NASA
Maj. James W. Wood, USAF

Boeing's full-size mockup of the X-20 Dyna-Soar with its nominated pilots. Front to rear: Capt. William Knight, Capt. Russell Rogers, Milton Thompson, Maj. James Wood, Capt. Henry Gordon and Capt. Albert Crews, Jr. (Photo: USAF)

Armstrong and Dana would leave the Dyna-Soar program in 1962 and be replaced by just one pilot, Air Force Capt. Albert H. Crews. The pilots began mission training at Edwards and Wright-Patterson AFBs, but the program was destined never to reach operational status. On 10 December 1963 it was abruptly canceled. The Secretary of Defense, Robert S. McNamara, told a Pentagon press conference that he had directed the Air Force to forgo any further development of the winged spaceplane and instead to develop the orbiting military space laboratory that would come to be known as the Manned Orbiting Laboratory.

If the Dyna-Soar program had not been canceled, it was anticipated that the first manned, single-orbit flight would occur in July 1966 under the command of senior X-20 test pilot, Maj. James Wood, who had previously been a finalist in the Mercury (Group 1) selection process.

X-15

Conceived in the early 1950s, the X-15 was a rocket-powered, winged spaceplane that eventually enabled a number of NASA, USAF and USN pilots to fly above the recognized boundary of space.* The X-15 was a continuation of the Research Airplane Program first approved by the government in 1944, which produced the X-series of rocket-powered aircraft. Prior to the first flight of an X-15, human-controlled flight had been restricted to altitudes below 130,000 feet and speeds less than 2,100 mph.

The first of three sleek X-15 airplanes constructed by North American Rockwell was delivered to Edwards AFB, California in 1958. The company was responsible for demonstrating the flight-worthiness of the innovative aircraft, and conducted the initial flight tests.

The first flight took place on 8 June 1959 when North American's chief test pilot Scott Crossfield was air-launched from beneath a B-52 in a planned powerless glide-flight lasting almost five minutes. Following completion of the contractor test flights, the aircraft was turned over to NASA for further research flights that continued to increase the speed and altitude capabilities of the craft. The X-15 reached its design speed on 9 November 1961 when Maj. Robert M. White (USAF) flew the first X-15 to a speed of 4,093 mph – a little over Mach 6. Then, on 30 April 1962, Joseph A. Walker (NASA) took the X-15 to almost its design altitude, reaching 243,700 feet.

On flight No. 191 of the program, on 15 November 1967, Maj. Michael J. Adams (USAF) guided the X-15 to a record 266,000 feet, but encountered a flight control system malfunction during re-entry which caused severe pitch oscillations and led to the vehicle breaking up and killing the pilot. Adams was posthumously awarded his astronaut wings for exceeding 264,000 feet (50 miles) in altitude.

The final flight in the test series, No. 199, took place eleven months later when

---

* At that time the U.S. military defined the boundary of space to be at an altitude of 50 statute miles (80 kilometers). The Fédération Aéronautique Internationale defined it to be at 100 km.

NASA pilot Bill Dana landed the X-15 after a flight to 255,000 feet on 24 October 1968, during which he achieved a maximum speed of 3,716 mph. In all, thirteen of the 199 research flights flew beyond the recognized boundary of space. Twelve test pilots flew the X-15 to the edge of the atmosphere and beyond:

Michael J. Adams, USAF *
Neil Armstrong, NASA
A. Scott Crossfield, North American
William H. Dana, NASA *
Joe H. Engle, USAF *
William J. Knight, USAF *
John B. McKay, NASA *
Forrest S. Petersen, USN
Robert A. Rushworth, USAF *
Milton O. Thompson, NASA
Joseph A. Walker, NASA *
Robert M. White, USAF *

Eight of them (marked with an asterisk) exceeded the 50-mile altitude recognized by the Air Force as penetrating space, and hence received astronaut wings. Of these, three were NASA pilots and five were USAF research pilots.[17]

Manned Orbiting Laboratory

In 1960 the U.S. Air Force began to study the feasibility of sending a manned space station into orbit, crewed by military astronauts. On 10 December 1963, Secretary of Defense Robert S. McNamara announced that the Air Force had been directed to develop such a space station. This program went by the functional name of Manned Orbiting Laboratory (MOL), and plans called for the first station to be launched into polar orbit in 1971 using a Titan III booster. The orbiting laboratory was to provide a shirt-sleeve environment in which two military astronauts could live and work for a month before returning home. Air Force General Bernard A. Schriever was given the post of First Director of the MOL program, directly responsible to the Secretary of the Air Force. Due to the sensitive nature of many of the military duties to be carried out on the orbiting laboratory, a tight blanket of security was thrown over the entire program once actual development work began. President Lyndon B. Johnson gave MOL the official go-ahead in August 1965.

For the launch from a new complex near Vandenberg AFB in California, it was envisaged that the two astronauts would ride in a modified Gemini spacecraft, known as Gemini B, mounted on top of the laboratory module. Once in orbit, the astronauts would pass through a transfer tunnel in the heat shield of the spacecraft and enter the 14-foot-long pressurized laboratory module. What the Air Force did not reveal was that MOL would provide a platform for observing military activities over sensitive areas of the world.

Any MOL astronaut applicant was required to be a United States citizen born

U.S. Air Force depiction of a Manned Orbiting Laboratory (MOL) launch. (Photo: USAF)

after 1 December 1931, no more than six feet tall, and able to pass an appropriate military physical examination. They had also to be a graduate of a service academy or possess a bachelor's in engineering, natural science, physical science, or biological science. Unlike NASA's Gemini astronauts, no civilians were permitted to apply.

Three cadres of MOL astronauts were selected between 1965 and 1967, and while some would eventually fly into orbit, this would be with NASA, not a MOL mission. The program was canceled on 10 June 1969 owing to its escalating costs and the huge economic drain of the nation's involvement in the Vietnam War. In any case, by then it had become evident that unmanned reconnaissance satellites could achieve the desired objectives much more cost-effectively.[18]

The military astronauts selected for MOL were:

MOL Group 1 (18 November 1965)
Maj. Michael J. Adams, USAF
Maj. Albert H. Crews, USAF
Lt. John L. Finley, USN
Capt. Richard E. Lawyer, USAF
Capt. Lachlan Macleay, USAF
Maj. Francis G. Neubeck, USAF
Capt. James M. Taylor, USAF
Lt. Richard H. Truly, USN

MOL Group 1. From left: Crews, Truly, Lawyer, Taylor, Neubeck, Adams, Macleay and Finley. (Photo: USAF)

MOL Group 2: Crippen, Overmyer, Bobko, Fullerton and Hartsfield. (Photo: USAF)

MOL Group 3: Herres, Lawrence, Peterson and Abrahamson. (Photo: USAF)

MOL Group 2 (17 June 1966)
Capt. Karol J. Bobko, USAF
Lt. Robert L. Crippen, USN
Capt. C. Gordon Fullerton, USAF
Capt. Henry ("Hank") W. Hartsfield, Jr., USAF
Capt. Robert F. Overmyer, USMC

MOL Group 3 (30 June 1967)
Maj. James A. Abrahamson, USAF
Lt. Col. Robert T. Herres, USAF
Maj. Robert H. Lawrence, Jr., USAF
Maj. Donald H. Peterson, USAF

REFERENCES

1. NASA *Space News Roundup*, issue April 4, 1962, *More Astronauts to be Picked Soon*, MSC, Houston, TX, pg. 1
2. North, Warren J., undated paper, *Astronaut Selection and Training*, NASA MSC, Houston, TX, pg. 1
3. NASA *Space News Roundup*, issue June 12, 1963, *Recruiting Opens for More Astronauts*, MSC, Houston, TX, pg. 1
4. Slayton, Donald K. and Michael Cassutt, *Deke! U.S. Manned Space: From Mercury to the Shuttle*, Forge Books, New York, NY, 1994, pp. 118-119
5. *Ibid*
6. *Ibid*
7. NASA *Space News Roundup*, issue June 13, 1962, *MSC Personality: Flt, Crew Ops Chief W.J. North Took Training With Astronauts*, MSC, Houston, TX, pg. 10
8. NASA *Space News Roundup*, issue September 19, 1962, *NASA Names Nine New Pilot Trainees*, MSC, Houston, TX, pg. 1
9. Smith, Robert W., *The Robert W. Smith Autobiography, NF-104.com* (incomplete). Website: http://www.nf104.com
10. *Ibid*
11. Stafford, Tom and Michael Cassutt, *We Have Capture: Tom Stafford and the Space Race*, Smithsonian Institution Press, Washington, DC, 2002, pg. 36
12. Hansen, James R., *First Man: The Life of Neil A. Armstrong*, Simon & Schuster, New York, NY, 2005, pg. 195
13. E-mail correspondence with M/Gen. Ken Weir, 10 November 2011 – 28 May 2012
14. Conrad, Nancy and Howard A. Klausner, *Rocket Man: Astronaut Pete Conrad's Incredible Ride to the Moon and Beyond*, New American Library, New York, NY, 2005, pg. 130
15. Swenson, Loyd S. Jr., James M. Grimwood and Charles C. Alexander, *This New*

Ocean: A History of Project Mercury. Chapter 4, *Man in Space Soonest?* NASA Special Publication 4201 in the NASA History Series, 1989
16. Wikipedia online encyclopedia, *U.S. Air Force Test Pilot School*. Website: http://en.wikipedia.org/wiki/U.S._Air_Force_Test_Pilot_School
17. Evans, Michelle, *The X-15 Rocket Plane: Flying the First Wings Into Space*, University of Nebraska Press, Lincoln, NE (for release Spring 2013)
18. Wikipedia online encyclopedia, *Manned Orbiting Laboratory*. Website: http://en.wikipedia.org/wiki/Manned_Orbiting_Laboratory

2

Screening the volunteers

Although NASA had the well-chronicled experiences of the Mercury astronauts to draw upon, by mid-1962 human space flight was still in its infancy. Many of those applying for the second astronaut group only had a remote concept of their long-term future within the astronaut corps, and thought they might only be involved in one or two space missions before resuming their interrupted service careers. Nevertheless, the prospect was well worth the cost of a few years.

Those candidates receiving the good news that they had reached the next phase of Group 2 selection knew that they faced the daunting prospect of full-on medical and psychological testing at the School of Aerospace Medicine at Brooks AFB, Texas.

Normally outspoken and gregarious people, pilots would readily discuss most things in order to get the answers they wanted – except when it came to physicians. They knew throughout their days of flying high-performance aircraft that any hint of a complaint relating to their well-being could attract the unwanted attention of the medics, and a temporary – or even worse, a permanent – grounding from flight duties could result. Historically then, there was something of an unspoken chasm between a military pilot and any medical practitioner. Nevertheless, it was a troublesome patch of turbulence that they had to negotiate in order to get through to the next stage of qualification.

As Marine Corps Capt. Ken Weir admits, he knew very little of space exploration or anything related to what was transpiring, so in the knowledge that he would be quizzed on this he tried to find out what he might be in for. He also went into a crash physical fitness program.

"We'd heard some reports about the tests that had been conducted at the Lovelace Clinic in Albuquerque, New Mexico in the previous selection process, but not much to hang our hats on. Another Marine at Patuxent River ordered to take the physical was Bill Geiger. Bill was a Marine Corps Officer Basic School classmate of mine at Quantico, Virginia. We flew together during our instrument training in flight school at Corpus Christi, Texas and he had graduated with USN TPS Class 26 a couple of years before my TPS class 29 and was one of the first project test pilots on the F4H-1/1F."

The two men were sent to Headquarters Marine Corps in Washington, D.C. to

Capt. Kenneth W. Weir, USMC. The large "X" on the fuselage was painted by his colleagues who felt there ought to have been a tenth astronaut chosen in Group 2, and that it should have been him. (Photo: USMC)

receive an early brief from a lieutenant colonel who happened to be the Corps' space projects officer. Unfortunately, he did not have much insight into what they should expect, or how they should prepare themselves. Instead he took them to see the Deputy Assistant Marine Corps Chief of Staff for Air, the legendary World War II hero and multiple "ace" test pilot, B/Gen. Marion Carl. The general didn't seem overly enthusiastic about the whole business and shrugged it off with a laconic, "Go down there and give it your best shot. If you don't make it, no big deal – we can use you two here with the Marines."[1]

HUMAN FACTORS IN SPACE FLIGHT

At the time the selection process began for NASA's second astronaut cadre, 34-year-old Dr. Robert B. Voas, a former Navy psychologist and author of more than thirty technical papers and articles in his field, was serving as assistant for human factors to the Director of the Manned Spacecraft Center, Dr. Robert R. Gilruth. He had been fully involved in the selection of the Mercury astronauts, and though he would have less input and association with the second group, Voas was nevertheless still a key figure in the process, as he explained:

"Human factors is a term that has developed in the last twenty years. It deals with that area involving man as an integral part of a complex mechanical system. It has to do, among many things, with the display, design and use of his instruments and controls for maximum efficiency."

Born in Evanston, Illinois, Voas graduated from high school in New Brunswick, Canada. He then received a bachelor of philosophy degree from the University of Chicago in 1946, and later his bachelor of arts, master of science and PhD degrees in psychology from the University of California in Los Angeles in 1948, 1951 and 1953 respectively. He spent his first year out of college at the U.S. Navy Electronics Laboratory in San Diego, working with human factors in the design of radar sets and sonar systems.

In January 1954 he joined the U.S. Navy "to see the world", but actually saw only a very limited part of it over the next three years by being stationed at the School of Aviation Medicine psychology laboratory in Pensacola, Florida. While there, he was involved in research on the selection and training of pilots, and determining the proper training methods to turn out successful pilots. In 1957 he transferred to the Naval Medical Research Institute in Bethesda, Maryland, conducting research on

Dr. Robert B. Voas. (Photo: NASA)

physiological responses of pilots during jet and high altitude flights. He also participated in ballistic rocket experiments involving animals.

Assigned by the Navy to NASA's Space Task Group at Langley Research Center in Hampton, Virginia on the establishment of the agency in October 1958, he assisted Project Mercury management in the selection of the first group of astronauts. He was then appointed the astronauts' training officer and coordinated the training program. He was also involved in human engineering work on the Mercury spacecraft and the spacecraft simulators.

Married to the former Carolyn Merry, and the father of a son and daughter, Dr. Voas separated from the Navy on 11 October 1961, remaining at MSC in the position of training officer. On 1 July 1962 he was named assistant for human factors, and it was in this role that he had a hand in the selection of the second group of NASA astronauts.[2]

THE FINAL 32

After trimming the considerable pool of applicants down to a manageable 32, which was the same number as the finalists for Project Mercury, those required to proceed to the next phase of the process were advised to make their way to Brooks AFB in San Antonio, Texas for medical and psychological testing at the School of Aerospace Medicine's Aerospace Medical Division.

Those who received this notification were:

Neil A. Armstrong
Lt. Cmdr. Roland E. Aslund, USN
Lt. Alan L. Bean, USN
Lt. Cmdr. Carl Birdwell, Jr., USN
Maj. Frank F. Borman II, USAF
Capt. Michael Collins, USAF
Lt. Charles Conrad, Jr., USN
Capt. Roy S. Dickey, USAF
Thomas E. Edmonds
Capt. William H. Fitch, USMC
John M. Fritz
Capt. William J. Geiger, USMC
Lt. David L. Glunt, Jr., USN
Lt. Richard F. Gordon, Jr., USN
Orville C. Johnson
Lt. Cmdr. William P. Kelly, Jr., USN
Lt. Cmdr. James A. Lovell, Jr., USN
Lt. Marvin G. McCanna, Jr., USN
Capt. James A. McDivitt, USAF
Lt. John R. C. Mitchell, USN
Capt. Francis G. Neubeck, USAF

Lt. Cmdr. William E. Ramsey, USN
Elliot M. See, Jr.
Capt. Robert M. Smith, USAF
Capt. Robert E. Solliday, USMC
Capt. Thomas P. Stafford, USAF
John L. Swigert, Jr.
Capt. Alfred H. Uhalt, Jr., USAF
Capt. Kenneth W. Weir, USMC
Capt. Edward H. White II, USAF
Lt. Cmdr. Richard L. Wright, USN
Lt. Cmdr. John W. Young, USN

A PHYSICIAN'S INSIGHT

Dr. Lawrence ("Larry") Lamb was a key scientist in the early days of America's man-in-space program. He helped to develop medical examinations specifically for

Dr. Lawrence Lamb (left, in light suit) photographed in 1957 with aviation medical researcher Dr. Hubertus Strughold and the commander of the School of Aerospace Medicine, General Otis Benson. (Photo: USAF)

the selection of NASA's pioneering astronauts. In 1959 the testing of the candidates for Project Mercury was conducted at the Lovelace Clinic in New Mexico and at the Aero Medical Laboratory at Wright Air Development Center in Ohio. But in 1962 NASA turned to a newly completed facility that specialized in such procedures. Now the School of Aerospace Medicine (SAM) would singly oversee all the test functions previously carried out at the other two centers.

Dr. Lamb, then working at SAM, became fully involved in the medical aspects of astronaut selection following a formal request from NASA for the school to conduct candidate medical examinations. The authorization of Dr. Hugh L. Dryden, Deputy Administrator at NASA Headquarters, was contained in a letter dated 19 June 1962, signed by Robert Gilruth, Director of the Manned Spacecraft Center.

By this time, as hopeful candidates for NASA's second astronaut group were busy filing their applications with the agency, SAM had been performing pilot evaluations on a routine basis.

"The medical examinations for the selection of new astronauts were well within the range and scope of the daily activities of a highly experienced, well trained team," Lamb wrote in his memoirs. "The facilities had been specifically developed for such purposes."[3]

All of the applications were initially screened by medical personnel at MSC, who worked diligently to reduce the vast number of volunteers to a manageable few. This time, three years on, there were more qualified applicants than for the Mercury selection process, as the training program at the USAF Test Pilot School at Edwards AFB had expanded. One other major development was the inclusion of Mercury astronauts on NASA's selection boards, designed to lend some first-hand experience and a sense of reassurance to the candidates, and to give them a better understanding of actual mission requirements.

Another significant difference in the selection of the second astronaut group was that NASA now required two-man crews for their Gemini series, introducing a specific need for men who were well-coordinated team players, rather than a solo, do-it-all-yourself pilot, experimenter and trouble-shooter as in the Mercury program. Flights of increasing duration undertaken in extremely close quarters with a second astronaut meant that they had to display compatibility and tolerance in abundance. In addition, more advanced and diverse science experiments could be undertaken by the crewmembers, plus complex space-related tasks that would require them to function as experimenter, observer and medical subject. Therefore engineering and any other specialized academic and technical skills in potential astronauts would come into play in the selection process.

In Project Gemini, the chosen men would have to become adept at performing intricate rendezvous techniques and conduct hazardous spacewalks, or EVA (Extra Vehicular Activity, in NASA-speak), which would be crucial in the later transition to Apollo. Moving on to the Apollo program, in which many of the Group 2 astronauts would obviously participate, a third person would be added to the crew mix; one that must possess a solid scientific and/or technical background. Piloting skills might not be quite as crucial a requirement, but this third person would have to be sufficiently skilled and trained to operate either the command module or lunar module in space if required.

INSTRUCTIONS ARRIVE

Prior to the arrival of the finalists at Brooks AFB, form letters notifying them of their appointment were sent to their base or place of occupation. This letter included instructions for filling out an extensive aeromedical questionnaire which they were to complete and return to the examining center prior to the date of evaluation. This questionnaire encompassed all the different specialties of medicine and sought, in an objective fashion, detailed information on the individual's past history, habits, and any symptomatology that they might have. The letter also included instructions to the individual's referring flight surgeon, plus information pertaining to their preparation prior to testing. This instruction sheet read:

- *First day*: Report to the Laboratory, Room 161, fasting where your blood samples and urine samples will be taken. After you have drunk your glucose water used for your sugar metabolism test you should report to Room 183 to meet Dr. Lamb, Chief of the Clinical Services Division. This will be a very short meeting. Thereafter you will return to the Laboratory for the first blood sample which should be drawn 30 minutes after you have finished drinking your glucose water. After this blood sample has been drawn you should report back to Room 184 where your picture will be taken. You should then return to the Laboratory for the next blood sample on your glucose tolerance. Following this blood sample (approximately 0915 hours) you should report to Dentistry, Room 152, where your dental x-rays will be taken. During the time you are having your sugar metabolism studies done you should not drink any coffee or smoke any tobacco. You should have reported at the Laboratory fasting and the only intake which is permitted until the test is entirely through is tap water. The rest of the day's schedule is as indicated on the Schedule Sheet that has been given to you.

  In the evening you should have no alcohol for any night preceding any of the five days of testing. The first evening you may follow your normal dietary habits with the omission of alcoholic beverages. Get a good night's sleep each night during the period of testing.
- *Second day*: You would be well-advised to eat a light breakfast if it is your custom to have breakfast and avoid an excessively heavy breakfast because you will be doing physical exertion and performance type tests. After completion of your day's testing, again, you should avoid alcohol and follow your normal dietary habit.
- *Third day*: Be sure to eat a normal breakfast and report to Building 100 for the day's schedule as outlined. During the day be certain to obtain your Lugol solution which must be taken prior to the time of your testing scheduled for the following day. This is indicated on your schedule. Also, you should be

certain to pick up some gallbladder dye pills from the Scheduling Desk in Building 100 where you reported in. At the end of this day your evening meal should consist only of tea and toast. Do not eat any eggs, cream, milk, butter or anything which has fat in it. Do not put butter on your toast. Beginning at 2000 hours, take the gallbladder dye tablets, taking one for each five-minute interval with a small amount of water. Do not smoke after midnight and remain fasting from 2000 hours on.

- *Fourth day*: Do not eat breakfast or drink coffee or water. Report to x-ray in the fasting state. Do not smoke. At 1000 hours return to your normal dietary habits and smoke if you so desire. After the completion of the day's schedule you may eat a normal meal, avoid alcohol and do not drink or eat anything after midnight.
- *Fifth day*: Report to Room 116, Building 100, to drink your heavy water to be used to determine the amount of water content in your body which enables calculation of percentage of body fat. You should report in the fasting state. Do not drink any water or eat. At the time of reporting you should urinate and evacuate your bowels prior to taking the heavy water. After you have drunk the heavy water in the presence of the technician, he will note the time it has been administered. Once the water has been drunk it is absolutely essential that you do not eat or drink anything and that you do not urinate or move your bowels until after your blood sample has been obtained in Room 116 four hours after the completion of administration of the dose. The technician will notify you of the time you should report back to his room for the blood sample. Be certain to keep this appointment accurately. After this blood sample has been withdrawn you may return to your normal dietary and beverage habits.[4]

PREPARING FOR SAN ANTONIO

In order to raise his fitness level ahead of the physical tests, Ken Weir began running three miles every day. Someone had helpfully mentioned hearing that the testing physicians turned the candidates upside-down for long periods of time, so when he came home he would lie on the stairs, feet up and head down.

Word also filtered through that physicians had the candidates place their hands in buckets of freezing ice water for several minutes while hooked up to a blood pressure cuff and an electrocardiograph machine. Every day, therefore, Weir emptied the refrigerator of ice cubes, tipped them into a bucket of water and shoved his hands in for 20 minutes. However, "I later discovered this test meant planting your feet, not your hands, in freezing cold water for seven minutes to test your tolerance to extreme temperatures."

When the candidates arrived at Brooks they were assigned billeting alphabetically. Weir shared a suite with Capt. Ed White, a 1952 West Point graduate and USAF test pilot who was in amazingly good physical condition, the result of being a track star at the academy. Every day, White donned his shorts and gym shoes and ran the entire perimeter of the base.[5]

"It was hot and humid, typical August weather for Houston," recalled Tom Stafford, "and I was assigned a room in the Ellington Air Force Base BOQ with a civilian test pilot named Jack Swigert. There were thirty-two of us by this time, including friends from Edwards like Frank Borman, Mike Collins, Greg Neubeck, Ed White, as well as some familiar Navy faces, such as Pete Conrad and John Young, and Jim Lovell, a classmate from Annapolis."[6]

Weir soon learned that among the 32 finalists were another two fellow Marines, Captains Bill Fitch and Bob Solliday, making four Marine Corps candidates in all. Solliday had been one of the 32 finalists in the Mercury selection process and was well versed on what was taking place. As Weir recollects, "He was well known by the Navy and Marine members of the original seven astronauts, and appeared to have an inside track to being selected. However he was very guarded and reluctant to share any knowledge or experience he possessed with other candidates.

"I also ran into Pete Conrad, a Navy test pilot who was leaving Patuxent River for a new F4H squadron just about the time I finished Test Pilot School. Pete had also been one of the 32 finalists for the first group of astronauts and I was eager to pick his brain to try and learn as much as I could to be prepared for the screening. He said, 'Don't sweat it; come on, let's go over to the swimming pool and have a few beers,' and so very casually we did. Pete told me they probably already made their selections and just needed to make the whole process look like it was a fair evaluation."[7]

Or perhaps the canny Conrad was simply trying to lull the competing candidates into a relaxed state of preparedness for what he knew from experience would be a particularly excruciating period of tests, tests and even more tests.

SCHOOL OF AEROSPACE MEDICINE

"The potential astronauts came in groups of two or three at a time, on a daily basis," recalled Dr. Lamb. "Their eyes were bright with hope, their faces well-scrubbed, and their fingernails clean. Each was the epitome of friendliness, with an outgoing personality. Clearly, each applicant regarded this as one of the most important events of his life."

He already knew several of the 32 candidates from previous visits to SAM. "Two of the prospective astronauts, who had just recently completed examinations, came into my outer office and peered around the door at me with mischievous looks on their faces. They wanted to eliminate part of their examination procedures – specifically those which they didn't like and had recently had. This amused me, but since their request was completely legitimate, I discussed their request with the MSC physicians, and they agreed it was not necessary to repeat those tests after such a short time interval."[8]

Since its creation, the School of Aerospace Medicine had accumulated volumes of data significant to the evaluation of flying personnel. This resulted from an extensive research program coupled with the development of a referral consultation service to evaluate aeromedical problems within the Air Force. Since July 1959 the majority

Group 2 Candidates' Arrival at the Aerospace Medical Center, San Antonio, Texas, July–August 1962

July 9	Stafford
July 10	McCanna, Mitchell
July 11	Borman, Dickey
July 12	Collins, Uhalt
July 13	Lovell
July 16	Johnson, Fitch
July 17	Geiger, Gordon
July 18	Glunt, McDivitt
July 19	Kelly
July 20	Neubeck
July 23	Edmonds
July 24	Ramsey, See
July 25	Smith, Swigert
July 26	Birdwell, Solliday
July 27	Aslund
July 30	Weir, White
July 31	Wright, Young
August 1	Armstrong, Conrad
August 2	Bean, Fritz

of graduates from the USAF Test Pilot School had undergone extensive evaluation at the school, with their findings being used specifically for the selection of Air Force space pilots. By the time that the 32 finalists arrived at Brooks, a prodigious amount of information had been obtained, based on medical findings in flying personnel as related to normal population values.

Basically the same aeromedical techniques as used for testing the finalists during the Mercury astronaut selection process would be applied for the second group at the SAM facility, instead of being split between the Lovelace Clinic in New Mexico and the Aero Medical Laboratory at Wright-Patterson AFB, Ohio. As before, many of the tests would involve closely monitoring subjects during maximum exertion exercises. This data was recorded on magnetic tape and later used in evaluating the individual physiological variations of each subject. The ultimate goal of this comprehensive evaluation was to identify those men most likely to complete astronaut training and participate in space missions on a long-term basis without any medical impairment, either physical or mental.

There were four phases in the candidate evaluation. The first phase involved the detection of any significant disease processes or medical abnormalities. The second phase revealed any predisposition to diseases, or other medical problems that might limit performance capability – such as obesity and borderline glucose intolerance, classified as probable diabetes. The third area under examination was concerned with the assessment of an individual's mental and dynamic characteristics. This phase of the tests included a complex evaluation of the candidate's motivation, dependability, judgment, intellectual capability, learning aptitude, emotional adaptability, and maturity. Then there was the fourth phase. This placed considerable emphasis on a subject's physiological capacity and was conducted using a number of procedures, as recorded in the July 1963 paper on 'Aeromedical Evaluation for Space Pilots', edited by Dr. Lamb.

> This is accomplished by different procedures such as the study of the individual during maximum exertion, the use of the tilt table studies for automatic control of the central nervous system, and combinations of physiological stresses including common respiratory maneuvers such as hyperventilation, breath holding combined with orthostatic influences. By maximum stress testing such as maximal physical effort combined with suitable measurements which record the function of the heart as a pump, the fluid dynamics of the circulatory system and the ventilation of the lungs, the physiological capacity of an individual subject can be expressed. This requires the simultaneous recording of multiple biological signals during periods of stress testing or a dynamic approach. This is in direct contradiction to the usual clinical situation in which an individual presenting with disease is studied in the resting or idling condition and presents a disability that is apparent at rest or during idling circumstances.[9]

The SAM examiners came to the conclusion that while none of the candidates was considered a perfect specimen of health, as a group they were healthy enough. The principal objective of the physicians was to ensure that each of the men had good eyesight and hearing, and had no difficulty in maintaining their balance – important at a time when so little was known about the effects of weightlessness on a person's sense of position and balance. Some of the candidates were also carrying a few extra pounds, and many did not have ideal cholesterol levels, although more than half of the men did have levels below 200. However, those who were a little overweight or whose cholesterol levels were high – or both – were informed this was not altogether serious and could be overcome with a little diet and exercise. The physicians were fully aware that excess body weight did not necessarily correlate with cholesterol levels.

"One striking example was a topnotch test pilot I had seen several years earlier," Lamb wrote. "Originally, his cholesterol level was 300, even though he only weighed 155 pounds. We talked, and he began a better diet and a daily exercise program. Two years later, although he still weighed 153 pounds, his cholesterol had decreased to 173 .... Much later, at age 50, he was the command pilot for one of the nation's important space flights."[10] The pilot alluded to by Dr. Lamb was Gemini and Apollo astronaut John Young, who commanded the first space shuttle mission in 1981, at 50 years of age.

What the physicians did discover, was that a number of candidates had somehow slipped through the MSC screening by concealing certain health problem areas in the desperate hope that these would not be discovered by the expanded testing process in San Antonio. According to Dr. Lamb one candidate, who otherwise seemed to be in excellent health, failed the radiology results when X-rays revealed anomalous opaque tracks in his thigh and groin areas. Further investigation revealed that he had earlier been treated for a malignant growth, and the opaque tracks were due to injections for chemotherapy.

There had been some relaxation in the required standards and qualifications from the first group of astronauts, much to the relief of the candidates. As an example, the slightly increased volume aboard NASA's second-generation and third-generation spacecraft meant that the Project Mercury size restriction could be slightly increased. This allowed for six-foot astronauts rather than the 5 foot 11 inch limit imposed on earlier candidates. But this still posed problems, as Dr. Lamb recalled.

"One candidate, who was barely over six feet, spent the day jogging and exercising in hopes that by the end of the day he would be shorter. People usually are taller when they first get up than they are at the end of the day – yet another influence of gravity. The body weight gradually compresses the small discs between the vertebrae, like compressing a pillow. A person can be as much as an inch shorter by bedtime compared to when he first got out of bed."[11]

A FEW FACTS AND FIGURES

It was interesting for the aerospace medical staff to note that in 28 of the 32 instances the candidates were the first-born male child, and 21 of the candidates were the first-born sibling. Three of the men were the only child in the family.

The total number of flying hours for individuals in the group ranged from 2,100 to 4,500 hours including military and civilian flying, with an average of 2,913 hours of total flying time. Four of the men flew fighters in combat during the Korean conflict.

With such a large accumulated flight time, a number of in-flight incidents would be expected. These were as follows:

Number of Accidents	Number of subjects
None	15
One minor accident	4
One major accident	8
One major, one minor accident	1
One major, two minor accidents	1
Two major, one minor accidents	1
Three major accidents	2

Relative to the accidents, five of the candidates had successfully ejected from crippled aircraft and one (Bill Geiger) had bailed out. The reasons for the emergency escapes included mid-air collisions and in-flight fires, and one loss of control due to a systems malfunction.

Twelve of the candidates were non-smokers. Eleven had never smoked, and one only smoked for six months in his late teens. The remaining 20 men smoked between 10 and 40 cigarettes daily, with an average of 20 per day. Two smoked cigars only; eight per month for one candidate and one to two per day for the other. Two men smoked two to three cigars and a pack of cigarettes daily. Another two smoked cigarettes and pipes; one smoked a quarter packet of cigarettes and four pipes daily, and the other smoked a half packet of cigarettes and two pipes daily. There were no candidates who smoked pipes exclusively.

The usual childhood diseases were commonly reported by the men. Four of them had experienced episodes of aerotitus media (inflammation of the ear caused by pressure changes) as a result of flying with a cold. Two had mild allergies during childhood. The majority had suffered episodes of spatial disorientation or "aviator's vertigo" while flying under a cockpit training hood or in instrument weather.[12]

THE TESTS BEGIN

Once the candidates had settled in, the testing phase swung into full operation. In addition to the usual cardiograms and other methods of recording physiological data, the candidates were subjected to hypoxia tests to determine their reaction to being starved of oxygen, spun in a barany chair in a dark room to test their tolerance to motion sickness, and told to drink a quart of glucose as a preliminary to having their blood taken every hour in order to monitor their blood sugar levels.[13] Some of the information Ken Weir had gleaned beforehand proved to be correct.

"They put us on a tilt table upside-down and all around. Put our feet – not our hands – into freezing ice water for several minutes that seemed like Arctic hours. Hooked us up to the electrocardiograph machine, the blood pressure cuffs, and placed us onto a treadmill which they could gradually raise at one end to constantly increase the incline. We were evaluated until they observed a problem, or someone's blood pressure or heart rate reached outlandish limits.

"One other applicant and I (not Ed White, the Army track star) maxed out the treadmill test, staying on the contraption up to the maximum angle for a total of 26 minutes. I was informed, rightly or wrongly, that only John Glenn among the original applicants had accomplished that feat, and I was suddenly very proud and felt somewhat competitive. Little did I know."[14]

John Fritz, who underwent testing along with fellow General Electric test pilot Elliot See, said he "treasured" the opportunity to compete with Gemini aspirants such as his good friend from Edwards AFB, Neil Armstrong, "and to meet the original Mercury 7 astronauts including Alan Shepard and John Glenn". He said he also took great pride in eclipsing the time on the treadmill said to have been previously held by Glenn.[15]

A later astronaut once wrote that one of the tests was not to anyone's liking. "We were also subjected to the pretty unpleasant experience of having ice water poured in one ear to see how our inner ears reacted to the imbalance of the inner canals in one ear being warm, while those in the other ear were so cold. It was very disorienting. The brain didn't know how to cope with these mixed messages. It made your eyeballs spin pretty wildly."[16]

Yet another test was a carryover from the Mercury examinations, involving a vexatious device that was officially known as the Complex Behavior Simulator and nicknamed the "idiot box" by earlier subjects. Each candidate was seated in front of

Five-Day Schedule

	DAY 1	DAY 2	DAY 3	DAY 4	DAY 5
0800	Lab. Rm 161	Tilt Table Treadmill Rm 114	EEG Rm 171	X-ray Rm 145	Tot. Body Water Rm 114, 0800 Dentistry Rm 152, 0815 Pul. Func. Rm 114, 0830
0900		Valsalva Cold Pressor Rm 114	CBS Rm 171, 0930	Sigmoidoscopy Rm 114 Blood Vol. Rm 114, 0930	
1000	Int. Med – History & Px Rm 114	X-ray Rm 146	Aptitude Testing Rm 171, 1030	Ophthalmology Rm 136	Dr. Flinn Rm 171, 1000
1100		Ophthalmology Rm 136	Aptitude Testing Rm 171	Neurology Rm 171	Avn Med. Rm 103
1200	LUNCH	LUNCH	LUNCH	LUNCH	LUNCH
1245	Int. Med/EEG Precordial Map Rm 114	Ophthalmology Rm 136	Psychiatric Interview Rm 171	Clinical Psychological Testing Rm 171	
1345	Special Studies Double Master Rm 114	As Above	As Above	As Above	
1445	Vectorcardiogram Phonocardiogram Rm 114	As Above	ENT Rm 131	As Above	
1545	Plethysmogram	As Above	ENT Rm 131	As Above (to 6:00 P.M.)	

the multi-faceted device with a group of technicians looking on, pens poised over their memo pads. The candidates were required to initiate different responses to each of fourteen signals which appeared in random order at increasing rates of speed and complexity. Towards the end, the tasks became so rapid and complex that it was almost impossible to keep up. However this was the whole point of the exercise – the physicians were interested in a candidate's ability to organize their behavior and maintain emotional stability while under increasing stress. The simulator totally frustrated some candidates, but others recognized what it was all about and simply did their best without getting flustered. As Frank Borman recalled, "It would have fuddled an intelligent octopus."

A CANDIDATE DEPARTS

On 10 August, NASA Administrator James E. Webb informed a subcommittee of the Senate Committee on Appropriations that 32 candidates for astronaut positions had been selected, and that by 1 September ten of those men would have been chosen to enter NASA astronaut training.

Ken Weir's good Navy friend and mentor "Tex" Birdwell of the USAF Aerospace Research Pilot School was one of the 32 finalists. All too soon, however, and despite his best efforts to fall within the NASA-imposed height limitation, the tests at SAM revealed he was actually a little over six feet tall, which immediately brought about his disqualification from further testing. "He was sent back to Edwards the second day because he exceeded the height limits," Weir recalled. "He tried everything he could to scrunch down to the limit but they would have none of it, and the list was down to 31."[17]

Birdwell's son Bob revealed that his father was actually two inches too tall to qualify. "After he retired I asked him about it and he told me he was kind of glad he didn't make the cut because many of the astronauts had family problems, not to mention all the public relations nightmares. It was the same reason he twice turned down joining the Blue Angels demonstration team."[18]

CARL BIRDWELL, JR., USN

Carl ("Tex") Birdwell, Jr., was born on 6 January 1928 in Stephenville, Erath County, Texas, the son of Emmett ("Carl") and the former Oneta ("Peggy") Wall, and was an older brother to John and Mary Ann.

After graduating from Texas A&M College, Birdwell joined the Navy in 1945 as part of the V-5 Aviation Cadet Program and was commissioned as an aviator in July 1948. He was stationed aboard four aircraft carriers from 1948 to 1952, assigned to squadrons VC-21 and VS-21.

He was married on the morning of 21 December 1949 in Rancho Santa Fe, California, to Jean Mary Hadley, a 1946 graduate of Coronado High School.

Birdwell served a tour in Advanced Training Command as Flight Officer, Power Plants Officer and Flight Instructor at NAS Corpus Christi, Texas. In August 1952 he began a three-year bachelor's in aeronautics at the U.S. Naval Postgraduate School in Monterey, California. He then furthered his education with a degree in aeronautical engineering from the California Institute of Technology.

After completing jet transition training in July 1955, he was assigned to VF/VA-122 where he performed duties as Maintenance Officer and Operations Officer. In

Capt. Carl "Tex" Birdwell, Jr., USN. (Photo: USN)

February 1959 he graduated with distinction from the Navy Test Pilot School and was assigned as project test pilot to the Flight Test Division, Naval Air Test Center (NATC), Patuxent River, Maryland.

In June 1962 Birdwell was selected as one of eight experienced test pilots to attend the first Military Astronaut Course of the USAF Aerospace Research Pilot School (the former USAF Test Pilot School) at Edwards AFB, California, becoming the first Navy aviator in an ARPS class. On entering the school, Birdwell declared, "I'm vitally interested in further space research, which I've been thinking about most of my life. This new course seems to be a natural progression of the test business that holds my closest interest."

From February to May 1963 he was under instruction in VF-124 and VA-125, and then was assigned as Operations Officer to Commander, Carrier Air Wing 2.

From December 1964 to August 1968 he served as Executive Officer and Commanding Officer of Attack Squadrons 216 and 122. During two tours in Vietnam he led 177 combat airstrikes from the aircraft carrier USS *Hancock* (CV-19), including the first air attack on Dong Hoi in February 1965. In July 1966 he led an assault on oil storage facilities in Haiphong. His next assignment was as Project Manager for the A7 in the Naval Air Systems Command. He was commanding officer and chief test pilot of VX-5 at China Lake from September 1970 to August 1972, overseeing the flight testing of new air-to-ground weapons and electronic countermeasure systems. He also directed a team of engineers and technicians in the restoration of an F4U Corsair aircraft to flight readiness and later exhibited it at the Reno Air Races and at Tailhook Association reunions.

In September 1972, Capt. Birdwell became Director of the Test Pilot School at Patuxent River, where he oversaw the testing of 36 different types of aircraft. His final assignment came in January 1975 with the NATC Staff to fill in the new Plans and Programs Officer billet.

Birdwell was certified in 150 types of aircraft during his 30-year military career, and he logged more than 7,000 hours of flying time and 800 aircraft carrier landings. His many commendations included a Silver Star, Legion of Merit, the Distinguished Flying Cross with three gold stars, a Bronze Star and sixteen Air Medals.

For three years after leaving the military and moving to Bradyville, Tennessee in 1976, he instructed flight mechanics at the University of Tennessee Space Institute in

A later portrait of Captain Birdwell. (Photo: USN)

Tullahoma. He then operated a milk goat farm in Cannon County, Tennessee. During his last 20 years, Capt. Birdwell frequently visited Coronado, California, where the family owns property. He was also active in Alcoholics Anonymous, serving as the organization's Tennessee Assembly chairman.

After a long illness, Capt. Birdwell died on 12 April 1996 aged 68 at his home in Bradyville. He was survived by his wife and three sons, Carl, Hugh and Robert, sister Barbara and her husband Tex Earp, brother John and his wife Joyce, and four grandchildren. Following a military memorial service at NAS North Island his ashes were scattered at sea from the carrier USS *Kitty Hawk* (CV-63).[19]

FACING THE PSYCHIATRISTS

As part of their final psychiatric and psychological evaluation, each of the remaining 31 candidates was interviewed by two psychiatrists who had earlier scrutinized the men's records and written questionnaires. The first interview was two hours in duration and the second lasted for one hour, during which the psychiatrists pursued any areas warranting further attention. Although the interviews were nondirective to some extent, in that no formal list of questions was used, each examiner attempted to assess certain areas which had earlier been agreed on. These were:

1. **Review of flying career and experiences:** Original and current motivation; adaptability during training; major goals and reasons for changes; evidence of outstanding or ineffective performance; reaction to frustrating experiences; quality of relationships with co-workers and supervisors; reaction to competition and failure; adaptability to combat, test flying and other stressful experiences.
2. **Motivation for space flight:** Expectations; realistic versus unrealistic; pros and cons considered; quality and quantity of motivation; current job satisfactions; alternate goals.
3. **Marital history:** Marital adjustment; wife's attitude toward job; current situational problems; family adaptability to past transfers and separations; causes of marital discord, and response to them.
4. **Developmental history:** Early relationship with parents and siblings; causes of intrafamily tensions and applicant's response or participation; early educational history, academic achievements; social and sexual adjustment during and after adolescence; non-vocational interests.
5. **Psychiatric history:** Hospitalizations or consultations; symptom review; use of alcohol.
6. **Current situation:** Family relationships; social and recreational interests; interpersonal relationships.[20]

The testing psychiatrist asked all of the finalists a battery of probing and often highly personal questions, after which they did the entire Rorschach inkblot test. As Frank Borman relates, "We had to endure a battery of psychological tests, my first full encounter with the mysteries of psychiatry. There were the usual ink-blot tests, which implanted in us the fear that if something looked like a tree, it was a definite indication that we were sexual deviants."[21]

Ken Weir also remembers the Rorschach test, as well as being shown a blank piece of white 8 inch by 11 inch paper as part of the MMPI (Minnesota Multiphasic Personality Inventory) test. Like Borman and the others, he was asked what he saw in it. He had earlier been informed that when Alan Shepard was shown the piece of paper during his Mercury tests he said it looked to him like a snow storm. However, Weir took a far more pragmatic approach and told the psychiatrist that it was just a white piece of paper, a response which he said seemed to dumbfound the examiner. "After what seemed like an eternity, as he scribbled on his record sheet, he once again asked what I envisioned it to be. I gave him some story about a horse-drawn buggy at a railroad train depot in the Midwest with a steam locomotive and passenger cars as he was feverishly writing down everything I said. He then wanted to know what kind of sports I was interested in, and ones I had played. I listed them all and when I mentioned golf he stopped me and asked if I was any good at it. I said about bogey-like. At this he asked if I liked to score well, or hit the ball far. Next up, he wanted to know if I liked to hit the ball hard. When I responded 'Yes, very much so,' he suddenly asked me who I was hitting hard – my mother, father, sister, boss or anyone else. Altogether it was a confusing, outrageous, bewildering and often tedious ordeal."[22]

DREAMS AT AN END

As part of the test regime, candidates had to present themselves for otolaryngology tests, more commonly known as ear, nose and throat, or simply ENT examinations. These were considerably more detailed and comprehensive than the routine type of examination. First of all, the ENT specialists would peruse each candidate's survey form to check any medical history pertaining to their ears, nose or throat. This was followed by an acutely thorough physical examination. Finally, based on the results, the otorhinolaryngological examiners would have to place each candidate into one of four categories:

1. Recommended, best qualified
2. Qualified with reservation
3. Qualified
4. Not recommended

At the debriefing end of his physical, the flight surgeon showed Ken Weir an X-ray of his spine and said he was a little concerned about what might happen to a couple of his lower vertebrae in the event of an ejection. As far as Weir knew, NASA had no plans to include ejection seats in its spacecraft, so he could not correlate the significance. (In fact, the two-man Gemini spacecraft would include ejection seats. Fortunately, apart from one close call at launch, these were never used.) Then, when Weir was shown X-rays of his sinuses the flight surgeon did not indicate there was anything noteworthy about them. "And that was it: no comments about my inverted T-waves, hypertension or psychiatric evaluation."[23]

With the completion of the ENT tests and a great deal of deliberation, four of the candidates were placed in the fourth category, with a negative recommendation for their participation in the astronaut program. This recommendation was made because of physical findings which were disqualifying for Class I flying, according to the standards of Air Force Manual 160-1. Based on the best judgment of the examiners, the selection of those candidates would have compromised future monetary and training investment.

The following table demonstrates the examiners' findings:

ENT category	Examination findings	Number in group
I	Negative physical findings	18
II	Mild, high frequency deafness; mild, asymptomatic deflection of nasal septum	6
III	Moderate, high frequency deafness; minor, asymptomatic physical findings	4
IV	Moderate to severe deafness, paranasal sinusitis, otosclerosis	4

In the Aerospace Medical Division's closing summary, the following points and recommendations were made:

> The 32 space pilot candidates examined impressed the professional staff of the Otolaryngology Department as being the most eager, highly motivated individuals they have had the pleasure of evaluating. An ear, nose and throat rating device based on the overall examination was utilized in classifying the candidates and of the 32 examined, only 4 or 12.8 per cent of them could not be recommended for continued space pilot training.[24]

Those four candidates – not named in publicly accessible reports – were quietly told they could proceed no further. For them, their dream of one day flying into space had come to an abrupt end.

As Dr. Lamb later observed about these tests, the School of Aviation Medicine did not select the astronauts for NASA – that was not their job. Their function, as he stated, "was to provide a competent medical examination and evaluation. The staff did, in each instance, make an assessment of the significance of any observations they had made, but this was simply to provide information. The selection of future astronauts was made by the managerial staff at MSC after interviews and evaluation of the candidates' abilities as pilots and engineers."[25]

BACKGROUND CHECKS

The physical and psychological tests were not the only factors in determining a candidate's suitability. Like the Mercury astronaut candidates, they also had to pass stringent security checks. Ken Weir continues:

"All the time NASA security and FBI agents were conducting background investigations, the likes of which I never ever imagined. I already had a top-secret nuclear weapons codename clearance, but they were much more thorough than ever before in searching out any skeletons in my family closet. They were digging everywhere to find out everything they could about us.

"My neighbor across the street in Town Creek, Maryland was a Naval Academy graduate and the outstanding test pilot graduate in our Test Pilot School Class 29. He was a couple of years senior to me and we were very good friends, both having been assigned to the Flying Qualities and Performance Branch of Flight Test following our graduation. I was assigned to the F4H and he had been assigned to the OV-1. He was so cerebral that the USN Test Pilot School prevailed and had him reassigned back to the TPS staff as an instructor. When his wife was interrogated about me as a possible astronaut, the only comment she had was that, 'He really likes to entertain folks by always telling big stories that eventually turn out to be pure BS.' End of her interview – and probably me as well."

OFF TO HOUSTON

For an anxious Ken Weir, it seemed like he had to wait forever for the next step in the selection process: the examinations and personal interviews at NASA's Manned Spacecraft Center in Houston, Texas. He kept checking with the commander of the Marine Air Detachment at Patuxent River to find out when and what came next. But time dragged on and on without any news. Headquarters Marine Corps was queried almost every other day on behalf of the four applicants. They checked with John Glenn, but he did not know and could not find out either. Weir said they were told he was anxious to take his family on leave but could not do so because he wanted to be involved in the selection process and interviews and hence kept putting his leave off. Finally Glenn decided to go ahead and take leave and requested that he be notified of the date of the selection interviews and exams, saying he would return immediately to Houston to participate.

Curiously, almost as soon as Glenn departed on leave, notices went out to all 27 remaining finalists to arrive at Ellington AFB outside Houston on a certain date. Apparently he was not notified as he had requested. The finalists were instructed to wear civilian clothes and so, because he did not have anything that even resembled suitable civilian attire, Weir went to a small clothing store in Lexington Park outside the NAS Patuxent River main gate and purchased the cheapest suit he could find that came closest to fitting him. He was not the only one to have to do this, as most of the service applicants hadn't had any need of civilian business attire for many years. As Weir had just joined the Society of Experimental Test Pilots, he proudly placed his silver associate member's pin on the lapel. Although he now felt he was ready for anything, he was nevertheless filled with apprehension. "I had no advice or guidance on what to expect, and was not smart enough to figure any of it out," he reflected.

On the flight to Dallas he ran into and sat next to Hank Lankford, a highly-respected Chance Vought F8U test pilot he knew and had admired from his days flying Crusaders at El Toro, the Marine air base in California. Lankford pressed hard to learn what Weir was up to and eventually the truth slipped out. His friend did not even know NASA was looking for more astronauts.

At Dallas-Fort Worth airport, where he changed planes for Houston, Weir's father and mother, who had retired and built a farm in East Texas, were there to meet him. They had attended his Test Pilot School graduation the previous October and were truly surprised when he revealed what he was trying to do. His father wasn't sure that the United States could afford expending all that money on the space program, but he was nevertheless proud and interested.[26]

FACING THE SELECTION PANEL

On arrival the 27 candidates were billeted in the Bachelor Officers Quarters (BOQ) at Ellington AFB, for lack of any better accommodation. These barracks also served to facilitate the task of NASA escorts and security personnel in keeping the men's presence under wraps. Each group of three or four candidates had one security escort and vehicle assigned to them, although, as Air Force candidate Bob Smith recalls, they did not have the place to themselves. "A strange sidelight for a deactivated and closed base was that the barracks and mess hall were opened for us plus the Houston Oilers professional football team, of the fledgling American Football League, who were in residence at the base for their spring training. The old adage that politics makes strange bedfellows sure held there, but I suspect the base was under state control at that time."[27]

Once they had settled in, the men attended briefings, went on orientation visits to various sites and took extensive written exams before their turn came to suffer the dreaded interview. They would have the occasional encounter with one or two of the original seven astronauts but several of the candidates, particularly the members of the Marine Corps and some of the Navy, were still wondering where famed astronaut John Glenn was. When they inquired, they were told to their disappointment that he was on leave. Unimpressed, the four Marine candidates, Fitch, Geiger, Solliday and Weir felt he ought really to have been present to participate in the selection process.

"Everyone was treated as equals and military protocol was seemingly not observed," according to Weir. "Except for one Air Force candidate who was a couple of years and one grade senior to the rest of us; very much impressed with himself, his status and position, and did not hesitate to throw his weight around.

Occasionally he even ordered junior Air Force officers to go and do such-and-such that he wanted done.

"He seemed to be very confident he was a shoo-in and he let the rest of us know about it. I thought he was pompous and found his constant probing into my background a bit tiring. In front of several candidates he asked me point-blank how I ever got included in the group with so little meaningful qualifications. I told him that it was a well-known fact that I was there because the Director of the NASA Manned Spacecraft Center, Dr. Robert Gilruth, was my uncle. It was totally untrue, but he didn't pester me anymore and watched me like a hawk from then on. It was very apparent the rest of the Air Force candidates did not have much respect for him, but he was indeed one of the nine picked."

Neil Armstrong was already an experienced NASA test pilot at Edwards and it appeared to the rest of the candidates that he was a slam-dunk, pre-selected astronaut for the program. They did not interface with him much during the process, except on a social basis. He was recognized as a very unassuming gentleman and became very well liked and admired by the rest of the candidates for years to come.

Without doubt, the most daunting part of the entire exercise for Weir was the day he had to undergo his interview. "I was waiting to go before the interview board when I bumped into Tom Stafford, who had finished his interview, and asked him how it went. He said it went quite well and he seemed very confident at the time.

"When I went into the interview room, Warren North, Deke Slayton and Alan Shepard were seated behind an elevated dais. I was offered a seat well below their vaulted level and very much forced to look up at all three of them from the seemingly short-legged chair centered before them. All three were puffing on cigars and the room was heavy with cigar smoke. I was wearing my lowly SETP associate member lapel pin to show them I was a true test pilot. Slayton and Shepard – both long time full SETP members – were wearing their astronaut lapel pins. I was more impressed than they were."

Slayton asked Weir why he had applied for the astronaut program. He replied that it was a great opportunity for him to accomplish something truly meaningful with his life, and in some small way to make a significant contribution to the exploration of space. From there on he says he really got grilled, mostly by Shepard, who wanted to know exactly what Weir understood about space operations and how he learned about them.

"I told him what little I knew and that I had learned about it by reading as much as I could in the papers, magazines and books. He asked me what books I had read and I could not recall which ones I had looked at. He then asked me twice who Jules Verne was and what he wrote. He asked me who wrote *Twenty Thousand Leagues Under the Sea*, *A Journey to the Center of the Earth*, and *Around the World in Eighty Days*. To each one I had to admit I had no idea. Shepard then asked me my class standing upon graduation from the Naval Academy. I did not know the exact number but I assumed he must have it in front of him, so I admitted to being in the lower quarter of the class. He wanted to know who authorized my being the first test pilot to spin the F8U-2NE airplane. I admitted that no one had authorized the spin. I was performing some tests in the vicinity of NAS Oceana-Norfolk and had been jumped by an A4D. I promptly got the better of the A4D by raising the wing and using afterburner in the rat race until my F8U stalled and entered a spin from which I was able to recover by lowering the wing and letting the airplane arrow itself back into controlled flight. While Shepard appeared to be unimpressed with my aeronautical skills, Slayton did acknowledge my treadmill performance. Warren North seemed to be out of his element and just sat there listening to all of it."

Weir was finally excused and told to wait outside. After a short wait he was recalled and told that they had no further questions; did he have any for them? Having none, he was again dismissed. "When I left that interview I knew I did not have a prayer of being selected by those three folks to be one of the next generation of astronauts."[28]

When all the tests and interviews of all the candidates were completed, NASA had a "graduation" reception get-together with Dr. Gilruth and other NASA notables. A few astronauts were in town, but still no John Glenn. Bob Smith said he had one unsettling conversation at this time. "I remember, before we left Ellington after the final events finished, NASA gave us a party [at which] Gus Grissom told me how little they got to fly, except in travel in a T-38 jet trainer, how much time they spent in engineering reviews of the spacecraft and that he had been home only five days in a year. That didn't deter him but it certainly was not a lifestyle for everyone, and not for many families, which is one of my consolations."[29]

A LIFE-CHANGING PHONE CALL

On returning home, Weir was convinced that he had been through one of the most harrowing experiences of his life up to that point, and was certain that he would be eliminated from the future astronaut group.

"After several weeks waiting for some sort of notification, the phone rang. When I answered, a voice on the other end identified himself as a NASA employee whose name I did not recognize. First of all he said, 'You have been selected,' and then, as my heart momentarily skipped a beat, he paused and uttered the dreaded words, 'You have not been selected for the follow-on astronaut program.'"

Presumably the caller was Deputy Associate Administrator Charles Donlan, who normally made the "unsuccessful" calls. Deke Slayton rang those who had been selected. Weir thanked the person for the call and told him it was an honor to have been considered. The caller said he had other notifications to make, said 'Goodbye' and hung up.

"I was indeed disappointed but I knew when I departed Houston that I had not made a strong, positive impression and did not have any chance unless everyone else encountered the same experience I did. In discussions with others, few would admit to having not done well."[30]

John Fritz says he still feels the acute disappointment of missing out. "September 17, 1962 was the day NASA was scheduled to announce the successful applicants for the Gemini program, and [it] proved to be the most painfully traumatic of my life. Everyone at Edwards was awaiting word, since the candidates included military,

NASA and civilian pilots from that community. Early that afternoon, friend and fellow GE test pilot Bill Todd had a call from Col. Walt ("Danny") Daniels, Chief of Fighter Ops, who advised he had seen the list and 'Fritz was on it.' Bill rightfully expressed some skepticism since I did not have an engineering degree, but rather a BA in life science. Danny reiterated that he'd seen the list and once again said 'Fritz was on it!' Bill naturally shared this exciting news with me.

"Bill, Elliot See and I were in the GE pilots' room when the phone rang and our secretary advised Elliot that NASA Houston was calling. He picked up the phone and said, 'Oh, hi Deke!' He listened intently, and then said, to my amazement, 'Well, gosh, I'll have to think about it!' I darn near grabbed the phone to tell Deke that I *wouldn't* have to think about it!

"NASA advised the finalists that those chosen as Gemini astronauts would be called by one of the Mercury astronauts, and those who didn't make the cut would be notified by letter. Elliot had received his call. As I drove home an hour later I was an extremely disappointed astronaut 'wannabe'.

"At dinner that night at home in Lancaster with wife Dottie and toddler sons Greg and Jim, the phone rang. It was Houston, calling for me! I figured my notifying astronaut had been delayed for some reason, and this was my invitation to join the program. Unfortunately I was wrong. Instead, it was a member of the Houston astronaut office who said they decided it would not be fair to those not selected to wait for a letter, so they called everybody that evening. With that last cruel twist from elation to wrenching disappointment I went out into our backyard and looked east towards the rising full Moon just above the horizon, knowing my dream of ever walking there was dead forever. Some dreams die hard.

"I learned later that I was indeed on the list of selectees from NASA Houston, but was scratched by NASA Administrator James Webb, for lack of an engineering degree.

"Tragically, Elliot See and his assigned Gemini partner, Charlie Bassett, were killed while attempting a circling approach to St. Louis' Lambert Field under a 300-foot indefinite ceiling. They impacted the McDonnell Aircraft Company factory where the Gemini capsules were being assembled. Another cruel twist!"[31]

Orville Johnson also received the dreaded phone call. "I was deeply disappointed when informed of the outcome," he reflected. "The physical and mental testing at Brooks AFB was rigorous, but during the course of this process, including the later sessions at Houston involving interviews and exposure to the current Group 1, I felt confident that I would reach the finals. Later I would rationalize that my size and height was a contributing factor, since so much emphasis was being placed on the space and weight available in the Gemini spacecraft. I suppose I will never know the exact reasons, and maybe I shouldn't, or wouldn't, want to." He could not apply for the third astronaut group. "Unfortunately I was born in 1928, and that meant I was then too old to meet the requirements."[32]

To the surprise of many, Bob Solliday also missed out. Ken Weir had seen him as a shoo-in on his second attempt, but it has been suggested that Solliday and Deke Slayton had earlier been involved in a verbal altercation during the Mercury selection process. Slayton had allegedly stated that at 27 years of age the Marine captain was too young and inexperienced to have applied and ought not to have even been there. This appears to have caused resentment between the two men, and may ultimately have resulted in Solliday being excluded from consideration for the second astronaut group.

After the dust had settled, Ken Weir was able to devote all of his attention to the various interesting and demanding flight test activities at the Naval Air Test Center at Patuxent River.

As he told the author, there was not a single Marine selected in the second group of astronauts. He did an extensive analysis of the nine astronauts who were selected, examined their education, personalities, military and flight test experience, physical conditioning, etc., as best he could to determine just exactly what NASA was really looking for in their astronauts. A very strong common denominator jumped right out at him – *superior academic achievement!* In hindsight, he says, it should have been no surprise. NASA was not looking for another bunch of accomplished red-hot stick-and-rudder test pilots with limited academic credentials. They wanted intellectual giants to help to solve the complex problems of space exploration. All nine of those new astronauts had excelled academically at whatever institution they attended.

"They were all intellectually superior to the rest of us, and particularly me, by a long shot. They were head and shoulders above the original seven astronauts when it came to excelling academically. Tom Stafford and Jim Lovell were academic giants in our 1952 class at the Naval Academy and were the respective honor grads in their Test Pilot School classes. John Young was at the top of his Georgia Tech engineering class and his Test Pilot School class. Pete Conrad was a stellar aeronautical engineering student at both Princeton and Test Pilot School. Jim McDivitt had a master's degree in aeronautical engineering from the University of Michigan, one of the premiere aeronautical schools around. Both Neil Armstrong and Elliot See had long exhibited their intellectual prowess. Frank Borman had a master of science degree in aeronautical engineering from the California Institute of Technology, one of the two top engineering schools in the country. He and Ed White both excelled at West Point and in their Air Force Test Pilot School classes. White also had a master of science degree in aeronautical engineering from the University of Michigan. Tom Stafford had it exactly right when he said NASA was going to get some really smart test pilots this second time around."

A few weeks after receiving his call of rejection from Houston, Weir wrote to the Commandant of the Marine Corps, attaching his assessment of the selection process and recommending that if the Marine Corps wanted to be included in future NASA astronaut programs then it would have to concentrate on offering its aviators better education in the aerospace disciplines so they could excel academically and be more competitive. For starters, he recommended that the Corps request a student billet at the Aerospace Research Pilot School at Edwards AFB.[33]

NINE ARE CHOSEN

On Friday, 14 September, a NASA spokesman reported that the nine new astronauts would be named at the Manned Spacecraft Center in three days' time. MSC Director Robert R. Gilruth stressed that even though nine men would be named, it did not necessarily mean that they would all make space flights. "Assignment to flight crews will depend upon the continuing physical and technical status of the individuals concerned, and upon the future flight schedule requirements."[34]

As Deke Slayton explained: "By mid-September Al [Shepard], Warren North, the doctors, and I had agreed upon nine candidates. We passed the list through Gilruth to headquarters, where it was approved for public announcement on September 17. I spent the Friday before, the fourteenth, phoning the candidates to tell them they'd been selected. They were supposed to be in Houston for the public announcement. They would have until October 1 to report."[35]

That same day, the formal announcement was made of Deke Slayton's new title as coordinator of astronaut activities, even though he'd been performing that role for several months. This position entailed taking on such administrative duties as the scheduling of astronaut training activities, their public appearances and news media interviews, and visits to contractor facilities. But the most important responsibility Slayton had taken on, and one he administered totally by himself, was the selection of crewmembers for upcoming missions.

Maj. Frank Borman, USAF. (Photo: NASA)

Slayton would have to take many factors into consideration in setting out his crew recommendations. In particular, there was the issue of crew compatibility, with two – and later three – astronauts crammed into the sterile confines of a spacecraft for up to a fortnight at a time, with little or no privacy in such normally intimate matters as hygiene and regular bodily functions. He alone made the decisions, and his decision was final. In fact, in all the years in which Slayton performed this task, there was only one occasion (Apollo 17) when his crew listing was overruled by headquarters. To his credit, all of the crews that he assigned went on to function well together.

Frank Borman received his call from Slayton, asking if he still wanted to come to Houston. He accepted, after which he had one difficult task to perform. Still thrilled by the appointment he walked into the office of Chuck Yeager, then the commanding officer of the Test Pilot School at Edwards, and mentioned that he had received some good news. When Yeager asked what it was, Borman said he'd received a call from Deke Slayton and had been accepted into the NASA program. Borman would later write that his boss was not as excited about the news. "Yeager gave me a look that should have been reserved for someone who had just announced he was being audited by the IRS. 'Well, Borman,' he growled, 'you can kiss your goddamned Air Force career goodbye.'" He added that – in a way – Yeager's words would prove to be totally correct.[36]

PURE OXYGEN AND A FIRE WARNING

In November 1962, after Ken Weir's astronaut candidate experience, NASA contracted the U.S. Navy through Republic Aviation to conduct centrifuge testing for the Gemini program at NAS Johnsville, Pennsylvania, as well as 100 per cent oxygen mission profile tests in the pressure chamber of the Air Crew Equipment Laboratory at the Philadelphia Naval Shipyard. These tests involved donning pressure suits, experiencing launch accelerations by means of the Johnsville centrifuge, then being transported to the Bioastronautical Test Facility pressure chamber for two weeks of confinement in shirt-sleeve conditions breathing pure oxygen at simulated altitudes, and finally returning to Johnsville, donning a pressure suit and experiencing re-entry accelerations through to the recovery phase.

The pressure chamber was normally maintained at considerably less than sea level atmospheric pressure, approximately equivalent to 10,000 feet above mean sea level. All meals were freeze-dried food that required reconstitution by means of electrically heated water. About every third day, the resident in-chamber flight surgeon used a huge needle to draw blood from each volunteer's leg artery in the groin area without the use of any anesthesia to determine the oxygen content (PO2) of each subject's arterial blood. Six Navy and Marine volunteer subjects were involved, including Ken Weir, Al Bean and Bill Geiger, all of whom had participated in the second NASA astronaut selection program and still had hopes of someday making it to Houston.

According to Bill Geiger, "It was all meant to simulate a 14-day Gemini mission, and we had three guys right out of flight school with us. The testers wanted to see if there was any difference in the response of the younger fellows; they were roughly ten years younger than the three of us."

The subjects entered the chamber one day at a time, stayed two weeks and left one day at a time. Bean went first, Geiger second and Weir was third. "The three younger guys then went in one day after each other, 4th, 5th and 6th," Geiger continued, "so each of the six of us were in this spacecraft simulation for fourteen days, and we would come out on our fifteenth day. After we came out it was straight back to the Johnsville centrifuge and then they'd say the exercise was over and we were free to go."[37] The subjects would turn up at eight o'clock the following day for interviews and medical checks.

Ken Weir recalls that their experience ended in a bad accident, and sadly foretold a calamitous event for NASA.

"During the two-week experience at reduced atmospheric pressure, the special voltage incandescent light bulbs burned out rapidly for some unexplained reason, and the supply of bulbs was depleted. There was a shortage of bulbs and we moved them around as required in order to have light where needed. I transferred several bulbs around during the course of the two weeks. Some light sockets were left empty after a while.

"The day I finished my two-week tour and about an hour after I had departed the chamber, the insulation on one of the empty sockets in the berthing compartment sockets ignited and began burning. One of the three remaining test subjects attempted to extinguish the flames with an oxygen-saturated towel that immediately flashed with flames. The stellar resident flight surgeon who was in the chamber at the time smashed the glass window, causing an explosive decompression that sucked all of the oxygen out of the chamber. The occupants, though badly burned, were able to evacuate the charred chamber. It was a grim demonstration of what was at stake using a 100 per cent oxygen environment in a spacecraft and foretold the Apollo launch pad disaster that took the lives of astronauts Grissom, White and Chaffee."[38]

All three men (1st. Lt. Gabo, USMC, and Navy Ensigns Schlager and Marshall) managed to escape the chamber within 40 seconds, but suffered burns, two seriously. Lt. Cmdr. (MC) Francis M. Highly, Jr., the flight surgeon and inside supervisor-observer when the fire broke out, would later be awarded the Navy and Marine Corps Medal for his actions.

"It must have been late afternoon [on 17 November]," Highly later recalled for Florida's *St. Petersburg Evening Independent* newspaper. "One of the men was standing near an electric light outlet. We later learned there had been a spark from a faulty circuit, but all we knew was that suddenly his legs were covered in flames.

"Our first reaction was to put out the fire. But we found that we couldn't smother the flames with a blanket. When we tried, our own clothes caught fire .... A matter of some 15 to 20 seconds had elapsed." The men quickly began shedding their burning clothes. "I made the decision that we had to get out and punched the alarm button," Highly explained. "I grabbed a hammer [and] smashed through a porthole on the sidewall [reducing] the pressure in the chamber to a simulated altitude of 60,000 feet – an atmosphere so rarefied that it could not support flame. I looked down and saw the hairs on my chest burning." After the four men had clambered into a tunnel between the chamber and the next interior section of the chamber cylinder, Highly closed the door. They made their way down to ground level and were clear of the chamber about 45 seconds after the fire started. Highly had suffered burns to around 20 per cent of his body and as a result was hospitalized for five months.[39]

Bill Geiger was aghast when he heard of the fire. "I can remember visiting the young men in hospital afterwards and this brave young fellow said to me 'This must be the pain threshold test.' Poor guy. But I think they all survived."[40]

MORE COSMONAUTS

Following the single-orbit flight of cosmonaut Yuri Gagarin on 12 April 1961, large numbers of Soviet citizens felt encouraged to write to officials volunteering their services for future space missions. These letters increased significantly after the day-long flight of Gherman Titov aboard Vostok 2 in August of that year.

By this time, Soviet officials were seriously contemplating sending a woman into space, and began reviewing letters from female volunteers – particularly those with any sort of flying and parachuting skills. Around 24 women, all veteran parachutists, were asked to report in strict secrecy to Moscow, and from them five candidates were selected. Eventually the choice came down to two women, Irina Solovyova and Valentina Tereshkova, and the latter would fly amid a frenzy of propaganda aboard the Vostok 6 spacecraft, launched on 16 June 1963. Among other publicity snipes at NASA was the fact that they had outright rejected recruiting women pilots for space flights. It was later revealed that Tereshkova had suffered a lengthy bout of illness during her almost three days in orbit.

Earlier, in March 1962, the Soviet space chiefs began recruiting a new group of male cosmonauts. This time, there was greater emphasis on experience and academic prowess. Whereas the first group of cosmonauts had been relatively young (Gagarin was 27 and Titov was 25 when they flew their missions), the age limit for this second group was raised from the previous under-30 requirement to up-to-40 years. And in a similar move to NASA, this recruitment was open not only to jet pilots but also to engineers and navigators. But the most significant change was for all candidates to have graduated from a military college or civilian university. It was apparently no longer enough for a young pilot to possess stamina and a brave heart; comprehensive technical knowledge related to space flight was essential.

This cosmonaut enrollment initially comprised 15 officers, including eight pilots. They reported for training on 11 January 1963 and were joined much later by two other appointees, Col. Georgi Beregovoy and Lt. Col. Vasily Lazarev.

As the late respected Soviet space historian Rex Hall pointed out, "The arrival of the second group caused some friction in the ranks. The young pilots of the first enrollment largely worked as a collective, unconcerned with rank and title. With the addition of senior pilots and engineers, the cosmonaut team assumed the structure of an Air Force squadron."[41]

Beginning in March 1965, the Soviet State Commission began the lengthy process of selecting additional cosmonauts, suggesting a large number would be chosen to fulfill overly-optimistic flight projections handed to the Commission by Soviet space officials. In this way, out of 3,500 applicants, the third Soviet cosmonaut enrollment comprised 22 pilots and engineers. In fact, only six of this group would ever fly into space, despite years of concentrated training.

Conversely, out of the second and third groups selected by NASA, only four of the total of 23 astronauts – one from Group 2 and three from Group 3 – would never fly into space, and in their cases it would be as a result of fatal accidents.

REFERENCES

1. E-mail correspondence with M/Gen. Ken Weir, 10 November 2011–28 May 2012
2. NASA *Space New Roundup*, issue October 3, 1962, *MSC Personality: Dr. Robert R. Voas Serves as Asst. for Human Factors*, MSC, Houston, TX, pg. 6
3. Lamb, Lawrence E., M.D., *Inside the Space Race: A Space Surgeon's Diary*, Synergy Books, Austin, TX, 2006, pg. 233
4. Lamb, Lawrence E., M.D., (Ed.), paper, *Aeromedical Evaluation for Space Pilots*, USAF School of Aerospace Medicine, Aerospace Medical Division (AFSC), Brooks Air Force Base, Texas, July 1963
5. E-mail correspondence with M/Gen. Ken Weir, 10 November 2011–28 May 2012
6. Stafford, Tom and Michael Cassutt, *We Have Capture: Tom Stafford and the Space Race*, Smithsonian Institution, Washington, D.C., 2002
7. E-mail correspondence with M/Gen. Ken Weir, 10 November 2011–28 May 2012
8. Lamb, Lawrence E., M.D., *Inside the Space Race: A Space Surgeon's Diary*, Synergy Books, Austin, TX, 2006, pg. 234-235
9. Lamb, Lawrence E., M.D., (Ed.), paper, *Aeromedical Evaluation for Space Pilots*, USAF School of Aerospace Medicine, Aerospace Medical Division (AFSC), Brooks Air Force Base, Texas, July 1963
10. Lamb, Lawrence E., M.D., *Inside the Space Race: A Space Surgeon's Diary*, Synergy Books, Austin, TX, 2006, pg. 236
11. *Ibid*, pg. 234
12. Lamb, Lawrence E., M.D., (Ed.), paper, *Aeromedical Evaluation for Space Pilots*, USAF School of Aerospace Medicine, Aerospace Medical Division (AFSC), Brooks Air Force Base, Texas, July 1963
13. *Ibid*
14. E-mail correspondence with M/Gen. Ken Weir, 10 November 2011–28 May 2012
15. E-mail correspondence with John and James Fritz, 22 December 2011–7 July 2012
16. Scott, David and Alexei Leonov, with Christine Toomey, *Two Sides of the Moon*, Simon & Schuster, London, U.K., 2004

17. E-mail correspondence with M/Gen. Ken Weir, 10 November 2011–28 May 2012
18. E-mail correspondence with Bob Birdwell, 12–18 December 2011
19. *Ibid*
20. Lamb, Lawrence E., M.D. (Ed.), paper, *Aeromedical Evaluation for Space Pilots*, USAF School of Aerospace Medicine, Aerospace Medical Division (AFSC), Brooks Air Force Base, Texas, July 1963
21. Borman, Frank, with Robert J. Sterling, *Countdown: An Autobiography*, Silver Arrow Books/William Morrow, New York, NY, 1988, pg. 88
22. E-mail correspondence with M/Gen. Ken Weir, 10 November 2011–28 May 2012
23. *Ibid*
24. Lamb, Lawrence E., M.D. (Ed.), paper, *Aeromedical Evaluation for Space Pilots*, USAF School of Aerospace Medicine, Aerospace Medical Division (AFSC), Brooks Air Force Base, Texas, July 1963
25. Lamb, Lawrence E., M.D., *Inside the Space Race: A Space Surgeon's Diary*, Synergy Books, Austin, TX, 2006, pg. 237
26. E-mail correspondence with M/Gen. Ken Weir, 10 November 2011–28 May 2012
27. Smith, Robert W., *The Robert W. Smith Autobiography, NF-104.com* (incomplete). Website: http://www.nf104.com
28. E-mail correspondence with M/Gen. Ken Weir, 10 November 2011–28 May 2012
29. Smith, Robert W., The Robert W. Smith Autobiography, NF-104.com (incomplete). Website: http://www.nf104.com
30. E-mail correspondence with M/Gen Ken Weir, 10 November 2011–28 May 2012
31. E-mail correspondence with John and James Fritz, 22 December 2011–7 July 2012
32. E-mail correspondence with Orville Johnson, 6 December 2011–9 June 2012
33. E-mail correspondence with M/Gen. Ken Weir, 10 November 2011–28 May 2012
34. NASA *Space News Roundup*, issue September 19, 1962, *MSC Names Nine New Astronaut Trainees*, MSC, Houston, TX, pg. 1
35. Slayton, Donald K. and Michael Cassett, *Deke! U.S. Manned Space: From Mercury to the Shuttle*, Forge Books, New York, NY, 1994, pg. 120
36. Borman, Frank, with Robert J. Serling, *Countdown: An Autobiography*, Silver Arrow Books/William Morrow, New York, NY, 1988, pg. 91
37. Author interview with William Geiger, Los Angeles, CA, 28 May 2012
38. E-mail correspondence with M/Gen. Ken Weir, 10 November 2011–28 May 2012
39. St. Petersburg Evening Independent (Florida) newspaper, Wednesday, February 1, 1967, *They Escaped in Space Fire*, pg. 14
40. Author interview with William Geiger, Los Angeles, CA, 28 May 2012
41. E-mail message to author from Rex Hall, 2 January 2010

3

The finalists

It was the expected follow-up letter to the earlier, deflating telephone call. For those who received it, this was written confirmation – a final, official declaration – that they had failed to make the grade. The letter was signed in person by Dr. Robert R. Gilruth. In part, it read:

> The impression that you made on our Selection Committee was favorable. Overall, however, we did not feel that your qualifications met the special requirements of the astronaut program as well as those of some of the other outstanding candidates.[1]

For some, it marked the end of a quixotic thrust for one of the most prized jobs in modern history; others shrugged their shoulders, knowing and understanding that their qualifications had somehow fallen short of what was required, and they would try again – but next time they would be a little better equipped and certainly more knowledgeable about what was expected of them. They already knew that NASA was looking at a further astronaut intake the following year and these men set out to do something to make it happen next time round. Several would succeed.

TRY, AND TRY AGAIN

For Dick Gordon, missing out on selection in the second astronaut group in 1962 was "a major disappointment", and it was not until years later that the reason for his non-selection in that group was revealed to him, as he told the author:

"In February 1957 I was in Class 18 at Navy Test Pilot School along with Hal Crandall, who was a later finalist for the Mercury group of astronauts. This meant we were a year ahead of Pete Conrad and Jim Lovell – Wally Schirra was also in that class – and [we were] two years ahead of John Young, which means that much more experience. Wally, Pete, Jim and Bob Solliday from Class 20 were all selected to try out for the Mercury group, as was Hal Crandall, but I didn't get that particular tap on my shoulder, and I still don't know the reason. Therein lies the reason for my initial disappointment.

"Anyway, back in the summer of 1962, Pete Conrad and I were both in Fighter Squadron VF-96, which was the second F4 Phantom squadron on the West Coast. That year, the two of us were also involved in testing for the second NASA astronaut group. Of course Pete had earlier been under consideration for the Mercury group – which I knew about – and then we were both chosen to go to the School of Aviation Medicine down in San Antonio during the second group selection process. We both made the final group of 32, but in the end Pete was chosen and I wasn't. Although I was obviously delighted for Pete, that was a *real* bitter pill for me to swallow.

"Why didn't I make it? Once again I didn't know and wasn't given any reason. After the new astronaut group was announced by NASA, Pete and I had a long conversation about what it meant for both of us. I have to admit it had probably been a bit of an ego trip for me back then, but when I missed out I was extremely bitter and even thought about resigning from the Navy. With this in mind I went around to Boeing to see if I was interested in working for them, and also Rockwell. But then Pete and I had our conversation, which was interesting, because we agreed neither of us knew what the hell the future held for us, or what we'd gotten ourselves into.

"I guess failures teach you more than anything else in life, and you grow up a little as a result, so I finally got over my disappointment. The interesting thing for me now was that VF-96 was shortly going to be deployed to the Far East with the Seventh Fleet on the [USS] *Ranger* and this had been part of our conversation on future prospects. We'd also been told there were indications another astronaut selection would take place the next year. So I swallowed my disappointment and waited to see what would happen next. Like they say, time heals all wounds.

"As it turned out, I was selected in 1963, but it wasn't until several years later, while at NASA, that I learned the reason I may have been rejected as unsuccessful the year before. It was all to do with a question mark about the curvature of my lower spine. During the early part of the selection process Jim Irwin and I were sent to see a specialist – an orthopedic surgeon. Now Jim had earlier been in a significant airplane accident and had been badly banged up, so that was understandable, but I was a little curious as to why I was there as well. I thought, 'What the hell is this about?' but convinced myself it must be part of the medical evaluation, and of course back then you had to be 100 per cent fit or you'd be rejected. As things turned out I was selected and Jim wasn't, at least not then anyway, but I was none the wiser.

"So I guess that visit to the orthopedic surgeon cleared up the question mark on the curvature of my lower spine, as it never entered into any conversation from then on during the selection process. But I strongly believe that was the reason I wasn't selected in the second group along with my good buddy, Pete Conrad."[2]

ALIAS "MAX PECK"

As instructed by NASA, the nine new astronaut trainees arrived separately in Houston on Saturday, 15 February wearing civilian clothing, caught airport cabs and made their way downtown to the historic Rice Hotel on Texas Avenue. Then, as further instructed by Deke Slayton, they all checked in under the same alias – Max Peck – which happened to be the name of the hotel's general manager at that time.

"It was completely quiet," Neil Armstrong recalled in his memoirs. "Nobody was to know that we were coming in or that it was going to be announced. Leaks weren't as common in those days as they are now."[3]

While it may not be thought of as a "leak", there was a report in the *Tampa Bay Times* on 14 September, the day before the unnamed astronauts arrived in Houston, that Jules Bergman of the American Broadcasting Company had confidently stated 31-year-old X-15 pilot Neil Armstrong was one of those selected.[4] Meanwhile, there was a lot of confusion as nine men independently checked in under the same name. Jim Lovell relates in his book *Lost Moon* that the young girl checking him in didn't seem to be party to the subterfuge. When he said, as instructed by Slayton, that his name was Max Peck she looked at him strangely, and said, "Uh, I don't think so." Moments later the confusion was sorted when another clerk named Wes Hooper hastily joined the girl at the reception desk and said he would take care of it. "Glad you could make it, Mr. Peck," he said with a smile. "We've been expecting you. Here's your key, and do let us know if anything is not to your satisfaction."[5]

Any uncertainties Lovell might have harbored were resolved a little later when he left his room and made his way down to the lounge, where he spotted two of his old Navy buddies, Pete Conrad and John Young, sitting with someone he did not know. At this stage of the process no one knew who else had been selected, so there was a joyous reunion, and then he was introduced to the fourth man, an affable Air Force captain named Ed White.

PRE-ANNOUNCEMENT BRIEFING

The following morning the nine men assembled at Ellington AFB, as directed by Slayton. This gave them the opportunity to greet and catch up with those they knew beforehand, or meet others they did not recognize. Walt Williams, as head of NASA Flight Operations, briefed the new astronauts on their job and what was expected of them. Then Bob Gilruth took over, bringing the men up to speed on NASA's manned space flight plans and the envisaged flight schedule over the next few years. He also reassured them that there would be plenty of space flight missions for everyone, and perhaps even a lunar landing mission within Project Apollo.

Next, Slayton provided basic information on how the men would be expected to conduct their lives, both in public and in private; on the unknown pressures of being a celebrated astronaut; on a wide variety of temptations they would soon face, and how these should be dealt with. They were warned to be circumspect and constantly wary of accepting gifts and other free goodies, especially from companies seeking endorsements for NASA contacts. Then, understanding he was talking to mature and experienced test pilots very much like himself, the grounded astronaut sized them all up and added, "If there is any question, just follow the old test pilot's creed;

anything you can eat, drink or screw within twenty-four hours is perfectly acceptable."[6]

Both Gilruth and Williams were noticeably taken aback by this relaxed attitude to particularly rigid rules, but the men had laughed at the old adage, and some of the stiff formality evaporated from the proceedings, which is probably what Slayton was aiming to achieve. Finally, John ("Shorty") Powers, NASA's public affairs officer, explained to the group what was expected of them at the forthcoming announcement and press conference and how they should respond to certain questions.

Each of the nine new astronauts understood that their lives were about to change forever, but for the most part, they were generally ready to face up to this new and exciting challenge.

"LADIES AND GENTLEMEN ..."

At 1:00 p.m. on Monday, 17 September 1962, the members of NASA's newest astronaut group made their first public appearance in a televised press conference at the University of Houston's overcrowded, 1,800-seat Cullen Auditorium. The media was present in force this time, having been caught somewhat short with the adulation that followed the announcement and naming of the original seven astronauts. This time the auditorium was packed to capacity with television, radio, newspaper and magazine reporters from across the United States and around the world, all jostling to gain the best position, with television cameras and photographers recording the scene for the waiting public.

The nine men were seated at a long, curtained table bearing a single NASA logo in the center. They were positioned in alphabetical order with name boards displayed in front of them, similar to the way the Mercury astronauts had been introduced back in 1959. But this time Dr. Gilruth was sitting near the middle of the table, between Lovell and McDivitt. Also present were Walt Williams and Warren North, chief of MSC's Crew Operations Division. At the far left, seated next to Neil Armstrong, was the center's Chief of Personnel, Stuart Clarke, ready to field questions from the floor. He had already handed out a news release giving background details on the selection process. This read:

> In April 1962, the Manned Spacecraft Center asked for volunteers for its forthcoming Flight Crew Training Program. The following minimum qualification standards were published and distributed to the press, aircraft companies, government agencies – civilian and military – and the Society of Experimental Test Pilots.
>
> a. The candidate must be an experienced jet test pilot and preferably be presently engaged in flying high-performance aircraft.
> b. He must have attained experimental flight test status through the military services, the aircraft industry or NASA, or must have graduated from a military test pilot school.

c. He must have earned a degree in physical or biological sciences or in engineering.
d. He must be a United States citizen under 35 years of age at the time of selection, and 6 feet or less in height.
e. He must be recommended by his present organization.

Over 200 applications were submitted to NASA. A preliminary selection committee met in June to consider 63 of the most highly-qualified applicants. The preliminary selection committee was composed of MSC management, including representatives from the present group of astronauts. Various quantitative criteria concerning the applicants were studied, such as flight test experience, academic achievement, and their present supervisor's evaluation. Thirty-two applicants were selected to continue in the selection program. During July and August, they were given medical examinations, and one was eliminated as being "too tall." During the week of August 12, the 31 applicants reported to Houston for four days of examinations and interviews. [During] the next few weeks, the selection committee carefully reviewed and evaluated the tests and interviews. Nine were selected to participate as Flight Test Personnel. From these nine and our present astronauts will come flight crews for future space flight missions. It is planned that in late stages of Apollo spacecraft development a third group of Flight Test Personnel will be selected to join those then available as the pool from which Apollo flight crews will be chosen.[7]

As one would expect, the nine men were a little nervous. They could handle the newest, fastest, most temperamental airplanes in the skies, but this was a whole new ball game to them. As directed, they had turned up wearing mostly off-the-peg dark suits, apart from non-conformist Pete Conrad, whose white linen suit was in obvious contrast to the more conservative attire of his fellow astronaut selectees.

Then came the announcement that the world's press had been awaiting. It was the culmination of more than six months of extensive evaluation of 253 volunteers. The group – all married with children – comprised four members of the Air Force, three from the Navy, and two civilian test pilots. In introducing the new astronauts, Gilruth said they had already been assigned to a comprehensive training program. Following this, the nine men were named:

Neil A. Armstrong (32), NASA X-15 test pilot
Maj. Frank F. Borman II, USAF (34), instructor at ARPS, Edwards AFB
Lt. Charles Conrad, Jr., USN (32), safety officer, VF-142, Oceana NAS, Virginia
Lt. Cmdr. James A. Lovell, Jr., USN (34), flight instructor/safety officer, Oceana NAS
Capt. James A. McDivitt, USAF (33), experimental flight test officer, ARPS, Edwards AFB
Elliot M. See, Jr. (35), experimental test pilot with the General Electric Company
Capt. Thomas P. Stafford, USAF (32), ARPS Experimental Flight Test Division
Capt. Edward H. White II, USAF (32), experimental test pilot, Wright-Patterson AFB

Lt. Cmdr. John W. Young, USN (31), maintenance officer, VF-53, Miramar NAS, California.

Then came the usual assortment of bland media questions, the responses to which were hardly packed with memorable quotations for the world's press. This time there was no John Glenn to provide stirring responses encompassing patriotism, challenges and family values.

Starting alphabetically with Neil Armstrong, each of the new astronauts stoically gave an answer to the first question, "Gentlemen, what drew you to apply for the job?" by mentioning the honor of representing their country, their long-held interest in participating in space work, and wanting to be a part of NASA's space team. Some chuckles came when John Young, at the end of the table, mischievously looked back along the line and simply stated that he agreed with what the others had said. He then added, in a more serious tone, "I couldn't turn down a challenge like that."[8]

Pete Conrad was asked about when he first decided to become an aviator. "The psychiatrist asked me that too," he said, prompting ripples of laughter around the room, then added that he could not recall when he first wanted to fly. When asked about an adhesive patch on his head, Pete gave a toothy smile and said that pre-flight inspections could be more dangerous than flying and he had bumped his head on the wing of an airplane.[9]

Compared to the near-hysteria of the Mercury press conference, the introduction of the new cadre of astronauts was a relatively low-key event, with generally dull questions getting unexciting responses, prompting Neil Armstrong to later state that, "The questions at the press conference were typical, fairly unsophisticated questions, with answers to match." There was one moment of serendipity, however, when one of the reporters noticed that it was Tom Stafford's thirty-second birthday and offered his congratulations. "I can honestly say I never had a birthday before or since quite like that one," Stafford later noted.[10]

At the news conference it was also announced that Deke Slayton had been named coordinator of astronaut activities, and as such he would be responsible for assigning the training and engineering duties of all the astronauts. It was further pointed out by Gilruth that Slayton still held the title of astronaut and might eventually make a two-man or three-man flight. But then he qualified this by saying, "Unless there is some change in the medical situation it is very doubtful about his flying."[11]

When asked whether John Glenn or any of the Mercury astronauts would make flights to the Moon, Gilruth explained that with the first lunar missions being six or seven years away, age might be working against the original astronauts. "Some of the original seven are highly motivated to make the Moon shot," he told the newsmen, "but the age factor very well could make it difficult for them to make the team as the space program advances."[12]

Warren North weighed in, saying that the overall lower age of the new astronauts was a long-term factor in selecting future crews. "They will be younger, so they can fly longer," he remarked.[13]

While there was jubilation for the nine test pilots selected as the next generation of NASA astronauts, disappointment lingered like a cloud over the 22 finalists who

The "New Nine" astronauts. Back row (from left): Elliot See, Jim McDivitt, Jim Lovell, Ed White and Tom Stafford. At front: Pete Conrad, Frank Borman, Neil Armstrong and John Young. (Photo: NASA)

were not chosen. Some of these men would set out to add more flight experience and academic qualifications to their résumés and try again the following year. Alan Bean, Mike Collins and Dick Gordon would succeed in this ambition, as members of the third astronaut group, while Jack Swigert had to bide his time until his selection with the fifth group in 1966. Other finalists realized they had missed out for one reason or another and simply did not feel the urge to go through the process again. They chose instead to return to test flying and whatever their career might hold for them.

Although they had not been selected, the unsuccessful candidates could still hold their heads high, for they had participated in one of the toughest of all elimination contests. As in the selection of the Mercury group there was very little between most of the finalists, but it is undeniable that the nine chosen would eventually become recognized as the *best* astronaut group ever chosen by NASA, so the competition had been fierce. As each of the unsuccessful candidates dutifully returned to his base or aviation company, he could take deep pride in knowing that as one of the nation's elite test pilots, he was still amongst the best of the very best.

ROLAND E. ASLUND, USN

Roland Erhard ("Auz") Aslund was born on 20 July 1927 at Bemidji, Minnesota. His father, Berger (generally known as "Bill") was an immigrant from Sweden, while his mother, Magda Rygg Aslund, was a first-generation American born shortly after her parents' arrival from Norway. Following his father's arrival in the United States, immigration officers decided that the original family name of Inglebrittson was too difficult to spell, and it was changed to Aslund, which was the county where Berger had lived in Sweden. Roland was Berger and Magda's second son, and the second of an eventual four children. He had an older brother Austin, and two younger sisters, Jacquelyn and Marlys.

During the Second World War, the family moved from Minnesota to Vancouver, Washington, and Berger Aslund found a job in the shipbuilding industry. During this time, young Roland developed a love of flying and joined the CAP (Civil Air Patrol) while still in high school. He saved diligently for flying lessons, doing odd jobs and delivering newspapers. On completing high school he joined the Navy, signing up for the aviation midshipmen program, which granted high school graduates a subsidized two-year college education with a scientific or engineering major in exchange for a commitment to serve in the service for three years. This allowed him both a college education and the opportunity to subsequently train as a naval aviator.

Aslund took his college studies at the California Institute of Technology, and in January 1949 moved straight into pre-flight school at NAS Pensacola (Class 4-48). On finishing primary flight instruction at Whiting Field and Carrier Qualification at Corry Field, he joined the Gunnery Flight 100 class at Saufley Field in Pensacola to learn armament skills and formation flying at the controls of a yellow, single-engine North American SNJ advanced trainer, known as the "pilot maker" for its vital role in preparing aviators for combat.

The future naval aviators formed groups of six for the formation exercises, and fellow student John Jenista recalls one brazen attempt to impress an instructor. Their team consisted of Lieutenants Jenista, Aslund, Brandenburg, Crowl, Gooding and Pollard. "We thought we were really hot stuff, and named ourselves 'The Blue 100' after the Blue Angels." They even got themselves blue scarves. "One day we decided that we would really show the instructor how fast we could get joined up. We took off on runway 9 .... We wanted to have all six planes in formation by the time we got over the fence at the end of the runway. We almost made it. I think everyone was in place about 1,000 feet after we passed over the fence."[14]

Fresh out of flight and gunnery school and sporting his brand-new Wings of Gold, Aslund's first assigned squadron was VF-112 ("Fighting Twelve"), based at NAS North Island in Coronado, San Diego.

On 5 October 1950, VF-112 commenced combat operations from the carrier USS *Philippine Sea* (CV-47) as part of Carrier Air Group 11. Their missions, flying F9F-2B Panther jets over North Korea, initially consisted of attacking land and sea targets of opportunity southeast of Kunsan. Later, they would launch attacks on such targets as oil refineries, railroad tunnels and rolling stock, enemy lines of

Lt. Roland E. ("Auz") Aslund, USN. (Photo: Nadine Wisely, Blue Angels Association)

communication and anti-aircraft sites, while close air support missions were flown to provide relief and protection to United Nations ground forces. The next month the squadron came in contact with the deadly MiG-15 fighter jet. On 10 November the commanding officer of VF-112, Lt. Cmdr. John Butts, Jr., teamed with Aslund in damaging one of the swept-wing jets over Sinuiju. Some pilots were finally relieved of duty in March 1951, and six from VF-112, including Aslund, returned to the U.S. mainland aboard the carrier USS *Valley Forge* (CV-45). From 7 April that year the squadron would be based at NAS Miramar, California. In December 1951, VF-112 conducted a second combat tour to WestPac/Korea, again operating from *Philippine Sea*.

During the early stages of the Korean conflict, the U.S. Navy's renowned "Blue Angels" jet flight exhibition team had been assigned to fleet operations on a combat-ready status, but after high-level discussions in late 1951 their role was re-evaluated. The primary mission of the reformed team was to demonstrate to naval personnel the precision techniques and tactics which had been developed in actual combat. Their ongoing appearance at air shows would help to stimulate interest in naval aviation. Now flying the 600-mph Grumman F9F-5 Panther, a new demonstration team was established at NAS Corpus Christi by its former commander, Lt. Cmdr. Roy Voris. Lt. Aslund joined this team, although not initially in the 'first' five-man

Lt. Aslund (far right) with the Blue Angels team, 1953. (Photo: Nadine Wisely, Blue Angels Association)

squad, which made its public debut at the Memphis Mid-South Navy Festival in May 1952.

By December, Aslund, whose VF-112 squadron had returned from Korea aboard *Philippine Sea* four months earlier, had become one of the primary pilots, flying left-wing with the Blue Angels and would remain with the feted "Blues" until 1954.[15]

In July 1957, while studying for an astrophysics degree at the University of California at Los Angeles, Aslund met 21-year-old Californian divorcée Joan Kinney (née Hester) who had a three-year-old daughter and another ten months old. They fell in love and were married just six weeks later on 6 September, making him an instant father of two stepdaughters. As Joan recalls, "He was called the 'Golden Boy' by his team mates and soon earned a reputation as one of the best stick and throttle men in the Navy. After the Blues, he settled into semi-civilian life while still in the Navy, working on his [degree] and got his monthly flight time in at NAS Los Alamitos, California."[16]

In October he was selected as one of three American pilots to attend the highly regarded Empire Test Pilot School in Farnborough, England on a ten-month posting and the family moved to Surrey in January of 1958. He received high grades and graduated in December of 1958. In September of that year he also became the father of a baby boy, which meant that in a little over a year he had gone from being a 30-year-old bachelor to a staid married man with three children.

The Blue Angels team appeared on the front cover of *Naval Aviation News* in February 1954. From left: Lt. Wallace Rich, Lt. Roland Aslund, Lt. Francis Murphy and Lt. Cmdr. Arthur Hawkins. (Photo and permission courtesy Colin Babb, *Naval Aviation News*)

On his return from England, Aslund was sent to the U.S. Naval Postgraduate School in Monterey, California to finally complete his astrophysics degree (which he achieved in 1960). Meanwhile he and Joan had added another son to their family, arriving on the same date one year after his first son was born. Later in 1960, he headed to VX-4 Squadron at Point Mugu, California to join the experimental test squadron working on the Sidewinder missile and as Chief Test Pilot on F8 Crusader projects.

"It was at this time he was taken into the astronaut selection program," Joan points out. "I have no doubt he would have been chosen if it hadn't been for an inoperable kidney stone that could have caused problems in space flight."[17]

From Point Mugu, Aslund returned to NAS North Island, now operating aboard the carrier USS *Oriskany* (CV-34) as Flight Deck Officer with Carrier Air Group 16. Next was a short tour of duty at the Armed Forces Staff College in Norfolk, Virginia. During that time the United States became involved in the Vietnam conflict and he was ordered to NAS Miramar, San Diego as Executive Officer and subsequently Commanding Officer (June 1966 to June 1967) of VF-24, known as the "Fighting Red Checkertails". Tragically, while Aslund was the squadron's Executive Officer he learned that his 42-year-old brother Austin (known as "Jake") had been killed in an automobile accident.

The squadron was engaged in two lengthy WestPac deployments to Vietnam. The first was aboard USS *Hancock* (CVA-19) from 10 November 1965 to 1 August 1966, and the second on USS *Bonne Homme Richard* (CVA-31) from 28 January 1967 to 25 August 1967. During these tours the squadron flew F8C Crusaders. Aslund came back from his second eight-month deployment exhausted, but happy to be home once again.

His next duty was on the staff of Admiral John S. McCain, Jr., Commander-in-Chief, Pacific Command (CINCPAC), based at Pearl Harbor. Following this tour he received high praise from the admiral, and from there was assigned the job of Air Boss aboard *Bonne Homme Richard*, based in Alameda, California. The Air Boss was responsible for controlling all carrier aircraft, both on the flight deck and within a five-mile radius of the ship. During this one-year tour of duty the family stayed in Del Mar, California. Next he flew a desk for two years as Navy Liaison Officer at the Air Force Space and Missile Headquarters in Los Angeles.

During this time his marriage of sixteen years ended. He and his two sons moved to the Guantanamo Bay Naval Base in Cuba, where he led carrier inspection tours throughout the Atlantic Fleet. Then it was back to Hawaii as Assistant Chief-of-Staff under Rear Admiral Ralph Strafford Wentworth, Jr., Commander of the 14th Naval District.

Following his retirement from active service at the end of September 1980, Capt. Aslund stayed on in Hawaii. The next year he developed the Naval Junior Reserve Officers Training Corps (NJROTC) program for the state, working out of Radford High School, Pearl City, on the island of Oahu. During this time he found his future second wife, Diane Hasey, also a divorcée. "We met in Hawaii in 1977," she told the author. "I had just moved there from Massachusetts, and he had been reassigned to Pearl Harbor in 1976 as his last duty station before retirement. We met through friends, were married on base in Pearl Harbor on 17 August 1979 and lived on Oahu in Kaneohe and Kailua our entire married life."

Aslund died in Kailua on 20 October 1987, exactly two months to the day from when he was diagnosed with metastatic prostate cancer. He had just turned 60 years old and remained deeply involved in the NJROTC program to the end. "He was still 'wearing the uniform' and teaching up until the morning he died," according to his widow Diane. "In fact, he collapsed as he was getting ready to go to work."[18] Roland ("Auz") Aslund was buried with military honors at the Riverside National Cemetery in Riverside, California, which is also the final resting place of Mercury astronaut candidate and test pilot James Wood.

had undergone maintenance and needed functional check flights, he worked in the System Project Office for the Mach 3 North American F-108 Rapier. When funding for this program was withdrawn in September 1959 he was assigned to the Boeing and Michigan Aeronautical Research Center (BOMARC) System Project Office at Wright Field as the Mobile Inspection Unit (MIU) project officer. "The MIU was an oversized semi-trailer that contained all the test equipment required to test an air defense missile through its complete air defense profile without firing the missile," Dickey explained. "The MIU was manufactured by Boeing and I spent a lot of time shuttling between Wright-Patterson Air Base and Boeing, Seattle."

Capt. Dickey was then selected for the USAF Experimental Test Pilot School at Edwards AFB as a member of Class 61-C (along with future X-15 pilot and NASA astronaut Joe Engle), graduating on 19 April 1962. From 2 May 1962 to July 1964, his next assignment took him to the 4750th Test Squadron at Tyndall AFB, Florida, initially as an operational test pilot and later as Special Assistant to the Commander. It was during this time that he was chosen as an astronaut candidate by the Air Force, but ultimately missed the final cut. For thirteen months from July 1964 he attended the Army Command and General Staff College at Fort Leavenworth, Kansas. While at Fort Leavenworth he briefly flew the Cessna U-3A "Blue Canoe" from Richard-Gebaur AFB, Kansas before being designated an Instructor Pilot there in the T-33. In June 1965 he was again assigned to the 4750th Test Squadron at Tyndall where, as chief of the Projects Branch, he flew the F-101B/C Voodoo.

Maj. Roy S. Dickey as an F-105D pilot with the 469th TFS. (Photo: Roy S. Dickey)

ROY S. DICKEY, USAF

On 31 March 1929, Roy Sterling Dickey was delivered by a coun Meade County, Kansas farm home. Like many others, his parents O and Myrtle Ethel (Mathews) Dickey lost their farm to the bank du Depression. The family moved to Ashland, Kansas when he was six ye lived there until joining the U.S. Marine Corps in August 1946 a seventeen, just after his third year of high school.

Completing boot camp at San Diego Marine Recruit Depot, h stationed at MCAS El Toro in Santa Ana, California, with Marine Figh Squadron VMF-323 ("Death Rattlers"). In 1949 the squadron was featu movie *Sands of Iwo Jima*. While at El Toro, Dickey became an aircraft ar an F4U-4 Corsair until one day his ordnance sergeant (NCOIC) asked if he to go to the U.S. Naval Academy (USNA) at Annapolis. As he hadn't co high school Dickey did not believe this was possible, but the sergeant said he take the competitive fleet examination. The Marine Corps allowed him to take tests, which he passed. Upon being selected as a fleet candidate, he was transferr the Naval Academy Preparatory School (NAPS) at Bainbridge Naval B Maryland. After competing again in the entrance examination in May 1948 was delighted to be selected for appointment to the Class of '52.

"Since I was about four or five I had always wanted to fly, and going to the Nav Academy afforded me the potential opportunity for flight training upon graduation I was strongly considering going Marine Corps on graduation and applying for flight training out of Quantico, Virginia. So during my last year in the academy I asked my Company Officer, a Marine, what the policy of the Corps was for Marine 2nd lieutenants applying for flight training. He replied that the requirement was for two years in the field as an infantry officer and after that it depended upon 'the needs of the Corps'. So I said, 'Semper Fi' and elected to go United States Air Force."

Dickey graduated from USNA with a bachelor's degree in applied sciences and engineering on 6 June 1952, having been ranked 315th in his class of 783. He then joined the Air Force. "My first flight was on the first day after signing in for flight training with Flying Class 53-E at Marana Air Base, Tucson, Arizona, on 12 August 1952." His primary training began in the T-6G Texan before being transferred to Williams AFB, Arizona for basic training in the T-28A Trojan, advancing to the T-33 Shooting Star. After being awarded his pilot wings he undertook Phase I gunnery in the T-33 at Laughlin AFB, Texas, and from October 1953 to January 1954 moved on to Phase II at Luke AFB, Arizona, now flying the F-84G Thunderjet. Assigned to McGhee Tyson Airport in Knoxville, Tennessee, he would spend the next year and a half with the 469th and 460th FIS (Fighter-Interceptor Squadrons) flying F-86A/D Sabre jets.

In July 1955, Dickey was posted to Naha Air Base on the island of Okinawa, where he flew F-86Ds with the 16th FIS for the next three years. In early 1958 he returned to the mainland to earn his master of science degree in electrical engineering from the Air Force Institute of Technology (AFIT) at Wright-Patterson AFB, Ohio. After an occasionally hazardous but enjoyable period of test-flying T-33 jets which

During this assignment he also attended worldwide survival school at Stead AFB, Nevada.

On 17 June 1966, Maj. Dickey and 16 other pilots undertook combat training for Southeast Asia in the fifth F-105 RTU Class (67CR) at McConnell AFB, Kansas. Nicknamed "The Thunderstuds", the class deployed 15 Republic F-105 Thunderchief ("Thud") aircraft with the 561st TFS (Tactical Fighter Squadron) to George AFB, California between 23 September and 17 October 1966 for conventional weapons delivery training. Fourteen of the original 17, including Maj. Dickey, graduated from McConnell a week later. He was then assigned a combat tour, based with the 469th TFS, flying F-105Ds out of Korat Royal Thai Air Force Base (RTAFB), Thailand from 29 November 1966. The day after his arrival at Korat, he flew his first combat mission.

Flying over North Vietnam on 2 December 1966, Maj. Dickey was on just his third combat mission in a group of four aircraft attacking Phuc Yen airfield in North Vietnam, when the F-105D of flight leader Capt. Monte Moorburg was hit by flak from the ground. Moorburg calmly continued the attack in his stricken aircraft, dropped his bombs on the target and jettisoned his fuel tanks before heading west to the Red River, trailing fire, which quickly engulfed his aircraft. "The last time I saw Monte's Thud, it was on fire from the tip of the pitot tube to the afterburner eyelids," Dickey recalled. Moorburg somehow managed to eject before his F-105 exploded, but was killed in the process. Altogether, seven Air Force and Navy aircraft were lost that day. "My first real baptism of fire was a very sobering experience," Dickey later recorded.

Then, on his fourth combat mission two days later, and still largely inexperienced in combat flying, Maj. Dickey shot down a MiG-17 while flying an F-105. His flight was one of several in a second wave assigned to strike a railroad yard approximately two miles north of Hanoi. As the flight rolled in on the target, the pilots sighted four MiG-17s directly over the target, several thousand feet below their flight level. As Dickey came off his bomb run, he noticed one of the MiGs at his 10 o'clock position, attacking another Thunderchief. He was then 2,000 feet behind and slightly above the target's 4 o'clock position, so he began to fire his 20-mm gun as he closed to within 700 feet. He ceased firing when the heavily damaged MiG burst into flames at the wing roots and spun into the ground, becoming more intent on evading another MiG which had closed on his tail, shooting at him. He would receive the first of his eventual three Silver Stars for this mission. His citation read, in part, "Major Dickey avoided the attacks of MiGs, surface-to-air missiles, and the exceptionally heavy flak and delivered his ordnance precisely on target. He then exposed himself to great danger when he shot down a MiG that was attacking the other members of his flight."

On 3 June 1967, Dickey flew his 100th mission with the 469th TFS at Korat. Then, following his first combat tour in the F-105 Thunderchief, he was assigned to the Pentagon, Headquarters USAF, and was on the team that wrote the requirements for the F-15 in the FX Concept Formulation. "I then volunteered for a second combat tour in the F-105 in 1969 to get out of the Pentagon," he revealed, wryly adding, "In combat you know who is shooting at you." He would fly an additional total of 152 missions over Southeast Asia on his second tour, this time operating out of Takhli RTAFB from 29 November 1969, first as the Operations Officer in the 333rd TFS (Tactical Fighter Squadron) and later as Commander of the 44th TFS. He flew his final sortie on 6 October 1970, the day operations ceased for the F-105D. During his two combat tours, Lt. Col. Dickey had flown a total of 274 missions without taking a single hit from enemy fire. This also extended to all the pilots on the flights that he led, with the exception of one pilot who took a small-arms round in his trailing edge flap.

Dickey's next assignment was to 401st Tactical Fighter Wing (TFW) at Torregón Air Base in Spain, where he was promoted to colonel. He was then selected for the Air War College at Maxwell AFB, Alabama. Col. Dickey's subsequent assignments included Chief of Maintenance, Deputy Commander for Logistics, and Deputy Commander for Maintenance at George AFB. However, he could sense his career as an active combat pilot slowly dwindling. "I could not fly while in maintenance and they would not give me a flying assignment, so I retired from the U.S. Air Force," he stated. His retirement took effect on 31 July 1975, after 23 years in the service. Over the next two years he worked for Hughes Aircraft, became a crop duster for a further two years, and then purchased and operated a 210-acre farm/ranch for several years. He then worked for the aircraft simulator company Singer-Link for around two years before fully retiring to become what he cheekily describes as "a professional vagrant full time. I enjoy my present profession."

A recent photo of Roy Dickey. (Photo: Roy S. Dickey)

Dickey and his late wife, the former Wilhelmina ("Billie") Weeks Griffin, had four children: Roy Douglas, Stephen James, Clare Ann and Laura Lee. He has since married Gloria Ann Thomas. "We are happy together," he notes, "and are actively in love."

Col. Dickey's service decorations include three Silver Stars, five Distinguished Flying Crosses, a Bronze Star for meritorious service, the National Defense Service Medal with a bronze star, and 21 Air Medals.[19]

THOMAS E. EDMONDS

Tom Edmonds was born during the Great Depression in a farmhouse northwest of Kansas City, Kansas. When he started first grade he often rode to school with his grandfather on the family pony until the home farm was lost because of a local bank failure. After relocating to a nearby town his parents found work in the government's ambitious Works Progress Administration (WPA) program, which, at its peak, hired millions of workers to carry out public works projects such as the Timberline Lodge in Oregon, roads like the Merritt Parkway in Connecticut, and the laying of 20,000 miles of water lines. Tom contributed to the family's finances by delivering grocery advertisements, selling magazines, and cleaning a doctor's office. As he grew older he worked in a wholesale book company, delivered newspapers, and worked with an uncle laying carpet. Later still, he was employed in a biology laboratory.

In his youth, as he recalled, "The town we lived in celebrated an Old Settlers Reunion each summer. In the summer when I was 11 years old, I won enough money at a game of chance to pay for an airplane ride in an early Piper [that was] flown into a local hay field by a friend's uncle. The ground temperature was in the nineties with relative humidity in the eighties, so the pleasant conditions along with the beautiful view of the countryside gave me a great interest and incentive to have a life in aviation." After the experience of his first airplane flight, model balsa wood airplanes became a large part of Edmonds' life. "World War Two also gave me an opportunity to see military fighter airplanes on the local runways, and this further piqued my interest in aviation."

With family members adept at fixing almost anything and a boyhood associated with tools and science subjects, Edmonds made what he called "a natural choice" of an engineering education. Graduating from high school with a heavy load of science and math classes and a scholarship to the local community college gave him a good start to an engineering degree. His education continued at the University of Kansas and he was able to obtain a private pilots' license as a part of the Aeronautical Engineering (AE) program.

With the Korean War under way, and no desire to be drafted into the U.S. Army, Edmonds enlisted in the Naval Aviation Training program. After graduating as a naval aviator he was assigned to carrier-based Grumman F9F Cougar jet fighters, and met his future wife Wilma while in Navy jet training at Kingsville, Texas. After four years in the Navy he returned to the University of Kansas, graduating in 1956

Tom Edmonds at Edwards AFB. (Photo: USAF, courtesy Tom Edmonds)

with a bachelor's in aeronautical engineering. One special program involved rebuilding a surplus Sikorsky HOS-1 helicopter and flying it with an ex-Navy classmate, but this was terminated when the tail rotor support was damaged by the main rotor during a hard landing.

Fortunately 1956 was a good year for AE graduates and Edmonds received offers from airlines and aircraft manufacturers. He joined the Boeing Aircraft Company in their Flight Test Department, located in Seattle, Washington as a flight test engineer on the first KC-135 tanker, and was later assigned as a B-52 Stratofortress co-pilot, conducting structural testing during low-level flight. This was hazardous test flying, as evidenced when a primary structural failure in the horizontal stabilizer resulted in the loss of a B-52 and its five crew members flying low level near Burns, Oregon on 23 June 1959. Edmonds and the other test pilot had taken turns flying that aircraft on alternate days, and it was not Edmonds' turn that particular day.

It was an exciting time to be working for the aircraft company; in 1959 Edmonds began development flying with Boeing Model 367-80, the company's sole prototype for what would become the 707, America's first commercially successful jet airliner. The swept-wing aircraft was soon nicknamed the Dash 80 within Boeing. He started as the test engineer and was later assigned as project pilot for the 367-80. The next year his assignments increased, and he was conducting aerodynamic and structural load evaluations on numerous aircraft types. In December of that year the company sponsored him to the U.S. Air Force Test Pilot School at Edwards AFB, California as a member of the 12-strong Class 61-A. At the graduation ceremony on 18 August 1961 he was presented with two of the three TPS recognition awards by Col. Chuck Yeager; the A. B. Honts Trophy for outstanding student and the H. M. Ekeren Award for outstanding pilot.

On returning to Seattle, Edmonds continued test work on the Dash 80 and other commercial models, which involved automatic pilot, aerodynamics, all systems and new instrumentation for landing and navigation. During this time, he was shown a letter from NASA to the Chief Executive Officer of Boeing, indicating the agency's desire to have civilians involved in the second group of astronauts. The Director of Flight Test told Edmonds he was the only one qualified to apply and they would hold his job open if he wanted to try out. Much as the title of "astronaut" appealed to him, he was reluctant to apply due to the interesting test work in which he was involved at the time. Another important consideration was his family, with one young son, Dale, and another on the way. After discussing the situation with Wilma he finally decided to apply, knowing that he could opt out if he felt it was not what he wanted.

Edmonds at the controls of Boeing's Dash 80. (Photo courtesy Tom Edmonds)

Eventually, having completed all the preliminary medical and psychological tests, Edmonds found himself in the final group of 32 candidates, which included five civilian test pilots. It was not long before he began to realize that he was not ticking many 'boxes' in his desire to join the astronaut corps. "They were not forthcoming as to our pay, moving expenses, and such things as where we would live. I became spokesman for several civilians and would not take 'I do not know' for an answer. So my reputation was established. I was not selected and, somewhat relieved, I returned to a very happy group of flight test assignments at Boeing."

Four months after NASA announced the nine successful Group 2 astronauts, the Edmonds family was completed with the birth of a second son, Alan. Over the next several years, the family would enjoy seaplane fishing vacations to remote regions in Alaska and Canada – something he now knows would have been out of the question had he become a NASA astronaut.

In the years following, as Boeing's chief test pilot, Edmonds had assignments as project pilot, including several models of commercial airplanes and a Canadian De Havilland C-8A "Buffalo" short-haul research aircraft which had been modified by Boeing for NASA augmenter-wing research. He also conducted the first flights of the E-3 Sentry AWACS (Airborne Warning and Control) aircraft and the 737-300, while continuing his work on the testing and development of the Dash 80. "The Supersonic Transport (SST) fly-by-wire simulation for the swept-wing versus the delta-wing configuration as conducted on the Dash 80 was very interesting, as was the soft-field landing research," he noted. In this program the Dash 80 was outfitted with multiple-wheel landing gear and twenty normal sized tires operated at low pressure. "We landed on the grass at Wright Field, Ohio and the dry lake beds in California, but they refused to let us land on the grass by the Washington Monument in Washington, D.C., which was a shame as it would have been great publicity for Boeing [which was] then locked in competition with Lockheed and their massive C-5A Galaxy."

Overall, Edmonds served as project pilot on the Dash 80 from 1961 to 1972 and was a pilot on its final flight from the Transpo '72 exposition at Dulles International Airport, Washington D.C., delivering the test jet to a desert aircraft storage site near Tucson, Arizona. In 1990 the sun-faded aircraft was retrieved by Boeing, lovingly restored by company employees, repainted in its former brown and yellow livery and put on public display at Dulles Airport's Udvar-Hazy Center, a companion facility of the Smithsonian Institution's Air and Space Museum.

With his background in preliminary design, Edmonds served as project pilot on Boeing proposals for aircraft that became the Lockheed C5A, Northrop B2, General Dynamics F16, and Boeing AWACS. He was also Boeing's project pilot on the development of remotely piloted vehicles, contributing to the console design for the control of these vehicles. He said that flying a Cessna over a thousand miles away through a satellite control link was a real highlight for the entire Boeing crew.

Another assignment was assisting in the development of a short-field commercial airplane that could land at Meigs Field in Chicago. Ten years later that program had evolved into the 767. "The 767 was a wonderful experience," he reminisced. "I had been through many development phases of commercial jets and we were given a clean slate to change anything that we thought would improve the airplane and make

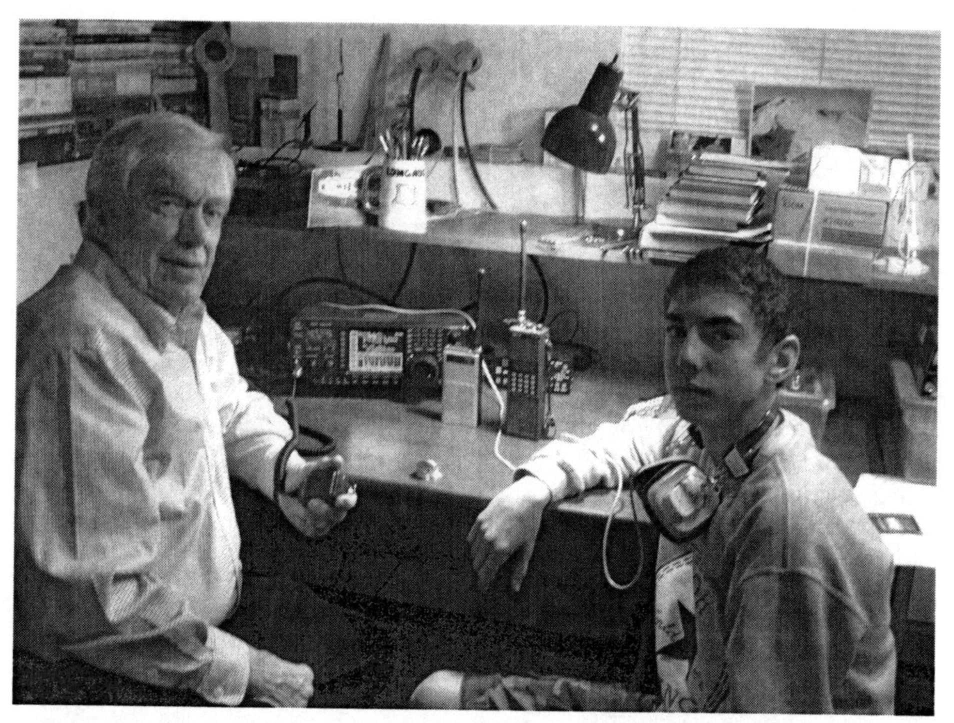

Tom Edmonds with fellow ham radio enthusiast, grandson Samuel Thomas. (Photo courtesy Tom Edmonds)

it first class from crew flight deck operations, including handling characteristics, as Boeing knew that these were critical items regarding safety. The flight instrument viewing, lighting, air flow, heating, control wheel size displacement and position, rain removal, seat comfort and automation of systems were all improved in the 767, as was the fly-by-wire primary flight and engine controls. It was the first commercial airliner to use color CRT (Cathode Ray Tube) primary flight instruments, which introduced a whole new era of information displays including navigation, systems diagrams and checklists. New navigation and automatic pilot computers, under pilot supervision, made lift-off to touchdown and stopping an all-automatic operation. The 767 design set off a new era in commercial airplane design."

Edmonds finished his Boeing test career in 1989 on the 747-400 series aircraft, flying the structural dynamic demonstration ("flutter") program. He continued in the industry for several years as a Federal Aviation Administration (FAA) Designated Engineering Representative (DER), which involved fire suppression, navigation and electromagnetic interference testing (EMI). Eventually he went into full retirement, operating ham radio with other volunteers including his grandson Samuel Thomas in emergency management, as well as vegetable gardening and an occasional fishing trip in his Cessna 185. Altogether, Edmonds had completed over 3,000 test flights and flown "85 models, 51 types not including 8 types of seaplanes and stick time in five different helicopters".

Asked if he regretted not entering NASA's space program, the response came quickly. "I have enjoyed seeing the beautiful pictures from space, but I've never regretted not being an astronaut because I ended up with more time with my family and had many satisfying experiences with the Boeing Company."[20]

WILLIAM H. FITCH, USMC

William Harold Fitch, who would one day be a highly decorated lieutenant general in the U.S. Marine Corps, was born in Chattanooga, Tennessee on 6 November 1929, the son of William E. and Lelia Fitch, and a one-year-younger brother to Beverly. Soon after his birth the family moved to Fort Meade, Florida.

Bill Fitch grew up in Fort Meade but attended school in Bartow, 11 miles away. He graduated from Bartow High School's Summerlin Institute – then the only public military school in Florida – in June 1947. Three years later, in June 1950, he would graduate with his bachelor's degree from the University of Florida. On 2 September of that year he entered the Naval Aviation Cadet Program and in November began flight training at NAS Pensacola, Florida flying the SNJ (a Navy-modified version of North American's T-6 Texan) and Grumman F6F Hellcat. He was designated a naval aviator, got his Wings of Gold, and was commissioned a 2nd lieutenant in the

Naval Cadet William Fitch, 1950 (Photo courtesy William Fitch)

U.S. Marine Corps Reserve on 1 April 1952. He then undertook advanced flight training at NAS Corpus Christi, Texas from April to June 1952, and was assigned to VMF-114 at MCAS Cherry Point in North Carolina. Over the following year he flew Vought F4U-5N Corsairs from USS *Wright* (CVL-49) and *Tarawa* (CV-40), logging over 100 carrier landings. His time aboard *Tarawa* included a two-month Caribbean cruise and a seven-month Mediterranean cruise.

In July 1953, Lt. Fitch joined VMA-324 at MCAS Miami, and three months later his squadron was deployed aboard USS *Saipan* (CVL-48), flying Douglas AD-4B Skyraider aircraft on an extended Far East tour. Up until November 1954 he flew operations in support of the Korean War armistice and ferried aircraft to the French in Vietnam in the first Indo-China War, once again logging over 100 carrier landings in the AD-4B. He was promoted to the rank of captain in December 1954 and given a regular commission in February 1955. On completing photo reconnaissance school, Capt. Fitch joined VMCJ-3 based at MCAS El Toro in California, where from June 1955 to April 1956 he flew the Grumman F9F-5P Panther. During this assignment, on 7 August 1955 he married Margaret Marie Williams from his old hometown of Bartow, Florida. He went on to serve another 15 months at El Toro as Operations Officer for the Tactical Air Command Center of the 3rd Marine Air Wing.[21]

The next step in Marine Capt. Fitch's service career came with his appointment to the U.S. Navy Test Pilot School at Patuxent River, Maryland. "At the time TPS was a six month school," he stated. "Before I finished the school [Class 19] in February 1958, I was told by the senior Marine Corps officer at Patuxent River that after TPS I would be going to China Lake and to VX-5. I had never heard of VX-5 until a few months prior to TPS graduation." After graduating, Fitch served the next two years as project test pilot with the Navy's aviation test squadron in China Lake, California, where he flew the A4C and FJ-4B on developmental flights for conventional and nuclear weapons tactics, as well as flying the FJ-4B aboard several aircraft carriers. "In 1958, as it had been for perhaps ten years, all the flying in VX-5 was connected with nuclear weapons delivery, whether it was low-level navigation, aerial refueling, mission profiles, loft bombing, or dive bombing. When I joined VX-5 in March 1958 we did every kind of nuclear weapons delivery that there was."[22]

It was while serving at China Lake that Capt. Fitch invented the Multiple Carriage Bomb Rack (MCBR) that revolutionized bomb carriage on jet aircraft. Fortunately, the commanding officer there, Cmdr. Dale W. Cox, Jr., saw the merit in the proposal and set things in motion under complete secrecy. As directed by Capt. Fitch and the squadron's Maj. Rice, special wiring harnesses were used to build the multiple bomb racks by the squadron's metallurgists and avionics personnel. On 30 September 1959, just three months after developing the concept, Fitch made the first high-speed, low-altitude test flight carrying the prototype MCBR. The equipment held up so well that the next day Fitch carried and dropped six MK-81 (250-pound) inert bombs. "It worked great," Fitch later recalled. Some years later he, Cox, and Rice attained a U.S. patent on the innovative bomb rack, which evolved into the multiple ejector and triple ejector racks still in use on jet attack and fighter aircraft today. For his work in developing the MCBR, and for operating the first thirteen

L/Gen. William Fitch, USMC. (Photo: USMC courtesy William Fitch)

test flights with the device on A4 aircraft, Fitch was awarded the Navy Commendation Medal.

From May 1960 to March 1962, Fitch was assigned as an A4B Skyhawk pilot with VMA-311 at Yokosuka, Japan. Promoted to the rank of major, and subsequent to his unsuccessful bid to become a Group 2 NASA astronaut, he served as a staff officer in the Bureau of Weapons and with the Research and Development Division in the Office of the Chief of Naval Operations at the Pentagon until July 1965. After finishing Marine Corps Command and Staff College, Maj. Fitch served as Logistics Officer of MAG-14 and then assumed command of VMA(AW)-225, an A6A all-weather attack squadron. Promoted to lieutenant colonel on 1 July 1967, he reported to Chu Lai in the Republic of Vietnam, where he took command of a second A6A squadron, VMA(AW)-533, also serving as Group Operations and Group Executive Officer. He flew 310 low-level combat missions in A6A and A4 aircraft, including 127 against targets in North Vietnam.

Fitch graduated from the National War College and received a master of science degree in international affairs from George Washington University. From April 1972 to September 1973 he led Marine Air Group 14, then commanded the 32nd Marine Expeditionary Unit in the Mediterranean Sea and flew a number of helicopter types from the amphibious assault ship USS *Iwo Jima* (LPH-2), completing 30 helicopter landings on a number of amphibious vessels. After serving

Bill Fitch addressing a 2007 meeting of the Golden Eagles. (Photo courtesy John R. ("Smoke") Wilson, Jr.)

for a year with the 2nd Marine Air Wing, he was assigned to Marine headquarters on his third Washington tour until July 1975. Advanced to brigadier general on 1 April 1976, he transferred to the 1st Marine Air Wing on Okinawa, where he served for a year as Assistant Wing Commander, then commanded the 9th Marine Amphibious Brigade from December 1976 to May 1977. After another tour at Marine Corps Headquarters, on 2 June 1980 he was appointed Commanding General, 1st Marine Air Wing, flying jet, turbo prop, and helicopter aircraft before relinquishing command on 2 June 1982. Further promoted to lieutenant general on 1 July 1982, Fitch assumed duty as Deputy Chief of Staff for Aviation until his retirement from the Marine Corps on 1 September 1984, at which time he was awarded the Navy Distinguished Service Medal.

During his 34 years of active duty, 32 of them as a Marine Corps officer, L/Gen. Fitch flew 6,895 flight hours in 121 models of aircraft, including more than 4,000 jet flight hours in fighters and attack aircraft, over 1,000 hours in propeller fighters and attack aircraft, more than 1,000 hours in various propeller trainers, proficiency and utility aircraft, and close to 600 hours in helicopters.[23]

His combat awards include the Silver Star, Distinguished Flying Cross, 29 Air Medals (4 single mission and 25 strike/flight), the Legion of Merit with Combat "V" and two awards of the Navy Commendation Medal. In March 1985, L/Gen. Fitch was appointed by President Ronald Reagan as one of fourteen new members of the National Committee on Space. Today, he and his wife Margaret live in McLean, Virginia.

JOHN M. FRITZ

Perhaps the biggest influence in John Fritz's life was a man that he understandably admired – his father. Lawrence George ("Larry") Fritz joined the Aviation Section of the U.S. Army Signal Corps in June 1917 and was attached to the Royal Air Force in England for flight training. Frustratingly, the war ended before he earned his wings and he was discharged in December 1918.

On his return to the United States, Larry Fritz became a marine engineer before rejoining the Army Air Corps to earn his wings and was discharged in 1924. He later found work as a barnstormer until he entered civil aviation and flew for a number of airlines including the one operated by the Ford Motor Company, flying Ford-built Stout 2-AT aircraft. Beginning in October 1925, Postmaster General Harry S. New awarded eight airmail routes to seven airmail carriers, including Ford Air Transport, which received two routes. Ford became the first company to commercially fly U.S. airmail under contract, starting on 15 February 1926. The pilot on that historic first airmail flight to Cleveland, Ohio was Larry Fritz. Later, he took on work as chief pilot for Maddux Air Lines and in September 1934 was appointed Vice President of Operations for Trans World Airlines (TWA). When the Second World War broke

John M. Fritz at Edwards AFB. (GE photo courtesy John Fritz)

out he was recalled to service, rising to the rank of brigadier (later major) general in the Air Transport Command's North Atlantic Division, and he was awarded the Army Distinguished Service Medal, Legion of Merit, Air Medal and Distinguished Flying Cross. Post-war he was appointed Vice President of Operations for American Airlines until his retirement in 1955.[24] Little wonder that young John wanted nothing more than to emulate his father.

John Merritt Fritz was born on 15 August 1930 in Tulsa, Oklahoma, although the family relocated to Kansas City, Missouri the following year. When he was 13 years old the family moved again, this time to Manchester, New Hampshire, where he took his early high school education at Manchester Central High School. In October 1945 the family moved yet again to Manhasset, New York, where he completed his high school studies in July 1948. He then took on two years of on-campus Air Force ROTC at Johns Hopkins University, Maryland, with a requirement that he join the service upon graduation. On gaining his bachelor of arts degree in 1952, he entered flight training with the Air Force as an aviation cadet in July of that year.

Fritz got his initial training at Columbus AFB in Mississippi flying T-6 Texans, and then received six months of basic single-engine training at Foster AFB, Texas in T-28 Trojan propeller aircraft and T-33 Shooting Star jet trainers. He received his wings and was commissioned a 2nd Lieutenant in the U.S. Air Force in August 1953, the day that the Korean War armistice was signed. Fritz then reported to Jet Combat Crew Training at Laughlin AFB, Texas where he was provided with basic bombing

The GE project team at Edwards, 1960. John Fritz fourth from right, kneeling in front row; fellow test pilot Elliot See extreme right, front row. (GE photo courtesy John Fritz)

and gunnery combat skills in the T-33. In September 1953 he transferred to Luke AFB for tactical fighter-bomber training in the F-84B. He reported to Hamilton AFB in November 1953, assigned to the 325th Fighter-Interceptor Squadron (FIS) which flew F-86D Sabres.

In December that year, Fritz was undergoing all-weather instrument training at Tyndall AFB, Florida where his older brother Jim was taking F-86D training. On 15 December, Lt. James D. Fritz was on a training flight when the turbine wheel of the J47 engine powering his F-86D failed after takeoff. In order to avoid crashing into a school house under his flight path, Fritz delayed his ejection until it was too late for his parachute to fully deploy and he was killed. For his gallantry he was awarded the Air Medal and buried with honors at Arlington National Cemetery, Virginia.

Still getting over his brother's death, John Fritz was assigned to ferry duties in July 1954, flying the all-new F-86D interceptor from North American to bases across the United States. In 1956 he completed his active duty career as a 1st lieutenant, married his sweetheart Dottie on 24 March 1956, and over the next twelve months flew F4D Skyray and A4D Skyhawk aircraft as a production test pilot for Douglas Aircraft. In July 1957 he was hired as a test pilot by the General Electric Company (GE) Jet Propulsion Division, based at Edwards AFB, California. There he worked on engine development, in particular the J79 (the company's variable-stator turbojet engine and the world's first production Mach 2 power plant), and made engineering test flights with the F-104 Starfighter. The Starfighter was Fritz's favorite airplane in which he zoomed without rocket assist to 92,000 feet and successfully achieved two dead-stick landings after engine failures. One of these dead-stick landings was at

John Fritz in his 'g' suit with the F-104 prior to his flight up to 92,000 feet. The name on his helmet reads "Fearless". (Photo: USAF courtesy John Fritz)

Mojave Marine Corp Air Station in which Fritz safely brought the F-104 to a stop in 4,752 feet on a runway that was less than 5,000 feet long. The F-104 Pilot Operating Handbook recommended using a runway no shorter than 15,000 feet for a dead-stick landing attempt. In recognition of his saving a highly valuable aircraft with its failed engine for technical investigation, Fritz received the Air Force Flight Test Center Flight Safety Award for March 1962. This award was marked by the presentation of a Zippo lighter with the AFFTC coat of arms affixed on the front, which he described to the author as "a really big deal".

Over the course of his six years with GE he would train pilots on the new power plant at U.S. Air Force bases, as well as nations in Europe, the Middle East and Asia that operated the F-104. In the course of developing the J79, CJ805, J85, CF700 and TF39 he flew the RB-66, XF4D Skyray, F11F-1F, F-101, F102, F-104, T-38, B-52 and F5B-21.

In 1962 Fritz was a finalist in the NASA Gemini selection along with fellow GE test pilot Elliot See. He did not make the final cut but See did.

Fritz left GE in July 1963 to take work over the next two years with the Northrop Corporation, test flying their T-38 Talon and F-5A, F-5B Freedom Fighter aircraft, and demonstrating the latter fighter jet to allied countries overseas. He demonstrated the F-5B at the Paris Air Show of 1965. He also flew Northrop's X-21A Laminar Flow Control experimental aircraft to explore the low drag benefits of this concept. In 1965 he returned to General Electric as chief test pilot and continued to work on the development of the J79 engine. It was during this time that he was involved in a

Moments from disaster; the XB-70A-2 flanked by (from left), a T-38, F4B, F-104 and John Fritz's YF-5A on the Valkyrie's final, fatal flight. (Photo: NASA)

tragic aircraft accident that took the lives of two top test pilots, including X-15 pilot Joe Walker, and led to the loss of an XB-70 experimental jet.

The delta-winged XB-70A Valkyrie was developed by North American as the prototype of a long-range strategic nuclear-armed bomber envisaged by the U.S. Air Force. A large aircraft crewed by only a pilot and co-pilot, it was capable of Mach 3 and altitudes exceeding 70,000 feet. Two XB-70A aircraft were built to undertake experimental test flights from 1964 to 1969, and were identified as Air Vehicle 1 and 2 (AV-1, AV-2).

On 6 June 1966, XB-70A-2, powered by six GE J93 engines, was scheduled to complete a flight test program out of Edwards AFB and then take part in an 8:30 a.m. airborne photo session organized by John Fritz, which would feature a formation of aircraft powered by GE engines. He would be flying a Northrop YF-5A, while at the controls of the XB-70A were North American chief test pilot Alvin White and Air Force Maj. Carl Cross, making his first flight in the aircraft. Also participating in the photo op was a Northrop T-38A flown by Capt. Peter Hoag along with Col. Joseph Cotton – another XB-70A pilot and XB-70A program director. The other two aircraft were a McDonnell F4B flown by USN Cmdr. Jerome Skyrud and E. J. Black, and a Lockheed F-104A chase plane flown by NASA's chief test pilot, Joseph Walker, who had previously set records in the supersonic X-15 rocket plane. A civilian Gates Lear Jet, chartered by GE and flown by H. Clay Lacy, was carrying several company photographers, and it was to follow the formation, documenting the event.

Towards the end of the 90-minute photo session the five aircraft were flying in a "V" formation, with XB-70A-2 in the lead. The F4B was neatly tucked in to its left, with the T-38A on its left wing. To the Valkyrie's right were in turn the F-104A and Fritz's YF-5B. At 9:26 a.m. it appears that Walker's aircraft drifted aft in such a way that its "T" tail became caught up in the XB-70A's right wingtip vortex, flipping the smaller aircraft violently to the left and causing it to hit the leading edge of the larger aircraft's right wing before rolling up and left across the back of the fuselage. It then sheared off most of the Valkyrie's right vertical stabilizer and about 60% of the left stabilizer. Fritz followed Walker's F-104 down to impact on the desert floor. Walker had evidently been killed by the impact with the XB-70.

Unaware for a few moments that they had been hit, White and Cross continued to fly the XB-70A straight and level for another 15 seconds. White later said, "I heard a good loud thump – an explosion – and I heard somebody yell 'Midair!' But with all that length and mass behind me I didn't know it was us."[25]

When the XB-70A finally reacted to aerodynamic forces it yawed abruptly and violently to the right, and then became inverted, nose down, before rolling again. The nose was then pointing skyward, almost vertical. A portion of the left wing broke off and the doomed aircraft fell in a flat spin. White activated his escape capsule, but it seems that Carl Cross had rapidly lapsed into unconscious due to the high g-forces being created by the out-of-control aircraft. White survived, although his capsule's protective air bag, designed to inflate when the main parachute deployed to cushion the shock of impact, failed to inflate. "Then I heard the XB-70 hit," he recalled. "It made a terrible explosion and an enormous plume of smoke came up."[26] The capsule hit the ground hard, and White was badly injured. Incredibly, there were no bones broken in the instantaneous 45 g force at impact, but the organs of his lower body were virtually paralyzed, and his life hung in the balance for several days before he began to slowly recover. Maj. Cross was killed when the XB-70A-2 crashed in the desert near Barstow, California.

John Fritz continued to work at GE as chief test pilot, conducting tests of the J85 power plant in Northrop's T-38 and YF5B-21. He was co-pilot on GE's seven engine B-52 flying test bed when the world's first high bypass engine – the TF39 – made its maiden flight. He then returned to the Air Force Reserves in 1968 as a troop carrier pilot, based at March AFB, California, prior to his honorable discharge with the rank of captain and retirement from professional flying in June 1969. By this time he and Dottie had two sons, Greg and Jim.

The next challenge came as GE's manager of Commercial Engines Programs in Washington, D.C., representing GE to the Department of Transportation, the Federal Aviation Administration and Congress on the Boeing SST (Supersonic Transport) and other commercial aircraft programs. When Congress canceled the SST in 1972, Fritz relocated to GE Engine headquarters in Cincinnati to manage post-production programs. In 1974 he transferred to Seattle, Washington to become District Manager of GE's Northwest Office, which provided sales and service support to Boeing. He played a key role in winning the competition for the National Emergency Airborne Command Post with the CF6-50 engine, GE's first application on a large Boeing aircraft. This led to its selection for the "Air Force One" fleet of 747 aircraft for U.S. Presidential use. Of perhaps greater significance was the selection of the CFM56 GE engine for the 737-300, displacing the Pratt & Whitney engine used on the original 737-200 series. The CFM56 engine, the result of a joint venture between SNECMA of France and GE, is now the most highly-produced jet engine of all time, with more than 7,000 orders for 737 aircraft (as of this writing) for all 737 series through to the 737-900.

In 1985 Fritz relocated to Melbourne, Australia, where for the next seven years he served as Vice President of GE's Australasia Aircraft Engines Division, responsible for all marketing and support of the company's engine products in the Pacific region. Through his work, GE engines were mounted on a number of military and civilian aircraft, including the locally based Qantas Airways. He retired from GE in October 1992.

Following retirement, John Fritz has been active in his church as President and Elder. He is also a member of the Society of Experimental Test Pilots, as well as the Ancient Order of the Quiet Birdmen, a fraternal organization of civilian and military aviators. Now residing with his wife Dottie in Kent, Washington, he enjoys playing golf, shooting skeet and hunting game birds with his sons Greg and Jim in eastern Washington state.[27]

WILLIAM J. GEIGER, USMC

Toledo is located in northwest Ohio, at the western end of Lake Erie. It was here, on 6 January 1931, that William John Geiger became the first-born child (and only son) of the former Marion Good and William Christian Geiger, a newspaper advertising salesman. Along with two younger sisters, Bill Geiger grew up in the middle-class family during the Depression years. He attended Renwood Elementary School and Scott High School in the Old West End neighborhood of the city, where he enjoyed the sporting pursuits of basketball, track, baseball and tennis.

The recipient of a U.S. Navy ROTC scholarship which gave him the opportunity to attend Vanderbilt University in Nashville, Tennessee, he studied there for two years before his father's failing health caused him to switch to Ohio State University in Columbus for the final two years of study. He graduated in 1952 with a bachelor's degree in biological sciences, and promptly married Adrianne Hawk. They were both graduates of Scott High School and students at Ohio State University. To serve his NROTC post-graduation obligation of a two-year period of military service, Geiger elected to go to the U.S. Marines and was commissioned a 2nd lieutenant. As this was the time of the "police action" in Korea he was then whisked off to the Quantico Marine Base in Virginia to become a platoon leader. He began to enjoy military life so much that he abandoned his earlier plans to attend medical school at the end of his military service.

While attending Basic School class he became friends with Ken Weir, with whom he would later apply for astronaut selection. After Basic School he applied for flight training, and in early 1953 went to Pensacola, Florida along with Weir. After training on North American SNJs they were assigned to Corpus Christi, Texas for advanced flight and instrument training, flying Grumman F6F Hellcats. "I remember the first day I got behind a Pratt & Whitney R-2800," Geiger reflected. "The power it had was just amazing." Then the training switched to jets, first flying the Northrop T-33 and later the Grumman F9F-2 Panther. "We were in Kingsville, Texas, which was so hot that the aircraft took every darned inch of the runway before lifting off."

Geiger was assigned to his first squadron at MCAS Cherry Point, North Carolina in the summer of 1954. He flew the F9F-6P Cougar over the next 18 months with Marine Photographic Squadron VMJ-2. Then the Marine Corps decided he needed to broaden his scope and sent him to Communications Officer School back in Quantico. By this time he and Adrianne had two young daughters, Gwynn and Christianne. On completing the training he was shipped to Atsugi, Japan. After serving with a photo squadron he became the Communications Officer for Marine Air Group 11. Towards the end of 1957 he returned to Cherry Point, where he flew the photo reconnaissance Vought F8F-2P Crusader, the F3D and the Douglas F3D Skyknight, a portly aircraft that had earned the nickname of 'Willie the Whale'. On 29 July 1958 Geiger had to bail out of an F3D through the downward escape chute when the aircraft caught fire.

"We were about 15,000 feet above Pamlico Sound, North Carolina when we suffered engine failure. The J34 engines on the F3D were in the fuselage and the fuel was in tanks between the engines. The J34s were produced by Westinghouse, which had graduated from making toasters to producing jet engines. At that time they had a nasty habit of shucking their turbine blades, which would go right through the side of the aluminum into the fuel tanks, allowing quantities of kerosene to run over hot

Capt. William Geiger, USMC. (Photo: USMC courtesy William Geiger)

engines. We actually lost a lot of flight crews trying to bring those airplanes back – but not this time. The pilot and the avionics operator were more or less seated side-by-side, and by pulling the panic handle I could blow the door on that chute, which also disassembled the sides of our seats and we could just swing around and go out feet first. The nylon letdown was completely uneventful; just the way they'd trained us to do it." Geiger and Sgt. Taylor landed in the Pamlico River and were rescued by an Army LCU (Landing Craft Utility) in the inland waterway. "Six months later I got my orders to Test Pilot School at Patuxent River, Maryland."

Following his graduation as part of USNTPS Class 26 in 1960, Capt. Geiger remained at Pax River in the Fighter Branch of Service Test Division, flying F8 and F4 aircraft doing zoom climb work – once making it to 96,000 feet – and did a lot of thunderstorm evaluation work on the F4 equipped with a steel radome and lightning attenuator. "We determined that in most cases the airplane could get through a thunderstorm and still be functional on the other side. After all, it was supposed to be an all-weather fighter." During his time at Pax River, Geiger and his wife had two more children – sons Greg and John, born 1960 and 1962 respectively.

In the spring of 1963, having missed out on NASA's second astronaut group selection the year before, Geiger received orders for MCAS El Toro, California, and

U.S. Navy Test Pilot School Class 26 on a tour of aircraft manufacturers following their graduation. In the back row: Bill Geiger fifth from left, Alan Bean seventh from left and instructor Pete Conrad third from right. (Photo: USN courtesy William Geiger)

joined the F4B all-weather squadron VMF(AW)-513. On being promoted to major soon thereafter, he was made the squadron's Executive Officer under Commanding Officer Tom Miller.

Having trained for a little over a year at El Toro, in the fall of 1964 VMF-513 flew their aircraft across the Pacific to Atsugi, Japan, led by Maj. Geiger, while at the same time the all-weather squadron they were replacing, VMF-314, flew from Atsugi to California. In doing so, the two squadrons chalked up another "first' in Marine Corps history. "Not only a first," according to Geiger, "but I don't think it was ever repeated." Adrianne would remain at Dana Point, California with their four children whilst he was overseas. Soon after his arrival in Atsugi, Geiger was pulled from the squadron and attached to helicopter group MAG-36 in Da Nang as a liaison officer for the arrival of fixed wing aircraft. He would later rejoin VMF-513, returning home at the end of 1965. Assigned as personnel officer to MAG-33, he then returned to the 3rd Marine Air Wing at MCAS El Toro. "This didn't exactly thrill me, but I tried to do a good job. However, it was pretty obvious to me that Vietnam was a political war and not a military war, and I re-evaluated a lot of things, deciding that when the opportunity came for me to do something else I would leave the Marine Corps." Maj. Geiger resigned and left active duty in August 1966.

Geiger opted to remain with the Marine Corps Reserves, initially at Los Alamitos, California and shortly thereafter was promoted to the rank of lieutenant colonel. His

Capt. Geiger (right) in 1963 with Chief Warrant Officer (CWO) Carl Meyers. (Photo: USMC courtesy William Geiger)

Lt. Col. William Geiger, USMCR (Photo courtesy William Geiger)

first command, which lasted eighteen months, was with Air Support Squadron 4. He remained on staff for the next six months before being given the opportunity not only to create the first Marine Reserve Cobra Helicopter Attack Squadron, HMM-764, but also to learn to fly helicopters. Meanwhile, his civilian job was as a test pilot with the Autonetics Division of North American which involved testing radar and fire control systems. He would remain there for almost five years before joining the faculty at the University of Southern California. But four years later Geiger decided he wasn't cut out for academia, and even though it would be the first time he had been involved in corporate aviation he took up a position as manager of aircraft operations with Owens-Illinois, a company specializing in packaging and container glass products situated in his hometown of Toledo, Ohio. He remained with Owens-Illinois from 1975 to 1986, and when they began to downsize, which involved selling several of their aircraft, he took a job with an oil exploration company in Louisiana. Meanwhile, in September 1979 he retired from the Marine Corps Reserve, and in 1981 he and Adrianne were divorced. After two years with the oil company Geiger was on the move again, this time to Federal Aviation Services, a management company in San Jose, California, which operated Challenger aircraft and later Gulfstream business jets in Los Angeles. In 1989 he met gifted Chinese-born artist Rulan Weng when he approached her with a commission to design artwork for the interior of the company's Gulfstream jets. They were married on 2 February 1992, and he became stepfather to her son Chien. Two years later he took up part-time work at Flight Safety International, instructing in Gulfstream programs and carrying out contract flying, giving him and Rulan more time to enjoy life and travel while she pursued her career in fine arts. Sadly, Rulan Geiger passed away on 2 March 2012 after fighting a brave battle with lung cancer.

These days, Bill Geiger still flies when he can, although he said his final flight in a Gulfstream was in 2011. Fortunately his son Greg owns three airplanes, including a Piper Lance that they fixed up, and he now flies that aircraft "quite frequently".[28]

DAVID L. GLUNT, JR., USN

Ann Glunt is understandably proud of her late husband and his many achievements in his all-too-short life. "All I can say personally, is that Dave loved flying and his career," she offered as a reflection on him. "I know he was honored to be considered as an astronaut but it was never his ultimate goal as he would have had to give up many years of flying and serving in the Navy for one flight in space. It wasn't even something that he talked about frequently once it was over."

David Latta Glunt, Jr., was born on 9 March 1930 in Charleroi, a borough in Washington County, Pennsylvania, situated along the Monongahela River and some 25 miles south of Pittsburgh. He grew up in a family that valued a good education. His father, David ("Chippy") Glunt, had studied at California State Teachers' College, later becoming a teacher and long-time principal at his *alma mater* Charleroi High School, a successful school football coach and basketball mentor, and Superintendent of Schools in 1956. His mother Henrietta was also a school teacher, a profession later followed by his younger sister, Ann, born in 1938. Unknown to him at the time, he lived just a few miles from another Navy man who later became a lifelong friend and fellow astronaut aspirant, John Yamnicky.

David Glunt attended the U.S. Naval Academy, Annapolis, in the Class of '54. Apart from his academic studies he also excelled in sports, particularly as a member of the USNA Varsity 150-pound football team that won the Eastern Championship in 1953. Whilst at the academy he began dating Ann Ely Buchanan, who had also attended Charleroi High School. "We were not high school sweethearts although we attended the same school," Ann reflected. "He was two years ahead of me. We dated his last three years at the academy and were married a week after his graduation."

After graduating from the academy in 1954, Glunt began his flight training that August, receiving his Wings of Gold as a naval aviator in November 1955. Assigned as an attack pilot and later attack instructor with VA-44 ("Hornets"), Lt. (jg) Glunt flew aircraft such as the F2H2 Banshee, F9F-8 Cougar and A4A Skyhawk while at NAS Jacksonville, Florida. Another member of his squadron at this time was Lt. (jg) Alan Bean, who would walk on the Moon two decades later. From August to October 1957 the squadron, led by Commander Tom Sedell, flew F9F-8 aircraft from the carrier USS *Wasp* (CVS-18) on a North Atlantic cruise that provided protection for the VS (Sea Patrol) squadrons operating anti-submarine patrols from that carrier. From December 1958 to January 1960 he served in VA-34 ("Blue Blasters") flying A4A and A4B carrier-capable attack aircraft, and as Assistant

Lt. David Glunt, USN. (Photo: USN courtesy Ann Glunt)

Maintenance Officer, deployed in the Sixth Fleet. The squadron was involved in two Mediterranean cruises during this time, flying A4D-1 and A4D-2 aircraft from USS *Saratoga* (CVA-60). In January 1960 he returned to VA-44 as an attack instructor, and from May to October of that year he served as the Attack and Instrument Training Officer on the staff of Commander Replacement Air Group 4, flying A4D-2N, F9F-8, T-28B Trojan and F8 Crusader aircraft.

From October 1960 to June 1962, Lt. Glunt attended the U.S. Navy Test Pilot School at the Naval Air Test Center (NATC), Patuxent River as a member of Class 28. On graduating from USNTPS he served in the Ordnance Branch of the Weapons Test Division of NATC, flying a variety of attack and fighter aircraft as an ordnance project pilot. He also took part in the Navy's preliminary evaluation of the Grumman A6 Intruder twin-jet engine aircraft. From July 1962 until June 1963 he attended the U.S. Naval Postgraduate School in Monterey, California, during which time he was an unsuccessful finalist for NASA's second group of astronauts. On graduating with his master's degree in Navy Management, from June 1963 to May 1965 he served on the staff of the Deputy Commander Operational Test and Evaluation Force, Pacific, as Assistant Radar and ECM (Electronic Countermeasures) Officer and Assistant Air Warfare Officer.

After A4D-5 refresher training with attack squadron VA-125, in September 1965 Lt. Glunt reported to VA-216 ("Black Diamonds"). One of his commanding officers at this time was Carl Birdwell, Jr., another NASA Group 2 astronaut finalist. Glunt subsequently served as the squadron's Administrative Maintenance, Operations and

Executive Officer and flew attack missions over North Vietnam during deployments in the Seventh and Sixth fleets aboard the carriers USS *Hancock* (CVA-19) and USS *Saratoga* (CVA-60). From December 1967 to April 1969 he served a second tour at NATC, Maryland as a flight instructor and Assistant Director for Administration and Maintenance.

His next assignment was to VA-93 ("Blue Blazers") aboard USS *Ranger* (CVA-61) during a combat deployment with the Seventh Fleet in October 1969, initially as Executive Officer prior to relieving Thomas W. Schaaf as Commanding Officer on 19 January 1970, a position he held until 19 November 1970, when was transferred to VA-153 ("Blue Tail Flies") as Executive Officer and flew the A7B attack aircraft.

His subsequent orders returned Cmdr. Glunt to Washington, D.C., along with his wife and two children, David and Linda. He served two years in the Naval Material Command (MAT 03) as Project Administrator of Advanced and Exploratory Aircraft Development. At the completion of this Pentagon assignment, he reported on 16 May 1974 as Commanding Officer of USS *Ponce* (LPD-15), an amphibious transport dock. Under his command *Ponce* completed an extended Mediterranean cruise from 25 September 1974 to 20 March 1975, participating in training exercises. Captain Glunt later served as Operations Officer on the Staff of Commander Carrier Group 4. Next was command of USS *Guadalcanal* (LPH-7), an amphibious assault

Capt. David L. Glunt, USN. (Photo: USN courtesy Ann Glunt)

helicopter carrier during a Mediterranean cruise from January to July 1976, followed by another Pentagon assignment, this time at the Navy's Human Resources Management Center. Glunt finally retired from the service on 1 July 1983 after a tour as Deputy Director of Commander, Norfolk Naval Base in Virginia, where he and Ann had decided to take up residence.

Capt. Glunt's many honors and awards include the Distinguished Flying Cross (four times), the Bronze Star, 26 Strike/Flight Air Medals, two Navy Commendation Medals with Combat "V", the Navy Unit Commendation, the Meritorious Unit Commendation, the National Defense Service Medal with Bronze Star, the Navy Expeditionary Medal and the Vietnam Service Medal (four stars).

In retirement, David Glunt became an active and long-time member of St. Aidan's Episcopal Church, situated on the nearby Little Neck Peninsula of Virginia Beach, Virginia. Those years of retirement with his beloved Ann after his 38 years of active service were both rich and fulfilling. However, the terrorist attacks on 11 September 2001 may have been responsible for yet another victim, far removed from the tragic events of that day. Horrified by the news unfolding before them all day on television, David and Ann watched live pictures as terrorist evil claimed the lives of thousands of innocent people. They heard about the American Airlines passenger jet slamming into the Pentagon, but David would never know that one of his good friends was on board Flight 77 that morning. The following day, saddened beyond belief at what had happened in his country, he suffered a fatal cardiac arrest.

Ann will never know if the two events were related. "Many people think his heart attack was caused by the events of September 11," she mused. "But we can never be sure. It was a difficult week." The names of many of the victims were not released for some time, so it is unlikely that he knew his old Navy buddy John Yamnicky was a victim that day. "I don't think Dave knew of his death," Ann added. "I don't think they announced the names until the next day after Dave's death."

A memorial service for the remarkable life of Capt. David Glunt was held at St. Aidan's Church on 16 September 2001. He was survived by Ann, his loving wife of 47 years, their son and daughter-in-law David and Lutricia Glunt of Chesapeake; a daughter and son-in-law Linda and Randy Adams of Peachtree City, Georgia; two grandchildren, Bridget Ann Adams and Matthew David Adams; and his sister Ann Shord of Richland, Washington, who died ten years later in November 2011.[29]

ORVILLE C. JOHNSON

Statistically, apart from its population declining in numbers over the decades, not much has changed in the small city of Lawton, North Dakota since Orville Christian Johnson was born on a small family farm four miles north of town on 24 December 1928, the son of Henry and Sophie Johnson. Back then, the population stood at around 230; today, only some 30 residents inhabit the city's one square mile. His grandfather, Christian Orville Johnson, had bought the farm for the princely sum of $5 under the Land Homestead Act in 1899 after he and his parents emigrated to the United States on a sailing ship from Norway in 1867.

K-13 Suwon, South Korea, 1953: Lt. Orville Johnson with the 36th FBS "Queen Flight" house boy. (Photo courtesy Orville Johnson)

Orville graduated from Lawton High School as one of a class of four students in the spring of 1946, then went to the University of North Dakota in order to major in mechanical engineering. He enrolled in the Air Force's ROTC scheme in 1948, and was honored to be chosen as a member of the National Society of the Scabbard and Blade, which recognizes exemplary ROTC cadets "of outstanding ethical and moral character". He was commissioned a 2nd lieutenant in the Air Force Reserves on 31 May 1950, and in January of the following year graduated with his bachelor's in mechanical engineering. He was employed by the General Electric (GE) Company of

Schenectady, New York, whose Test Engineer Program involved a number of three-month assignments in various engineering fields. "My first was in steam turbines and generators; my second was in gas turbines and generators; and the final was in a division called Aeronautic and Ordnance Systems for an extended period of six months. In that last assignment I was exposed to jet engine fuel controls, autopilots for jet airplanes and various ordnances. I think this sparked my first real interest associated with flying."

Johnson would remain with GE until March 1952, when the Air Force notified him that it was time to enter active duty and repay the two years he owed them for his training and 2nd lieutenant commission through the university's ROTC scheme. He had no hesitation in applying for pilot training, and entered active service with the Air Force on 7 April 1952, starting his training with Class 53-C at Malden Air Base, Missouri. The students went straight into primary skills training in the North American T-6G Texan, which was a tough assignment. "Without previous experience, it was indeed a handful with that big 550-hp radial engine up front," he commented. The next phase took the students to San Marcos Air Base, Texas for basic training in the T-6G/F, focusing on formation and instrument flying. Then it was on to James Connally AFB where they flew the Lockheed T-33 Shooting Star jet trainer from January to June 1953, when the class graduated. He then attended Fighter Weapons School at Nellis AFB, Nevada, graduating in September 1953. The following month he was assigned to the 36th Fighter-Bomber Squadron ("Flying Fiends") with the 8th Fighter-Bomber Group and deployed to Air Force Base K-13 in Suwon, South Korea, flying North American's F-86 Sabre jets.

1st Lt. Johnson commanded the squadron's Queen Flight until October 1954, then graduated from Instrument Flying School at the K-55 Base, Osan. Once back home, he was assigned to the USAF Armament Center at Eglin AFB, Florida, "available for less desirable tasks such as co-pilot on the B-29, B-47, target towing with the F-84G, parts pick-up in the C-45 and so on". He later became a project pilot on gun systems including the M-39 and Gatling gun installations. On 31 May 1956 he resigned from active duty in the Air Force, returned to the family farm, and bought a J-5 Piper Cub which he flew "using the county and state road as runways".

Wearing a suit instead of a uniform, Johnson joined the Westinghouse Electric Corporation's Air Arm Division in Baltimore, Maryland as an engineering test pilot. The flight test section operated airplanes on a lease-bailment status from the Navy or Air Force for the purpose of initial flight testing and concept proving of the systems developed in the company's engineering department. These projects usually involved changes to the standard configuration of the airplane which affected the performance and flight characteristics. On his arrival in Baltimore he also joined the Maryland Air National Guard (MANG) 104th Squadron, which was then flying F-86E Sabre jets, and in August 1958 was promoted to the rank of captain. The year before, he had met Betty Calligan, who was then working for American Airlines, and their romance had blossomed. As a result they were married in Pittsburgh, Pennsylvania on St. Patrick's Day in 1958. "Our five children were born in close succession," he said. "Christian in July 1959, Jeffrey in August 1960, Steven in November 1961, Laura in January 1963, and Lisa in February 1965."

Johnson (left) at Westinghouse Air Arm Flight Test in Maryland with a bailed USAF F-101B. (Photo courtesy Orville Johnson)

Following his requests, Westinghouse arranged for Johnson to attend the Navy Test Pilot School at Patuxent River, Maryland, where he joined Class 25 in the fall of 1959. There were fifteen students in his class, including two civilian trainees, and it was split into three "homerooms" of five students each. Each group was issued with a Pilots' Handbook for the propeller-driven and jet aircraft in the TPS hangar, and these were the airplanes they would fly over the next eight months. On the high end were the F4D Skyray, F11F Tiger and F8U Crusader. They studied hard, with most students up late each night writing and typing up flight reports. The TPS actually ran two classes in parallel. The students would fly for half a day, with academic studies for the rest of the day, and each week they would exchange mornings and afternoons with the other class.

"One of our flight instructors at TPS was astronaut-to-be Pete Conrad," Johnson recalled. "We became very good friends during the course, as did our wives. Pete and I flew together and shared cockpit duties on a number of school test flights. He was also the source of information which led to my future application for the Group 2 astronaut selection process, of which he was a successful applicant." Following their graduation, the class went on a two-week tour of aviation companies which included Grumman and Sikorsky (New York), McDonnell (St. Louis), Lockheed (Georgia) and Vought (Dallas). Johnson resumed his position at Westinghouse and the normal weekend status with the MANG. "At Westinghouse, we were developing and testing the tactical radar and weapons control system for the F4 Phantom series including

Maj. Johnson in an F-86H with the Maryland Air Guard, 1964. (Photo courtesy Orville Johnson)

the Navy F4J, and various infra-red detection systems, low-level TV, and terrain-following radar systems using our various bailment airplanes as test beds."

In April 1962, Johnson applied to NASA for consideration as a civilian astronaut applicant and reached the finalist group of 32 candidates, but was unsuccessful in making the final nine. In October 1962, still with Westinghouse, he was promoted to the rank of major in the MANG.

With tensions running high following hostile actions by the North Vietnamese in launching the infamous Tet Offensive in January 1968, several Air National Guard squadrons were called up, including the 104th Tactical Fighter Squadron. Recalled to active duty on 8 June 1968 and promoted to lieutenant colonel, Johnson was assigned to the 104th TFS, flying the F-86H at Cannon AFB, New Mexico. Their purpose was to take crew conversion people from bomber and other aircraft, carry them on T-33 check rides, and then take them aloft on some thirty flights in F-86H Sabres, thereby qualifying them as an "entry level fighter pilot, AFSC". Thus "trained", they could go to Vietnam as forward air controllers. But in June 1969 the program was canceled and Lt. Col. Johnson was released from both active duty and from the MANG. After some consideration, he informed Westinghouse that he would not be reclaiming his position in flight test.

"Betty and I would not be returning to Baltimore," he said of the decision. "We now had five young children along with attendant responsibilities. We'd made many good friends in New Mexico and elected to raise our children there." Along with a friend, he started a construction company in Albuquerque involved in residential and light commercial building. He continued flying, joining up with the New Mexico Air National Guard (NMANG) 188th TFS – nicknamed "The Tacos" and the "Enchilada Air Force" – attached to the 150th Tactical Fighter Group that had recently returned from peacekeeping duties at Tuy Hoa Air Base in the Republic of Vietnam. Although most of the NMANG personnel remained on active duty around the world, a small force kept the squadron's home base of Kirtland AFB operational. This force was busy recruiting officers and airmen such as Orv Johnson from those who had been released from active duty. He enjoyed it. "We flew the F-100C and F-100F airplanes until 1973 when we transitioned into the A7D Corsair and I flew that until retiring from the NMANG."

Meanwhile, Johnson's career in the construction industry had continued, with a move to Santa Fe in 1986 and more emphasis placed upon commercial development. On his retirement in 2005 he and Betty moved back to Albuquerque, although flying still remained one of the great joys of his life. "Until 1993 I continued to fly civilian airplanes, for business primarily, but often for sport or just for fun, at which time I allowed my medical to lapse for a period of about fifteen years. The urge to fly again surfaced in 2010 when a friend and I bought a little Grumman American Yankee AA-1. Currently, I own a small, home-built Vans RV-8A which is approved for full aerobatics of the casual – not competitive – kind."[30]

WILLIAM P. KELLY, JR., USN

William Perry Kelly, Jr. was born in Pittsburgh, Pennsylvania on 15 August 1928 to Beryl Rankin Kelly and William Perry Kelly, Sr. Although too young to participate in the Second World War, when he was about sixteen years of age he decided that he would serve his country as a naval aviator, and studied hard to achieve this. Once he was eighteen, his grandfather George Rankin, a former Pennsylvania state senator, managed to use his influence to have Bill sit the exams as a precursory requirement for entrance to the U.S. Naval Academy. He achieved impressively high scores, and became a midshipman in the USNA Class of '50.

Eager to begin his flying career, Kelly began taking private flight instruction in Buffalo, New York while on summer leave from the academy in 1948, later flying rented Piper J-3 Cubs and Cessna 140s. In 1949 he bought his own aircraft, a 1937 Taylorcraft A, which he sold the next year in order to update to a 1946 Cessna 140. As flying activity outside of their administration was contrary to academy rules, Bill Kelly, who didn't mind occasionally bending the rules, secretly stored his airplane at an airfield near the academy. One Christmas, while on leave from the USNA, he flew it from Buffalo to Florida and back.

After his 1950 graduation from the academy with a bachelor's in engineering, he was sent to Pensacola, Florida for basic flight training, and then took his advanced training at other bases. It was during his time that he met and married Barbara Jean Purifoy. With the Korean War now in full stride, he was posted to Fighter Squadron VF-51 ("Screaming Eagles"), whose members included a blond-haired ensign pilot

Lt. William P. Kelly, Jr., USN, next to his F-11F Tiger on board USS *Intrepid*. (Photo: USN courtesy Barbara Kelly)

named Neil Armstrong. In late December 1951 the squadron was deployed to Korea aboard the newly overhauled USS *Valley Forge* (CV-45). Their F9F Panther fighter-bomber jets would primarily attack freight yards, rail lines and rolling stock. *Valley Forge* returned to the U.S. mainland in the summer of 1952 and Kelly's next combat deployment was to Korea and Taiwan aboard USS *Philippine Sea* (CV-47). It was during this deployment, in 1954, that he and Barbara celebrated the birth of the first of three eventual sons. Also in that year, Lt. Kelly was stationed at Beeville, Texas in the Air Force Training Command, and from 1956 to 1957 was assigned to Williams AFB, Arizona under a Navy exchange program with the Air Force, enabling him to fly F-80 Shooting Star and F-86 Sabre jets.

Next he was assigned to VF-33 ("Fighting 33") at NAS Oceana, Virginia, flying F-11F Tigers, participating in non-combat carrier cruises along the Atlantic Coast of the United States and military exercises in the Caribbean. He would remain with the squadron until 1961, his posting concluding with a further Mediterranean cruise on USS *Intrepid* (CV-11). Upon his return, Kelly was faced with a life-changing decision – he could either attend the Navy's War College in Monterey, California, which was the option strongly recommended by his commanding officer, or he could enter Test Pilot School at Patuxent River, Maryland. He chose the latter option. After graduating from the USNTPS in 1962 he remained at the Naval Air Test Center as a

The modified Grumman S-2A with Fulton Skyhook equipment fitted. (Photo: courtesy David Shugarts)

flight instructor, and it was during this time that he unsuccessfully applied to NASA to become a Group 2 astronaut.

One amazing program in which Kelly was later involved took place on 24 March 1964. This innovative program, invented by Robert Fulton and an extension of Navy trials, was known as the Skyhook Covert Aerial Retriever System (SCARS). Military tests were being conducted of an airborne means of rescuing downed airmen. For this particular exercise, Lt. Cmdr. Kelly and R. J. Vasseur, the Skyhook project director, played the role of downed airmen. As envisaged in a real-life situation, a recovery kit would be dropped and the airmen would don an overall-type harness, then inflate a balloon using a helium canister and release it trailing a 500-foot nylon lift line.

With this done, Kelly and Vasseur reached the exciting part. The test involved a modified Grumman S-2A Tracker flown by test pilot Samuel ("Pete") Purvis. There was a V-shaped yoke extending beyond its nose, and the belly radar dome had been removed. With everything ready, Purvis swooped down low, aiming the S-2A's nose straight for the lift line. The yoke then snared the line well below the balloon, which was automatically cut free and jettisoned. The two test subjects were hoisted high into the air, then reeled in and assisted into the aircraft through the hole where the radar dome had been.

As Kelly's friend and biographer David A. Shugarts described the operation, "The pickup run was made at about 125 mph, but due to the catenary of the line, the first sensation was that of being lifted vertically in a smooth ascent. However, the tough part came during the 20 minutes of being hoisted up to the aircraft. With effort, they could 'duck fly' with their backs to the wind, but it was difficult to breathe with the 125 mile per hour wind roaring around them and they were

Left: Kelly, on the ground, foreground, is rigged with Vasseur for a two-man Skyhook test. Right: They begin their ascent. (Photo: courtesy David Shugarts)

exhausted when they finally struggled past the lip of the radar dome opening and into the plane."

Their work that day would help to form the basis for later Surface to Air Recovery System (STARS) Skyhook programs undertaken by the Air Force for the HC-130H, which was an extended-range Search and Rescue version of the Hercules transport aircraft. An example of this recovery system was seen to great effect in the 1965 James Bond movie *Thunderball*, with Bond and Domino Derval being plucked from the water at the end of the film.

After the excitement of flight test programs, Kelly served two years in the office of Commander-in-Chief, Atlantic Command (CINCLANT). In 1966 he returned to Pax River for test pilot duties in the NATC's Carrier Suitability Flight Test Division. During this time he was involved in the Automatic Carrier Landing System (ACLS) program, an early computerized system to provide safe automated carrier landings in low or zero visibility, or when the sea state and deck motion were deemed too severe for pilot-flown approaches. The tests were conducted using the F4G Phantom, A7 Corsair II and F8 Crusader, and it was a risky, trying time for any test pilot. Later he worked in the NATC's Board of Inspection and Survey.

By 1970 Cmdr. Kelly knew that if he stayed in the Navy he would have to endure yet another desk assignment, this time in Washington, D.C., so he made the decision to retire after 20 years of service. Now faced with a civilian lifestyle, Kelly elected to become a teacher in a new Navy Junior ROTC program in Maine. Although he had fully intended never to fly again, less than a year after his retirement from the Navy a neighbor asked Kelly if he could give him some flying lessons. He finally agreed, but had first to obtain his FAA license for Certified Flight Instructor, later adding that of Instrument and Multi-Engine Instructor as more and more people hired him to give them flying lessons. He was soon balancing his new teaching career at NJROTC with weekends instructing pilots, as well as engaging in church and Boy Scout activities with his and Barbara's three sons. After several years of this, Kelly received a call from Piper Aircraft, asking if he would consider working for them. This led to him becoming chief test pilot and flight test engineer for Piper Aircraft, initially in Lock Haven, Pennsylvania, and later in Lakeland, Florida from 1974 to 1978. Following his resignation he became a freelance test pilot and long-distance ferry pilot.

As David Shugarts recollects, "A huge variety of jobs came his way, from new-aircraft certification for small manufacturers, to experimental work for individuals and innovators." Bill Kelly also spent about five years delivering aircraft all over the world as a ferry pilot, crossing the oceans in single-engine and multi-engine planes to places such as Australia, New Zealand, Singapore, Europe, the Middle East, Africa, and Central and South America.[31]

Barbara Kelly agrees that it was an incredibly busy time for her husband. "Along with writing for *Aviation Consumer*, *Aviation Safety*, and *Light Plane Maintenance* – all for Dave Shugarts and Belvoir Publishing – Bill began working as a flight test analyst and engineering representative on various turboprop aircraft, both land and sea, crop dusters, development projects on new aircraft, and experimental home-builds. [In addition] he worked as a consultant on airplane crash and reconstruction investigations in lawsuits."[32]

In reflecting further upon his long-time friend and magazine article contributor, Shugarts was bemused by the thought that Bill Kelly could have actually given up flying after he left the Navy. "Bill's attempt to quit flying at 20 years in 1970 was very premature, to say the least," he mused. "Bill ended his flying career after more than 50 years, with about 15,000 total hours, including 5,300 military. About 9,000 hours had come in the 'second half' of his flying career."

In 2002, aged 74, some troubling symptoms took Bill Kelly to see a doctor, who immediately sent him to hospital where he was diagnosed with pancreatic cancer. He lapsed into a coma a few days later and passed away on 20 October 2002.

Remarking on her late husband's missing out on becoming an astronaut, Barbara Kelly noted that, "Though Bill suffered some disappointment at not having been one of the chosen few, ultimately he was more than pleased at what became his future with its many accomplishments."[33]

The last word on this incredible man goes to his friend David Shugarts. "He left behind a wife, three fine sons and three grandchildren – and thousands of admirers and friends. I have always felt privileged to be counted among them."[34]

MARVIN G. MCCANNA, JR., USN

Marvin George ("Marv") McCanna, Jr. was born 22 March 1929 in a tiny house on the banks of Spring Creek, a few miles north of the village of Hallton, Spring Creek Township, Elk County, Pennsylvania, the eldest son of Marvin George McCanna,

Sr. and Mary Wallace McCanna. His siblings were Charles L. ("Chuck") McCanna (also a Navy veteran) who died in 1989 at age 57, and James, who died after only eighteen months.

Until about 1950, Hallton was the site of a wood chemical factory. At that time it was served by the Tionesta Valley Railroad which brought wood to the factory, and the Clarion River Railroad which hauled tank cars of wood alcohol and acetic acid from the factory. Today, Hallton consists mainly of vacation and hunting camps.

"Dad told us that he worked on the railroad as a boy," Brad McCanna revealed. "He also told us that he had to ride the train or a railway motor car about 18 miles to go to school in nearby Ridgway, Pennsylvania. He was the Senior Class President and a star basketball player at Ridgway High School, from which he graduated on 6 June 1946."

On 17 July 1946, McCanna enlisted in the Navy under the V-5 program. After completing the Aviation Fundamentals School in Jacksonville, Florida and the Naval School of Photography in Pensacola, Florida he was assigned to VF-62 (at that time VF-6B) at NAS Norfolk, Virginia, which was operating the F4U-4 Corsair. He went on to attain the rate of Aviation Photographer's Mate Third Class (AF3). In October 1948 he reported to the U.S. Naval Academy Preparatory School in Bainbridge, Maryland. McCanna was honorably discharged from the Navy on 14 June 1949, and one day later was designated a midshipman at the Naval Academy, Annapolis, as a fleet appointee.

Lt. (jg) Marvin G. McCanna, Jr., USN. (Photo: USN courtesy McCanna family)

While at the academy, he was a letterman on the lacrosse team and a company commander, achieving the rank of midshipman lieutenant. As Brad McCanna relates, "Apparently he was a stickler for following rules and regulations, as his nickname at the academy was 'Reg Book Red'." He graduated 468th out of a class of 925 with a bachelor of science degree and was commissioned an ensign in June 1953. He stayed at the academy as a company officer until September 1953, when he reported to NAS Pensacola, Florida for flight training.

McCanna went on to complete his advanced jet training at NAS Corpus Christi, Texas and was designated a naval aviator in December 1954, along with a promotion to lieutenant (junior grade). Whilst there, he had met Emma Lou (always known as "E.Lou") Frandsen, and they were married on 7 January 1955. Shortly thereafter, he reported to VF-32 ("Fighting Swordsmen") at NAS Cecil Field, Florida for his first tour as a naval aviator. "It is debatable whether it occurred during flight training or while serving with VF-32," added Brad McCanna, "but early on he earned the call sign 'Red Dog' due to his curly, bright red hair."

While at VF-32 McCanna flew the F9F-6 Cougar, the Navy's first swept-wing jet fighter, and from 4 November 1955 to 2 August 1956 deployed to the Mediterranean aboard USS *Ticonderoga* (CV-14). During the cruise, all VF-32 pilots made 100 or more carrier landings, qualifying them for the Centurion Club. In all, they made a total of 1,875 landings and flew 2,788 hours in jets. In late 1956, VF-32 became the first fleet squadron to operate the supersonic F8U-1 Crusader. In preparation for this, six of its pilots had an indoctrination course at the Chance Vought Aircraft's Dallas Plant. They included Comdr. Gordon Buhrer, Lt. James Allen, Lt. Howard Rutledge, Lt. David Davidson and Lt. Marvin McCanna, Jr.

Thursday, 6 June 1957 was a bleak day for aircraft operating out of Cecil Field, Florida. A pair of McDonnell F-3H-2N Demons from VF-31 suffered engine failures after flying into a severe thunderstorm 150 miles off the coast of Georgia. Both pilots ejected, but only one was eventually recovered. That same day, President Dwight D. Eisenhower and members of his cabinet were on board USS *Saratoga* (CV-60) to observe air operations when an F8U-1 from VF-32 flown out of NAS Cecil Field by Lt. (jg) McCanna suffered an irrecoverable engine failure. He managed to bail out seven miles from the carrier and was rapidly rescued by one of its helicopters. Brad McCanna says his mother enjoyed telling the story that when she received word her husband had ejected over the Atlantic, her only concern was that the salt water would ruin his curly red hair. A month later, on 1 July 1957, he was promoted to full lieutenant. Then from February to July 1958 he was on another Mediterranean cruise, this time with VF-32 aboard *Saratoga*, on the ship's first deployment with the Sixth Fleet.

On 14 August 1958, Lt. McCanna reported to the U.S. Navy Test Pilot School at NAS Patuxent River, Maryland. After graduation with Class 21 in March 1959, he stayed at the Naval Air Test Center as a project pilot in the Armament Test Division (later Weapons System Test) until August 1961. At some point during this tour, he successfully landed an F8U-1 Crusader with its nose gear up, saving the aircraft.

In September 1961, McCanna reported to VF-174 ("Hellrazors") at NAS Oceana, Virginia for replacement pilot training. After F8U refresher training in December he

Lt. McCanna with VF-32 F8U-1 Crusader. (Photo courtesy McCanna family)

Lt. McCanna (left) receiving flight navigation briefing at NAS Patuxent River. (Photo courtesy McCanna family)

reported to VF-103 ("Sluggers"), also at NAS Oceana, as the squadron's Operations Officer. During this tour of duty, he completed a Mediterranean deployment from 3 August 1962 to 2 March 1963 aboard USS *Forrestal* (CV-59). He was promoted to lieutenant commander on 1 January 1963. He attended the U.S. Naval War College in Newport, Rhode Island, from August 1963 to June 1964. In addition to the regular course of instruction, he got his master's degree in International Affairs from George Washington University. In July 1964 he reported to NAS Alameda, California for duty as Aide and Flag Secretary to the Commander of Carrier Division 3. This tour included two combat deployments to the Western Pacific. The first was on USS *Ranger* (CV-61), which sailed on 6 August 1964 in the wake of the Gulf of Tonkin incident and returned to Alameda on 6 May 1965. After several months at home, he deployed aboard USS *Enterprise* (CVN-65) from 26 October 1965 to 21 June 1966. Beginning in August 1966 he served as a flight instructor in the

F-11A Tiger with VT-26 at NAS Chase Field, Texas. He was promoted to commander on 1 July 1967.

Cmdr. McCanna's next assignment came in August 1967 when he reported to VF-101 at NAS Oceana, Virginia for F4J Phantom transition training. After completing the training in February 1968, he was sent to VF-41 ("Black Aces"), also at Oceana, as Executive Officer. He deployed with the squadron to the Mediterranean aboard USS *Independence* (CV-62) from 30 April 1968 to 27 January 1969. He was given command of the squadron a month later, relinquishing it on 23 December 1969 with orders to join the Office of the Chief of Naval Operations as the Weapons Program Coordinator in OP-05 at the Pentagon from January to November 1970.

In November 1970, he commenced training in preparation to assume command of Carrier Air Wing 17 on USS *Forrestal* (CV-59), which he duly did in April 1971 in Barcelona, Spain. After the Mediterranean deployment the air wing returned to NAS Oceana in July 1971. Relinquishing command of the wing in April 1972, he reported to Commander, Naval Air Forces, Atlantic Fleet, where he served as Training Officer, Assistant Chief of Staff for Readiness, and Chief of Staff. He was promoted to the rank of captain on 1 July 1972.

Four years later, in August 1976, McCanna assumed command of NAS Oceana, and retired from that billet in October 1978 with 32 years of naval service. Following his retirement from the Navy he became a consultant to Maritime Associates, Inc., of Rosslyn, Virginia, and later an independent defense consultant. But he continued to reside in Virginia Beach, Virginia. He and E.Lou had four children: Brad Alan, born in 1951 by his mother's prior marriage and adopted by his stepfather in 1962; Marvin George ("Trey") McCanna III, born in 1955; Kimberly Anne, born in 1957; and Charles Galland, born in 1960. They were divorced in April 1982 after several years of separation.

Marvin George McCanna, Jr. died in Virginia Beach on 11 May 1986 (Mother's Day) of "acute coronary insufficiency" after a brief illness. Somewhat ironically, his former wife E.Lou also died on Mother's Day nine years later, in 1995.[35]

JOHN R. C. MITCHELL, USN

Born in Youngstown, Ohio, on 16 August 1929, John Robert Cummings Mitchell was a year old when his parents, John Henry and Maria (née Viera) Mitchell, moved to nearby Alliance. On later moving to the Navy town of Quincy, Massachusetts, they lived across Quincy Bay from NAS Squantum, and four-year-old John found the front row view of seaplane activity enthralling. Squantum also doubled as a private airport where Mitchell experienced his first flight around his thirteenth birthday. He attended Huntington Elementary School and later Central Junior High School. In June 1947 he graduated from Quincy High School and joined the Marine Corps, soon completing all the requirements as a Private First Class (PFC). He was enrolled as a midshipman at the US Naval Academy in June 1948.

Three months after his USNA graduation in June 1952 Mitchell undertook flight training. In December 1953 he received his Wings of Gold as a naval aviator and was sent to NAS Kingsville, Texas for jet training. Completing this in the spring of 1954, he was assigned to VF-193 ("Ghostriders") – a jet night-fighter squadron – at NAS Moffett Field in the San Francisco Bay area, flying F2H-3 Banshees off the straight-deck carrier USS *Oriskany* (CV-34).

The squadron was deployed on a WestPac tour when he arrived in California, so he received further orders to a fleet all-weather training unit at NAS Barbers Point in Hawaii for a jet instrument course and then extensive night-fighter training flights in the radar-equipped F6F-5N Hellcat propeller-driven night fighter and the jet-powered F3D-2T2 Skynights. He then rejoined VF-193 at NAS Moffett in June 1954, where he flew as a wingman to Lt. Alan Shepard in an informal aerobatics team called the Mangy Angels that had been established by the future astronaut.

During a squadron deployment on *Oriskany*, Mitchell had a close call on 22 June 1955 as he was coming in for a routine night landing after a night combat air patrol over the Sea of Japan to rehearse intercepts. In the last moments of the approach his F2H Banshee began to sink, even with full power, and he crashed onto the fantail at the stern of the ship. There was an explosion, and the rear half of the aircraft from the cockpit aft toppled into the sea in flames. But Mitchell had survived, suffering only minor cuts and bruises, whilst those who witnessed the incident were convinced he must have died. Promoted to lieutenant early 1957, he would depart VF-193 in Hong Kong near the end of his second cruise aboard the angled-deck carrier USS *Yorktown* (CV-10).

On 13 July 1957 Mitchell married Kaye Fratus (better known as "Billie") a girl he had known since their teens. They would go on to have five children: Joanne, twins John and Harold, Dana, and Ardis. Following the wedding they drove cross-country to Maryland, where he reported to NAS Patuxent River for Class 19 of the Test Pilot School. On finishing the 22-week course he was assigned to the Navy Test Center's Flight Test Division "as assistant to everything: Operations, Maintenance, Supply, Safety and Trial Lawyer". After a year he was transferred to the Carrier Suitability Branch as F8 Crusader project pilot.

In early 1959 he was invited to Washington to participate in the Mercury astronaut selection. Unsuccessful in his bid, he departed Flight Test for VF-74 ("Bedevilers") in the summer of 1960. The squadron was flying F4D Skyrays and had been chosen for the fleet introduction of the F4H Phantom. It would become the first squadron to deploy the F4 aboard USS *Forrestal* (CV-59). Promoted early, Lt. Cmdr. Mitchell was assigned to VF-101 ("Grim Reapers") at NAS Oceana, Virginia. In February 1961 he returned to VF-74. After extensive training, including seven weeks of operations in the Caribbean, the squadron made the first extended F4 deployment in August 1962.

On 18 April 1962, NASA announced that it would accept applications for a second group of astronaut trainees, and Mitchell was invited to apply. The reporting date happened to coincide with *Forrestal*'s departure from Norfolk, but he was given permission to attend. After completing the evaluation with NASA he returned to Oceana for a refresher instrument training course prior to rejoining his squadron in Naples, Italy. By the time he reached the carrier, his second letter of rejection from NASA had already arrived. Mitchell left VF-74 in February 1964 to serve as Flag

Lt. John R. C. Mitchell, USN, circa 1960. (Photo courtesy J.R.C. Mitchell)

Lieutenant and Aide to Commander, Naval Air Forces, Atlantic Fleet, VADM Paul H. Ramsey. He left this position in May 1965 and, after F4 refresher training with VF-101 in NAS Key West, Florida, was assigned to VF-33 ("Tarsiers") at NAS Oceana. Shortly after joining VF-33 he received a third invitation from NASA, this time to try out for selection in their fifth astronaut group – and this time he declined.

During the summer of 1967 he reported to the U.S. Army War College in Carlisle, Pennsylvania for a year of international studies. Aiming to return to research and development, he joined the staff of Task Force 77 for a one-year unaccompanied tour in the Gulf of Tonkin, which involved six months of carrier duty on USS *Kitty Hawk* (CV-63) and then six months on USS *Constellation* (CV-64) as the Anti-Air Warfare Officer. At the end of this tour in late summer 1969 Mitchell returned to San Diego. His next assignment was Commander, Carrier Air Wing 2, which would involve six months of travel and training to become qualified in the A6 Intruder and A7 Corsair II, prior to the F4's deployment on USS *Ranger* (CV-61). Then a Pacific Fleet staff member alerted him to expect a change of orders. The Commander, Naval Air Force Pacific Fleet confirmed the change of orders to Commander, Carrier Air Wing 9. After a brief period at NAS Fallon, Nevada there was a six-week Caribbean tour on USS *America* (CV-66), during which he was able to renew his acquaintance with fellow Mercury astronaut candidate Capt. Tom Hayward, and then a seven-month deployment to the Western Pacific. After a month of refresher tactical qualification in the F4 with VF-121 ("Pacemakers") at NAS Miramar, including day and night catapult launches and landings on an Atlantic Fleet carrier, he reported to Carrier Air Wing 9, which was scheduled to depart Miramar onboard *America* for six weeks of wartime exercises in the Guantanamo Bay training exercise operating area, returning to home base late February 1970. In April the various elements of the wing rejoined *America* to sail to Subic Bay in the Philippines on an eight-month deployment, where Mitchell conducted day and night combat sorties in the A7E and F4J and a number of day flights in the A6 from May to November 1970.

The carrier returned to San Diego in time for Christmas. In January 1971 Mitchell reported to Commander, Naval Air Force, Pacific Fleet as Air Wing Training Officer. After being promoted to the rank of captain in the spring of 1972, he was assigned to the Pentagon as F14A AWG-9/Phoenix Program Coordinator in the Office of the Vice Chief of Naval Operations (Air). In June 1974, Mitchell was given command of USS *Sylvania* (AFS-2), a *Mars*-class combat stores ship that departed Norfolk early in August for a seven-month Mediterranean tour with the Sixth Fleet and returned to Norfolk in April 1975. After dry dock work at the naval shipyard, *Sylvania* transited to Newport News, Virginia for a six-month overhaul. Meanwhile, Capt. Mitchell was given command of the carrier USS *John F. Kennedy* (CV-67), taking over from Capt. William Gureck on 29 November. Mitchell was in turn relieved on 14 May 1977 by Capt. Jerry Tuttle, allowing him to return home for a second Pentagon tour, this time as Deputy, Carrier Programs Office. That year he divorced his first wife and married Janellen McHugh, a close friend of many years who had three children of her own: John, Katherine and Marcella.

Capt. John Mitchell retired from the Navy in December 1978, his many awards including a Bronze Star, Meritorious Service Medal, Strike Warfare (with four stars), National Defense Service Medal (with a star), China Service Medal, Meritorious Unit Commendation (with a star), Armed Forces Expeditionary Medal (with two stars), and the Vietnam Service Medal (with five stars).

His early civilian employment was with companies that dealt in maritime training for the Navy. Eventually, he returned to the Virginia Beach area and joined a small family-owned shipyard that had won a contract for the phased maintenance of supply ships. After a decade of intense work, he joined his second wife Janey in a small real estate firm owned by a former naval aviator, where they worked until he passed away on 31 May 2011, aged 81.[36]

FRANCIS G. NEUBECK, USAF

Francis Gregory ("Greg") Neubeck is a third-generation Washingtonian, born in the nation's capital on 11 April 1932. He won early football recognition as quarterback of his Gonzaga High School team, the Eagles.[37]

After attending Sullivan's Naval Preparatory School, Neubeck entered the U.S. Naval Academy. He graduated with a bachelor of science degree as a midshipman on 3 June 1955, finishing 515th in a class of 742. When a number of the graduates were offered the choice of remaining with the Navy or joining the Air Force, he opted for the Air Force and was commissioned a 2nd lieutenant. After completing his flight training, Neubeck worked for a time on the development of weapons systems for jet fighters at Eglin AFB in Florida and also served as a flight instructor. By this time he and his wife, the former Margaret Bowlen, had become parents to what would prove to be their only child, a son named after his father, born on 20 July 1956.

Promoted to the rank of captain, Neubeck arrived at the USAF Experimental Test Pilot School at Edwards AFB on 29 August 1960 as a member of Class 60-C, with a 32-week program ahead of them. This class included future NASA astronauts Frank Borman, Mike Collins, Jim Irwin and Tom Stafford. The fourteen class members were taught the extreme parameters of test flying, beginning with basic instruction in a dated Northrop T-28 Trojan, before moving on to more high-performance aircraft such as the F-86, the B-57 jet bomber, the T-33 (in which Neubeck had already notched up 3,000 hours), and concluding with the supersonic F-104 Starfighter. After graduating on 21 April 1961 he remained at the pilot school as an instructor, along with fellow captain and classmate Frank Borman.

On 23 October 1962, it was announced that Capt. Neubeck, then stationed at Eglin AFB, had been selected to begin training for future military space missions with the Aerospace Research Pilot School (ARPS), also located at Edwards. This eight-month ARPS III course was designed to train future space pilots, engineers and program managers for the planned X-20 (Dyna-Soar) project and other Air Force programs then under consideration. The class included Mike Collins, as well as future NASA astronauts Charlie Bassett, Joe Engle and Ed Givens. It was during this period that Neubeck was tested for NASA's Group 2 astronaut selection,

Francis G. Neubeck, USAF. (Photo: USAF)

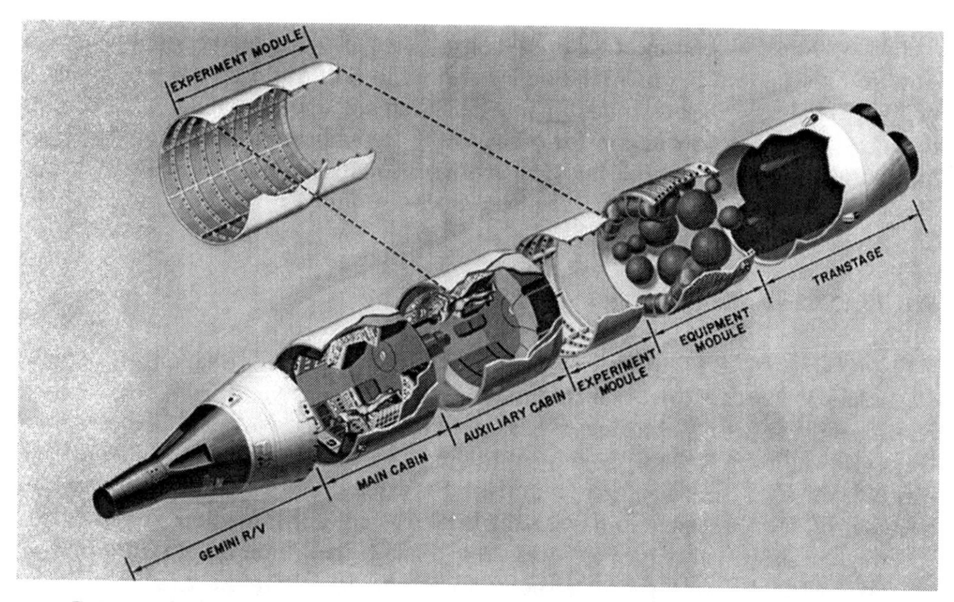

Cutaway illustration of the Manned Orbiting Laboratory (MOL). (Credit: USAF)

missing the final cut. The ARPS class graduated in May 1963, and on 5 June NASA announced that it was seeking a third group of astronauts. Class members Bassett, Collins and Alfred Uhalt submitted applications, but this time Neubeck opted to remain with the Air Force and its proposed space programs.[38]

The Defense Department's Manned Orbiting Laboratory (MOL) program looked to be going ahead on 12 November 1965 when Neubeck, now with 4,600 hours of jet flying time, was selected as one of the eight proposed aerospace research pilots who would train for the projected month-long Earth orbital missions. "Greg always got a lot done on his own," his fellow MOL pilot Lachlan Macleay once stated. "He was the quietest of all of us and maybe the most independent, and that worked for him because he really knew his stuff."[39]

The eight men underwent far more extensive ARPS training at Edwards prior to graduating in July 1966. Almost three years later, in June 1969 the critically over-budget MOL program was abruptly terminated before any of the pilots had a chance to fly into space. Acutely disappointed, Neubeck resumed regular flight duties with the Air Force.

In 1972 he earned a master's in business administration from Auburn University in Alabama, Georgia. After a combat tour with the 7th Air Force involving flying F4 Phantoms out of Ubon Royal Thai AFB in Thailand, Neubeck was appointed Deputy Commander for Test and Evaluation at the Air Defense Weapons Center at Tyndall AFB, Florida in 1973. He subsequently served as Assistant Deputy Chief of Staff for Requirements at Headquarters Tactical Air Command, stationed at Langley AFB, Virginia. Finally, he was Vice Commander of the Tactical Air War Center at Eglin AFB until his retirement from the service with the rank of colonel in 1982, having notched up over 7,000 hours flight time.

In 1985 Neubeck became a technical consulting aerospace engineer in his resident state of Florida, and began attending law school in order to pursue a new career in politics. He also wrote a textbook on missile design and another book on the nation's economy. The next year he ran for office as the Republican nominee for the House of Representatives from Florida's 1st Congressional District, but wasn't elected. Today, he remains an aerospace engineering consultant.[40]

WILLIAM E. RAMSEY, USN

William Edward ("Bill") Ramsey was born in San Diego, California, on 7 September 1931. As the son of a three-star admiral, he spent his formative years principally in Hawaii and Southern California. He attended the Schofield Barracks High School in Oahu, Hawaii prior to his appointment to the U.S. Naval Academy in July 1949. He was a member of the swimming team there for three years. After graduating with a bachelor of science degree in naval science he was commissioned on 5 June 1953.

Ensign Ramsey subsequently served his "black shoe" time during an initial sea tour as Assistant Boat Group Division Officer aboard USS *Glynn* (APA-239) from June 1953 to February 1954. He then received flight training at Naval Air Stations Pensacola, Corpus Christi, and Kingsville, advancing his flying skills from March 1954 to April 1955 and becoming a designated naval aviator on 13 April 1955. He served two months with Fighter Squadron VF-61 ("Jolly Rogers") at NAS Oceana, Virginia, followed by a four-year assignment to VF-84 ("Vagabonds"), operating the FJ-3 Fury from USS *Forrestal* (CV-59), *Randolph* (CV-38), and *Lake Champlain* (CV-39) as a pilot, safety officer and landing signal officer. He later joined Attack Squadron VA-43 at NAS Oceana as an instructor pilot for the A4D Skyhawk from June 1959 to June 1960.

In the summer of 1960 Ramsey was selected for the U.S. Navy Test Pilot School in Patuxent River. On graduation he remained assigned to the Naval Air Test Center as a project pilot in the Carrier Suitability Branch of the Flight Test Division. During this period he became the first pilot to complete a nose-tow catapult launch from an aircraft carrier. He was also the first pilot to conduct a single-engine landing aboard ship in both the F4 Phantom and E-2C Hawkeye; to take off and recover aboard ship in the E-2C; and to perform minimum air-speed catapult launches in both aircraft. He admitted there were some harrowing moments, such as the time that he was testing the minimum speed at which an E-2C twin turboprop could be flown off the carrier, and the airplane took off 10 knots below the planned speed. "We sank to forty feet," he recalled. "We're sixty feet off the water. Needless to say, it got our attention! That was probably the closest we've really ever been."

In 1962, Ramsey was one of only five test pilots in the country nominated for the Iven C. Kincheloe Award as the outstanding test pilot of the year by the Society of Engineering Test Pilots.

Towards the end of 1962, Lt. Cmdr. Ramsey was selected by NASA as one of the final 32 pilots interviewed in Houston as an astronaut for the Gemini program, but did not make the final nine. In the fall of 1963 he traveled to Brooks AFB in San

Lt. William Ramsey at the U.S. Naval Test Center, Patuxent River, as a project pilot in the Carrier Suitability Branch of Flight Test. Shortly after this photo was taken in 1962 he entered the NASA astronaut selection process. (Photo courtesy VADM W.E. Ramsey)

Antonio, Texas to complete medical examinations as one of 34 final contenders for NASA's third group of astronauts, but this time failed to make the final fourteen due to hypervascular limitations.

Late in 1963 Ramsey became Flag Secretary to Commander, Carrier Division 6, and in 1965 attended the Royal Air Force Staff College in Bracknell, England. After his graduation from the college a year later, and refresher training in the F4 and A4, he was assigned as Operations Officer for Carrier Air Wing 6, which operated from USS *America* (CV-66) in 1966 and 1967. Following a tour as Executive Officer for the A4 Attack Squadron VA-66 ("Waldos"), with deployments aboard *America* and USS *Intrepid* (CVS-11), he was given command of the squadron at the height of the Vietnam War in May 1968. He took the squadron on a Vietnam deployment aboard *Intrepid*, personally leading – and completing – 138 combat sorties. It is worth noting that from the time he assumed command, VA-66 flew more combat sorties and flight hours over North Vietnam than any other Navy squadron.

In March 1969, having been personally selected by Admiral Hyman G. Rickover, he attended Nuclear Power School at the Mare Island Naval Shipyard in Vallejo, California and in 1970 the Nuclear Power Training Unit in Idaho Falls, Idaho. Upon completing the course, Cmdr. Ramsey was assigned to USS *Independence* (CVA-62) as Operations Officer and promoted some two years early to the rank of captain, and then made Executive Officer in December 1970. After a Pentagon tour in the Office of Program Appraisal as an advisor to the Secretary of the Navy on Tactical Aircraft (TACAIR) and F-14 Tomcat matters, he was assigned command of the versatile amphibious landing ship dock USS *Pensacola* (LSD-38) from July 1972 to October 1973. Ramsey then joined the faculty of the Naval Air College, where he held the Admiral Halsey Chair for Air Warfare, concentrating on tactics and strike warfare.

In January 1975, Capt. Ramsey assumed the responsibilities and duties as the first commanding officer of the nuclear-powered carrier USS *Dwight D. Eisenhower* (CVN-69). By the time he relinquished this command in March 1979 he had taken "Ike" through the carrier's launch and christening, initial reactor criticality, dockside trials, builder's sea trials, acceptance trials, final contract trials, shakedown training, presidential weapons exercise, post-shakedown availability and initial deployment to the Sixth Fleet in January 1979. Two months later, now advanced to rear admiral, he was assigned command of Carrier Group 1, based in San Diego. He would remain in this position until October 1980, when he became First Director of the Naval Space Systems Division in the Office of the Chief of Naval Operations (OPNAV). Further promoted to vice admiral in September 1985, he served until March 1989 as the First Deputy Commander-in-Chief of U.S. Space Command at Peterson AFB, Colorado, managing and planning for the operational tasking of

Vice Admiral William E. Ramsey, USN. (Photo courtesy VADM W.E. Ramsey)

worldwide missile warning, satellite tracking, and satellite control facilities. During this time, in 1986, he was the first naval recipient of the American Astronautical Society's Military Astronautics Trophy, awarded for his "outstanding leadership in the application of astronautics to the development of space systems for national defense".

By the time he retired from active duty with the Navy on 1 March 1989, Ramsey had accumulated over 4,800 total flight hours and completed 912 fixed-wing carrier landings, with 258 night landings. In August, he was made Vice President for Space Systems Division of Computer Technologies Associates Inc., a company involved in aerospace information systems and engineering, where he remained until his final retirement in 1996.

In his long and distinguished career, Vice Admiral Ramsey was the recipient of many awards, medals and honors. He received the Defense Distinguished Service Medal, the Legion of Merit (twice), Bronze Star (2, Combat "V"), Air Medal (3), and Strike/Flight Air Medal (11). Among other responsibilities, he is currently a Trustee of the U.S. Naval Academy Foundation, the National Naval Aviation Museum, and Central Flight Leader of the Early and Pioneer Naval Aviators (Golden Eagles) Associations. In addition, he received the Defense Distinguished Service Medal, and the accompanying citation is worth reproducing:

> Vice Admiral William E. Ramsey, United States Navy, distinguished himself by exceptional service as Deputy Commander in Chief and Chief of Staff, United States Space Command, and Vice Commander in Chief, North American Aerospace Defense Command, from October 1985 to February 1989.
>
> Under Admiral Ramsey's personal direction and leadership, the roles, missions, and operational relationships of this command were defined; the organizational support foundation and operational force structure were established; and key operational planning initiatives in the areas of space-based wide-area surveillance, ballistic missile defense, and deep space surveillance were analyzed and developed.
>
> The distinctive accomplishments of Admiral Ramsey culminate a distinguished career in the service of his country and reflect great credit upon himself, the United States Navy, and the Department of Defense.

Married to the former Peggy Scott ("Scottie") Booth of Coronado, California, who also had a three-star admiral for a father – Charles Thomas Booth II – the Ramseys had three sons; Tim, Blake (USNA Class of '83, who sadly died in an automobile accident in 1986) and naval aviator Chris. They reside in Pensacola, Florida.[41]

ROBERT W. SMITH, USAF

As one of the outstanding, record-breaking test pilots of that era, it is something of a surprise that U.S. Air Force Capt. Robert W. Smith was not selected to be a NASA

Group 2 astronaut. It does reinforce, however, the depth of that incredible pool of talent from which the selection panel had to make its final choices.

Robert Wilson Smith was born in Washington, D.C., on 11 December 1928, although the Depression would later prompt a relocation to Seneca, Maryland, from where his father would commute to his job with the Southern Railway System in the capital. Young Bob was eight years old when his parents divorced and he was sent to live with an aunt and uncle in York, Pennsylvania, where he attended school from fourth to ninth grade. One major influence at this time was his uncle George Luckett, a naval aviator who had received Naval Aviation Pilot's License No. 18 in 1919 from the Department of the Navy. Bob was entranced by the stories his uncle told, and decided he too would like to fly for his country one day. At the age of 21 he enlisted in the Aviation Cadet Program of the Air Force on 20 June 1949. Just prior to this, however, he had met and married Martha Yacko.

After a year's training, first in the propeller-driven T-6 Texan and then moving to advanced training in the F-80A Shooting Star jet fighter, he was commissioned a 2nd lieutenant and awarded his pilot wings at Williams AFB, Arizona on 23 June 1950, coincidentally the day on which the Korean conflict broke out. By this time, he and Martha had celebrated the birth of their first child, a daughter named Lane.

Lt. Smith prepares for action with the 4th Fighter Wing, 335th Squadron at Kimpo Air Base, Korea. (Photo: USAF)

Assigned to the 27th Fighter Squadron of the 1st Fighter Wing in July 1950 for a year of training, Lt. Smith flew F-86A Sabre jets at Griffiss Field, New York, prior to flying for the 335th Fighter-Interceptor Squadron of the 4th Fighter Wing at Kimpo Air Base in South Korea from August 1951 until April 1952. Before embarking on this deployment, he and Martha had a son named Bobby. The primary combat zone for his squadron was a sprawling area in the northwest corner of North Korea which became known as "MiG Alley". He was credited with destroying two MiG aircraft in aerial combat, plus one probable and three damaged.

On his return to the U.S., Lt. Smith served with the 93rd Air Defense Squadron in New Mexico from April 1952 to August 1953, then took a course at the Air Force Institute of Technology (AFIT) at Wight-Patterson AFB, Ohio, graduating two years later with a bachelor of science degree in aeronautical engineering. He was then sent to Edwards AFB to complete Test Pilot School as a member of the thirteen-member Class 56-A. Upon graduating he was ordered to the Air Force Armament Center near Ft. Walton Beach, Florida as a flight test officer, where he served for two years. He was then transferred to Eglin AFB, Florida until January 1960. His next assignment was a project officer on the Titan II ballistic missile at Vandenberg AFB, California until June 1962, at which time he was sent to the Aerospace Research Pilot School at Edwards AFB.

As he recalled in his partially completed memoirs, "I had heard about a course of

The Smith family at home: Bob, Bobby, Lane and Martha. (Photo: USAF)

Capt. Robert Smith at Edwards AFB with the NF-104. (Photo: USAF)

training for future Air Force astronauts that had been formed at the Air Force Flight Test Center, where I had gone to Test Pilot School, and checked it out. The first class for the new Aerospace Research Pilot School [was already in progress] and I made application for the second." Smith's application was endorsed by none other than General Jimmy Doolittle. "A wonderful directive arrived in our headquarters that I was one of eight accepted for that training .... I joined my mates in Class II of ARPS. A six-month graduate course for selected military pilots to gain special flight and ground training for space flight."

The class had only just begun its studies when NASA announced it was beginning the process of selecting a second group of astronauts. "That took one of our primary instructors, Frank Borman, and me away from the ARPS for a period of time when we were designated by the Air Force as candidates." Maj. Borman was chosen as one of the nine new astronauts but Capt. Smith missed out in the final selection. When applications were invited for the third group of astronauts the following year, Smith was over the age limit.

At the conclusion of their ARPS course, four of the graduates, including Smith, remained at Edwards to join the Fighter Test Section of Test Operations, where he served until September 1964. Newly promoted Lt. Col. Smith was then assigned to the staff of Headquarters, Air Force Systems Command at Andrews AFB, Maryland. However, in March 1967 he decided that he wanted to qualify for combat flying in Vietnam, specifically in the F-105 Thunderchief supersonic fighter-bomber. "I went to Personnel and requested the first assignment specifically to an

F-105 combat slot, which meant F-105 training to check out and sharpen my dive bombing skills, the part of air combat where I had the least experience. I had the assignment in a flash." Following F-105 combat crew training at McConnell AFB, Kansas, he deployed to Southeast Asia and flew combat missions over North Vietnam as Commander of the 34th Tactical Fighter Squadron, based at Korat Royal Thai AFB, Thailand, between 5 July 1967 and 22 April 1968. His leadership and audacity under fire led to him becoming a recipient of the Air Force Cross. His citation reads:

> The President of the United States takes pleasure in presenting the Air Force Cross to Robert W. Smith, Lieutenant Colonel, U.S. Air Force, for extraordinary heroism in military operations against an opposing armed force while serving as an F-105 pilot of the 34th Tactical Fighter Wing, Korat Royal Thai Air Base, Thailand, 7th Air Force, in action in Southeast Asia on 19 November 1967. On that date, Colonel Smith gallantly led his force through a hostile aircraft attack, an awesome and extended attack by missiles which downed two aircraft, and into the heavy anti-aircraft defenses to strike crippling blows to the assigned target and to a large active surface-to-air missile site. Colonel Smith never wavered from his goal, and with complete disregard for his life, displayed great courage and determined leadership to accomplish an extremely hazardous and difficult mission. Through his extraordinary heroism, superb airmanship, and aggressiveness in the face of hostile forces, Lieutenant Colonel Smith reflected the highest credit upon himself and the United States Air Force.

With the end of his combat tour in June 1968, Smith returned to the United States and a posting to Air Force Headquarters in the Pentagon as the F-111 Project Officer, but he soon found this work unsatisfying and frustrating. "I was back once more to the city where I was born – but never wished to be – working in the Pentagon and assigned to DCS Procurement on the F-111 swept-wing fighter program. That project was prominent in a childish game between the Air Force and the Senate Armed Services Committee .... Extensive and costly paper evaluations passed back and forth, serving no real purpose. It was nothing like the Air Force that I had joined 20 years before, and which became increasingly political from without, and moving to within." Disillusioned, he retired from the service on 1 August 1969.

Putting his Air Force career behind him, Smith now began life as an engineering manager in the aerospace industry, which he pursued for the next 23 years. In those first four years he was an engineering manager on NASA's Skylab program, working directly with the astronauts. His next assignment was in New Orleans, Louisiana, as Systems Design Manager for the Space Shuttle External Tank, on which he worked from its inception. Following this, his company became involved in the U.S. Army's Apache helicopter program as contractor for the Target Acquisition and Designation Systems and Pilot Night Vision System. Later, when Operation Desert Storm began, Smith took over the LANTIRN project to accelerate delivery of the navigation and targeting pods which were used so effectively by the Air Force in attacking ground targets.

In 1998, at the age of 70, Bob Smith gave up flying and instead took to golf. He also began to write his memoirs. Although he came close to recording his entire life, it was a task he would never quite complete. His last few years in Montverde, Florida were spent in profound sadness when he and Martha lost their only son Bobby to a major heart attack. This was a terrible and unexpected blow from which Smith never fully recovered. He died from a massive stroke on 19 August 2010.[42]

ROBERT E. SOLLIDAY, USMC

Robert Edwin Solliday was born in Philadelphia, Pennsylvania on 4 December 1931, the son of Grace and Harry Solliday, and went on to be an older brother to Gordon. He attended Frankford High School and later was a varsity letterman in high school, earning a football scholarship to Lafayette College in Easton, Pennsylvania. He got his private pilot's license at the age of 17 and for a time worked at a flying school operated by the Flying Dutchman Air Service in nearby Somerton before joining the U.S. Navy in January 1952 to undertake flight training, which he completed at NAS Pensacola, Florida in May 1953. It was during that time that he met his future wife Charlene on the Atlantic's Jersey shore. On receiving his naval aviator wings he was commissioned into the U.S. Marine Corps. Three days later he and Charlene were married, and the following day he was transferred to NAS Corpus Christi, Texas for advanced jet training.

Lt. Solliday's first posting was to Marine Night-Fighter Squadron VMF(N)-531 ("Gray Ghosts") at Cherry Point, North Carolina. After training, the squadron was

Capt. Robert E. Solliday, USMC. (Photo courtesy Robert Solliday)

deployed to Korea, stationed at Air Stations K-6 Pyongteak and K-3 Pohang. At the end of the war in July 1953, Solliday accepted a regular Marine Corps commission, was promoted to 1st lieutenant, and transferred to VMA-212 ("Devil Cats") based at MCAS Kaneohe, Hawaii. His next assignment was Aircraft Engineering Squadron 12 at MCAS Quantico, Virginia, which was attached to Aviation Engineering and Demonstration Squadron "Topper 2", a close air support demonstration squadron.

In 1958, promoted to the rank of captain, Solliday attended Class 20 of the Navy Test Pilot School, which would graduate on 1 August 1958. His classmates included future astronauts Wally Schirra, Pete Conrad and Jim Lovell. Afterwards, Solliday remained there to serve at the Naval Air Station's Flight Test Division in the Flying Qualities and Performance Branch. Among other aircraft that he was assigned to fly at this time was the F4J Phantom II supersonic, long-range all-weather interceptor. He was there in late January 1959 when orders came to report in civilian clothing to Washington, D.C. for a top-secret briefing as a potential astronaut for the Mercury project. At the age of 27 he was the youngest of the 32 finalists, but failed to make the final selection. After missing out with NASA he remained at the Patuxent River Flight Test Center as Flight Test Project Officer for the Lockheed Martin KC-130F. He also participated in flight-testing the Grumman YAO-1 twin turboprop Mohawk, and was project officer for the innovative single-seat Goodyear AO-1 Inflatoplane.

Although not selected as an astronaut, Solliday still had a deep involvement in NASA's space program. He served as a CapCom (Capsule Communicator) Assistant on two Mercury missions at Vandenberg AFB in California, as well as in Guaymas, Mexico for Scott Carpenter's MA-7 mission in May 1962.

Solliday attended the Naval Postgraduate School in Monterey, California from 1961 to 1964. It was during this time that he applied for NASA's second astronaut group in 1962, a process that he described as "easier, and more sensible". As before, he reached the final group of 32 finalists but again missed out on the final selection. He next served as Executive Officer of Headquarters and Maintenance Squadron 11 at NAS Atsugi on the Japanese island of Honshu. In April 1965 the squadron was deployed to Da Nang in Vietnam as part of Marine Air Group 11. On his return he was stationed at the Naval Missile Center at Point Mugu in California. From August 1965 to April 1969 he served as the Air-to-Surface Program Officer, then as head of the Fleet Weapons Engineering Department. After that he was assigned to MCAS Iwakuni in Japan, where he commanded Marine Fighter/Attack Squadron VMF-232 ("Red Devils") from 4 March to 14 October 1970, operating Phantom F4Js. His next duty station was MCAS El Toro, California as Operations Officer of the Marine Air Reserve Training Detachment, flying aircraft such as the A4, MOV-10, SNB-5 and OE-1.

On 31 January 1972, Lt. Col. Solliday retired from the service and took on several major projects as an experimental test pilot for Hughes Aircraft Corporation, based at the Naval Missile Center, Point Mugu, California. This included the F-14A/Phoenix long-range, air-to-air missile project. On retiring from active test flying he remained with Hughes as its Marketing Manager for the Angle Rate Bombing

A recent photo of Charlene and Bob Solliday. (Photo courtesy Robert Solliday)

Systems. When Solliday left the company he started teaching at the Florida Institute of Technology and he and Charlene bought a boat so that they could cruise the Bahamas. He also served as National Vice President of the Society of Experimental Test Pilots and later as its Treasurer. Today, he and his wife reside near Nashville, Tennessee.[43]

ALFRED H. UHALT, JR., USAF

Alfred Hunt Uhalt, Jr. admits that his last name may be somewhat uncommon, but its meaning is serendipitous; Uhalt, translated from the Basque, means "waterside". In his case this was something of a misnomer, because in his youth growing up in New Orleans, Louisiana and later Jackson, Mississippi, Al Uhalt always had his eyes very firmly fixed on the skies.

Born on 3 June 1931 to Mary Ann (née Levy) and Alfred Hunt Uhalt, Sr., the younger Al Uhalt graduated from Central High School in Jackson, Mississippi a day before his eighteenth birthday. But he had begun civilian flight training at age fifteen and soloed a Piper Cub the day he turned sixteen. He worked weekends as a "soda-jerk" in a drug store and delivered magazines to pay for his flying lessons. Exactly a year later he gained his private license, long before he drove an automobile. He rode his bicycle to and from the airport until aged eighteen. Such was his fascination with aviation.

Three weeks after turning seventeen, Uhalt enlisted in the U.S. Air Force Reserve, enrolling in every officer-training course that the Extension Course Institute would provide. His goal was to learn as much as he could about the Air Force whilst also earning the college credits needed to become eligible for Air Force flight training and simultaneously ward off Army draft officials. After graduating from high school he studied physics at Millsaps College in Jackson for two years, then "shifted" to full-time work at the airport in June 1951, awaiting his "call up" from the Reserve to Air Force active duty. Meanwhile, he used his earnings to obtain his Commercial Pilot License and Flight Instructor Rating. The call finally came in March 1952 with an assignment as an aviation cadet. He graduated from flight training "with distinction" and was commissioned a 2nd lieutenant on 16 March 1953. He attended the USAF Instrument Flight Training School and was assigned to the 497th FIS, an all-weather squadron in Portland, Oregon. Next was a tour with the NATO Iceland Air Defense Forces in Keflavik, Iceland. On 18 August 1957 he married Deborah Ann Hurst of Biloxi, Mississippi, and they went on to have two sons: Alfred Hunt III and John Scott Uhalt.

On 11 August 1958 Uhalt graduated with honors from the University of Illinois with a bachelor's in aeronautical engineering under the auspices of the Air Force

Al Uhalt's ARPS Class III posing in front of an F-104 at Edwards AFB. Kneeling, from left: Ed Givens, Tommie Benefield, Charlie Bassett, Greg Neubeck and Mike Collins. Standing, from left: Alfred Atwell, Neil Garland, Jim Roman, Uhalt and Joe Engle. (Photo courtesy AFFTC History Office)

Capt. Uhalt at Holloman AFB, New Mexico in 1964 where he was performing lateral dynamic stability testing during air-to-air missile launches on the TF-102A. (Photo: USAF courtesy A. Uhalt)

Institute of Technology (AFIT) program. He was then assigned to Class 58-C of the Air Force Experimental Test Pilot School at Edwards AFB, where he undertook 32 weeks of courses in aircraft flight test performance, stability and control. All sixteen members of his class were USAF captains. Next, he was assigned to Tyndall AFB, Florida, where, as a project test pilot, he flight tested electronic sub-systems in all of the fighter-interceptor jet aircraft operated by the Air Defense Command.

On 22 October 1962 the Air Force announced the selection of a third group of ten officers, including Uhalt, for eight months' training at the Aerospace Research Pilot School at Edwards AFB. By this time he had qualified as Pilot in Command (PIC) in 86 different military and civilian aircraft, as well as owning and flying his own light airplanes. Having failed in his bid to be selected for NASA's second astronaut group, Uhalt trained hard at ARPS, sometimes relieving the rigorous classroom and flight training by flying a glider out of a nearby soaring operation simply to relax over the desert landscape. On graduating in May 1963 he was a qualified astronaut-designee with the Air Force.

As Project Chief Test Pilot, Uhalt flew all the qualitative and quantitative flight tests on the original Air Force evaluation of the Navy F4 Phantom II, later including hot and cold weather testing of the F4 at Yuma, Arizona and Fairbanks, Alaska. He also worked on a number of other projects. After graduating from the Air Command and Staff College in 1966, Uhalt gained a master's in public administration from George Washington University on 4 June 1967. He then served as a tactical fighter pilot in Vietnam, flying F4s out of Cam Ranh Air Base and earning the Silver Star, Distinguished Flying Cross, Air Medal with 17 Oak Leaf Clusters, and the Air Force Commendation Medal with an Oak Leaf Cluster. On returning to the United States in 1968, Maj. Uhalt resumed flight test work at Wright-Patterson AFB, Ohio. In 1969 he was assigned as an Executive Assistant in the Office of the Secretary of Defense in the Pentagon and then had a year at the National Defense College. With his tour at the Pentagon behind him, and now promoted to the rank of colonel, Uhalt joined the Air Defense Command Staff in Colorado Springs, Colorado. After commanding first the 14th Missile Warning Squadron and later the 46th Aerospace Defense Wing at Peterson AFB, Colorado, he retired from active duty in September 1979.

Additional decorations earned by Uhalt include the Legion of Merit with an Oak Leaf Cluster, and the Air Force Meritorious Service Medal with an Oak Leaf Cluster.

"As pilot-in-command," Al ("Trip") Uhalt III wrote about his father, "[he flew] at least 38 different military airplanes including the B-25, T-6, T-28, T-33, T-39, F-80, F-86, F-89, F-94, F-100, F-101, F-102, F-104, F-106 and F4. In addition, he has flown 118 different civilian airplanes as PIC including an initial 'first flight' in an experimental light sport aircraft."

These days, Al Uhalt is still very active in aviation, flying a number of different aircraft including the Piper Cub, ACA Citabria, Cessna 172, 182 and 210, Beechcraft Bonanza, Duke, Twin Beech, Thunder Mustang, Learjet, and Aero Vodochody L39 (a Russian jet fighter trainer). As a contract pilot operating out of Colorado Springs and flight instructor, he travels frequently to conduct initial checkouts, proficiency training and flight reviews.[44]

KENNETH W. WEIR, USMC

Founded a decade after the fall of the Alamo, the city of Sherman soon became a merchandising center in the northeast corner of the new State of Texas near the Red River boundary with Oklahoma and next door to the rail center of Dennison, which was the birthplace in 1890 of President Dwight D. Eisenhower. During and after the Civil War, Sherman saw the outlaw bands led by Jesse James and William Quantrill, and was later home to such notables as country and western star Buck Owens. It was here that Kenneth Wynn Weir was born on 20 October 1930 to U.S. Marine Corps aviator 2nd Lt. Kenneth H. and Bertye Lee Wynn Weir, a former U.S. Navy nurse. During the early 1930s Weir, his sister and his mother resided in Dennison while his father flew close air support in single-engine biplanes against Augusto Sandino and his followers in Nicaragua.

After graduating from Severn Prep School, Severna Park, Maryland in 1948, Weir received a presidential appointment to the U.S. Naval Academy as a member of the Class of '52. There he made his first ever flight in the N3N biplane single-float sea plane, with later additional flight training in the N3N and PBY seaplanes, and SNJ trainer. On graduating he was commissioned a 2nd lieutenant in the Marine Corps on 6 June 1952, and was the Honor Graduate in the 18th Special Basic School Class in 1953. He began his initial flight training as a student naval aviator in May 1953, and received his Wings of Gold on 30 July 1954. He was assigned to VMF-533 of the 2nd Marine Air Wing, Cherry Point, North Carolina. The squadron had acquired the F2H-4 Banshee the previous year, and he served a number of carrier deployments on USS *Lake Champlain, Bennington* and *Ticonderoga.* He flew the F2H-4 as a "special weapons" delivery pilot, remaining with the squadron until January 1956.

From February 1956 until April 1957 Weir served with the Tactical Air Control Center, 1st Marine Air Wing, at K-3, Pohang, Korea. He would later be assigned as Assistant Officer-in-Charge of Special Weapons Delivery Unit VMA-251 at NAS Atsugi in Japan. This was a detachment of about twelve pilots of the main squadron which flew AD-4 and AD-6 Douglas Skyraiders out of Iwakuni and was tasked with preparing to deliver nuclear weapons if required during the Cold War. Weir would also serve at bases on the Japanese island of Okinawa and in the Philippines.

On his return to the United States he was assigned for two years to the staff of the Basic School as Platoon Leader, Forward Air Controller, Infantry Tactics Instructor and Helicopter Operations Instructor at Quantico, Virginia, and then he undertook jet refresher training in Grumman F9F-8 Cougars at NAS Olathe, Kansas.

Weir's next assignment in 1959 was to VMF-334, flying F8U-1/2/2NE Crusaders with the 3rd Marine Air Wing based at MCAS El Toro, California. In February 1961 he was selected for the Navy Test Pilot School at Patuxent River, Maryland. On his graduation from USNTPS in 1961 he remained at the NATC's Flight Test Division as a Flying Qualities and Performance Branch project test pilot, flying such aircraft as the F4H-1/1F, F4A/B, F8U-2NE/F8D, OV-1 and A3J/A-5A until May 1963. It was during this period that Capt. Weir received a letter of commendation from the Commandant of the Marine Corps for noteworthy achievement, having

A Marine Corps family tradition. From left: Lt. Robert Weir, M/Gen. Frank Weir, B/Gen. Kenneth H. Weir and his son, Capt. Kenneth W. Weir. (Photo: USMC courtesy K.W. Weir)

been selected as one of the 32 finalists from which the second group of NASA astronauts would be chosen, but he did not make the final cut. Setting his disappointment aside, Weir next attended ARPS at Edwards AFB in May 1963 and graduated in December. A month later he returned overseas, serving as Staff Secretary with the 1st Marine Air Wing, and later as Assistant G-3, 9th Marine Expeditionary Brigade for the duration of his overseas tour from November 1964 to March 1965 covering the Philippines, South Vietnam, Malaysia, Thailand, and afloat. On his return, he served as Director of Test Operations for the Navy's highly classified part of the Manned Orbiting Laboratory (MOL), and as Assistant Chief of the Flight Crew Division for the Air Force in this program.

Maj. Weir resigned from active service in September 1966 and transferred to the Marine Air Reserve Training Detachment, joining VMA-241, a reserve squadron based in Los Alamitos, California. He also became an engineering test pilot with the Lockheed Aircraft Corporation, and in 1968 gained a master of science degree from the University of Southern California. From April 1970 he served as Commanding Officer, VMO-8 until September 1972. After several command and staff positions, in March 1977 he was assigned as Assistant Wing Commander, 4th Marine Air Wing. He was advanced in rank to brigadier general on 1 June 1977 and continued to serve in numerous positions including command of the 65th Marine Amphibious Brigade, the 4th Marine Air Wing (1985) and Deputy Commanding General, Fleet Marine

Ken Weir flew the U-2 high altitude reconnaissance aircraft for Lockheed during his 27 years with the company. (Photo courtesy K.W. Weir)

Force Pacific until his final designated mobilization assignment at Quantico, Virginia as Deputy Commanding General, USMC Development and Education Command. He served two terms as a member of the Secretary of Defense Reserve Forces Policy Board. He was assigned to inactive reserve status on 30 June 1986 and retired on 1 November 1990.

During the 27 years following his retirement from active duty in 1966, Gen. Weir worked as an engineering/experimental test pilot with the Lockheed Corporation at their Palmdale, California plant, including some advanced development project work at the company facility known as the "Skunk Works". This involved research and development of supersonic fighters and high-altitude surveillance spy planes. Eventually he became Lockheed's chief experimental/engineering test pilot for the U-2, TR-1 and ER-2 reconnaissance programs.

On 1 September 1993, at age 63 and after 31 years as an active experimental test pilot, with more than 19,700 accident-free hours of flight time (over half of which were accumulated flying jet aircraft and approximately 4,650 hours in pressure suits and single-piloted aircraft above an altitude of 60,000 feet), M/Gen. Weir retired as Lockheed's U-2/TR-1 chief test pilot. Over his combined military and civilian career, he flew virtually every Navy and Air Force fighter attack aircraft made, including the F2H Banshee, F9F Panther/Cougar, F8U Crusader, F4 Phantom, F-100 Super

U.S. Marine Corps Reserve M/Gen. Kenneth W. Weir. (Photo: USMC courtesy K.W. Weir)

Sabre, F-101 Voodoo, F-104 Starfighter, F-106 Delta Dart, F-18 Hornet, A4 Skyhawk, A5 Vigilante, A6 Intruder and A7 Corsair, as well as making the first test flights of the TR-1A, U-2S and A4S airplanes.

He is a Fellow in the International Society of Experimental Test Pilots and served as its president from 1980 to 1981. He is also one of the two hundred members of the Early and Pioneer Naval Aviators Association, better known as the Golden Eagles, which honors former Navy and Marine Corps pilots and astronauts who have made significant contributions to naval aviation. In 2008 M/Gen. Weir was designated a Severn School Distinguished Graduate by the school's Board of Trustees.

Today, he and his wife Nancy have three sons and five grandchildren, reside in Southern California and fly their Cessna 210 into the Idaho-Wyoming-Montana back country for camping and fishing.[45]

RICHARD L. WRIGHT, USN

Over his lengthy flying career, Richard ("Dick") Wright has accumulated more than 10,000 accident-free pilot hours among many and varied achievements in aviation, and is still taking to the air.

Richard Lee Wright was born in Denver, Colorado on Christmas Day, 1929, the son of Joseph and Grace Wright. He attended Teller Elementary and Gove Junior High prior to graduating from East High School. As a small boy, he had a fascination with airplanes and a yearning to fly. "I loved making model airplanes from the time I could glue two sticks together."

In June 1952 Wright graduated with a degree in electrical engineering from the Alabama Polytechnic Institute (Auburn University as of 1962), and was immediately commissioned in the Navy as an ensign. He served in the engineering department of the *Sumner*-class destroyer USS *Lofberg* (DD-759) until receiving his orders to pilot training in April 1953, while the ship was stationed off the coast of Korea. He began initial pilot training that same month at Pensacola, Florida, flying the naval variant of the North American AT-6 (SNJ) Texan. Then, keen to become a carrier-based attack pilot, he received carrier qualification training in the same aircraft type. In February the ensuing year he began advanced training at Corpus Christi, Texas. This included all-weather flight school operating a twin-engine Beech SNB Navigator, along with a fellow student and instructor. On gaining his instrument rating, a student was asked what type of flying he would like to tackle. Wright's choice was the Douglas AD-1 Skyraider, a propeller-driven, single-seat attack aircraft. "I was selected for that and did my advanced training in the AD-1 at Cabaniss Field near Corpus Christi."

In June 1954, having completed his advanced training, Lt. (jg) Wright received his wings as a naval aviator and was assigned to VA-55 ("Warhorses") at NAS Miramar, California. At the Fleet Air Gunnery Unit at NAS El Centro, California he practiced weapons delivery techniques using bombs, rockets and the 20-mm guns of the AD-4 Skyraider. There were also Grumman F9F-5 Panthers and F9F-6 Cougars available. The daily routine comprised two weapons delivery flights commencing at first light, plus afternoon classroom sessions. In April 1955, as part of Air Task Group 2, the squadron had an extended WestPac deployment on USS *Philippine Sea* (CV-47).

In 1956, while undertaking a second training cycle at NAS Miramar, Wright went on a blind date and was introduced to elementary school teacher Vivian Cates. They hit it off straight away and started dating. Before embarking on a second deployment in April 1957 aboard USS *Hancock*, he presented her with an engagement ring. They were married in October. In December 1957, VA-55 became the first attack squadron to transition to the swept-wing North American FJ-4B Fury fighter-bomber. It began another extended WestPac deployment in August 1958 on USS *Bennington* (CVA-20), flying the FJ-4B in the Formosa Straits during the Chinese Communist shelling of the islands of Quemoy and Matsu. While on this deployment their first son, Rick, was born.

In January 1959, Wright was assigned to the Navy Test Pilot School at Patuxent River as a member of Class 23. "We had a wonderful stable of airplanes to fly at the school," he reflected. "We had delta-wing airplanes, afterburner airplanes; we had reciprocating engine and twin-engine airplanes. Up until then I'd only flown three different types of airplane, but at USNTPS I had the opportunity to fly maybe eight or nine different types. John Young was in my class and we were good friends with him and his first wife, Barbara. Pete Conrad was one of the instructors in the school and we frequently played bridge with him and his first wife, Jane. Al and Sue Bean were also good friends and bridge players in those days."

After graduation, Wright remained at the Navy's Flight Test Center in the Carrier Suitability Division, where he became involved in testing multiple bomb racks on the A4 Skyhawk, checking the aircraft's capacity to withstand carrier catapult launches with the additional drag from the bomb racks. This, he said, was one of the scariest flight tests he ever did, "because the A4 weighed 22,500 pounds, and with the Curtis-Wright J65 engine it was a heavy bird with little delta wings. At that weight, with drag added, it had all the characteristics and performance of a flying rock." He then became project test pilot for the North American A3J-1Vigilante (later redesignated the A-5A), which became the only Mach 2 bomber to serve aboard a U.S. carrier. It was powered by two large GE J79 engines, giving it much greater thrust than the A4. Wright was able to carry out an initial series of carrier suitability tests aboard USS *Saratoga* (CVA-60) in July 1960, operating catapult minimum trials on the Vigilante "without feeling like we were a rock going off the end of the ship".

In 1962, Wright was posted to the first of ten eventual Vigilante squadrons based at NAS Sanford, Florida, and served two extended deployments flying the A-5A off USS *Enterprise* (CVAN-65). After further assignments at the U.S. Naval Academy, Air Force Command and Staff College, and U.S. European Command he was sent to the Naval Weapons Center in California's Western Mojave Desert, where he flew the A4 Skyhawk and A7 Corsair II. In 1965 he received his master's degree in business administration from George Washington University in Washington, D.C.

Cmdr. Wright retired from the Navy in 1974 and took up a position as a flight test engineer with Cessna, but decided there was far too much paperwork involved and left after three months. Eventually he secured a job with Flight Systems, Inc., private

Lt. Dick Wright seated in the North American A3J-1 (A-5A) Vigilante. (Photo courtesy R. Wright)

sector specialists in flight test design and implementation, for which he flew the T-33 at Mojave until 1981 during tests of the Pershing II medium-range solid-propellant ballistic missile. After five years with Flight Systems, he was offered a job as a civil service pilot for the Navy at China Lake, now flying a number of QF-86F Sabres and QF4s (a reusable full-scale target drone version of the Phantom). These converted airplanes, once owned and operated by the Japanese Defense Forces, were used for developmental testing by the U.S. military, including many of the Navy's new air-to-air missiles. "They were beautiful airplanes," Wright acknowledged. "The Japanese had maintained them well; there was no corrosion, a relatively low flight time, and the Navy picked them up for the cost of shipping. That really was the best flying of my career; I did twelve years at China Lake because I flew the F-86s and F4s just about every day." As well as test flying these aircraft, Wright operated them from the ground as full-scale, radio-controlled unmanned target drones. It was exciting work because, as he suggests, "The best test for a missile is to actually go against a real airplane, and we had real airplanes."[46]

In his 2007 book, *Too Far From Home* (Vintage Books, London), author Chris Jones describes how Navy pilot and later astronaut Ken Bowersox "would chase these doomed planes and watch unarmed missiles punch holes through their wings .... But every now and then, just often enough to keep things interesting, Bowersox would drop in behind one of Wright's drones, lock it in his sights, and launch the latest, greatest warhead straight up its ass."[47]

Dick Wright at China Lake remotely operating old drone antisubmarine helicopters used as missile targets. (Photo courtesy R. Wright)

Test-flying at China Lake. (Photo courtesy R. Wright)

Dick Wright retired from China Lake in 1993 and undertook part-time test flying for Avtel Services, Inc., and the Honda Aircraft Company. He now restricts himself to his personal Piper Cherokee 180 Challenger. He and Vivian have three children: Rick, the oldest, enjoyed an Air Force career flying the F4 and F-117 and these days is a commercial A320 pilot; younger son Bob is a defense systems engineer working in Tucson, Arizona; and daughter Michelle is married to a petroleum geologist who is a Marine reservist currently serving in Afghanistan. There are also a total of eleven grandchildren.

When asked if he was disappointed about not being selected as an astronaut, Dick Wright remains philosophical. "At the time it was a big disappointment to me," he mused. "Looking back, however, it was probably for the better. The guys that they selected were all smarter than me – I just had to acknowledge that. But also a lot of their marriages fell apart. My lovely lady and I have been happily married for some 54 years. Who knows what would've happened if I'd become an astronaut? Perhaps I wouldn't have had the wholesome family life. So I'm perfectly content, and I also had a wonderful career which I loved."[48]

REFERENCES

1. Collins, Michael, *Carrying the Fire: An Astronaut's Journeys*, Farrar, Straus and Giroux, New York, NY, 1974, pg. 33

2. Interview with Dick Gordon, Tucson, Arizona, 1 June 2012
3. Hansen, James R., *First Man: The Life of Neil A. Armstrong*, Simon & Schuster, New York, NY, 2005, pg. 207
4. St. Petersburg Times (Florida) newspaper, Saturday, September 15, 1962, *Nation Will Soon Meet Nine Newest Astronauts*, pg. 2
5. Lovell, Jim and Jeffrey Kluger, *Lost Moon: The Perilous Voyage of Apollo 13*, Houghton Mifflin Company, New York, NY, pg. 185
6. Hansen, James R., *First Man: The Life of Neil A. Armstrong*, Simon & Schuster, New York, NY, 2005, pp. 207,208
7. NASA News Release, September 17, 1962
8. Hansen, James R., *First Man: The Life of Neil A. Armstrong*, Simon & Schuster, New York, NY, 2005, pg. 211
9. St. Petersburg Times (Florida) newspaper, Tuesday, September 18, 1962, *Nine New Astronauts Picked for Moon Shot*, pg. 2
10. Stafford, Tom and Michael Cassett, *We Have Capture: Tom Stafford and the Space Race*, Smithsonian Institution Press, Washington, DC, 2002, pg. 41
11. St. Petersburg Times (Florida) newspaper, Tuesday, September 18, 1962, *Nine New Astronauts Picked for Moon Shot*, pg. 2
12. Rome News-Tribune (Georgia) newspaper, Tuesday, September 18, 1962, *Director Introduces 9 New Astronauts*, pg. 3
13. Bonham Daily Favorite (Texas) newspaper, Monday, September 17, *9 New Astronauts Plan Moon Flight*, pg. 1
14. Jenista, John E., blog extract from *The Brown Shoes: United States Naval Aviation History*, 13 November 1988. Website: http://thebrownshoes.org
15. Blue Angels article, published in the August 1955 issue of *Naval Aviation News*, author unknown.
16. E-mail correspondence with Joan Cudeback, 1 December 2011–12 March 2012
17. *Ibid*
18. E-mail correspondence with Diane Aslund, 29 November 2011–17 April 2012
19. E-mail correspondence with Roy Dickey, 21 December 2011–11 March 2012
20. E-mail correspondence with Tom Edmonds, 9 November 2011–2 April 2012
21. Official veteran biography, *Lt. Gen. William H. Fitch, U.S. Marine Corps*. Website: http://www.veterantributes.org/TributeDetail.asp?ID = 1484
22. Oral interview with General William Fitch (unidentified interviewer) for the A-4 Skyhawk Association. Website: http://www.a4skyhawk.org/2c/a4parts/mcbr.html
23. Golden Eagles biography: *William H. Fitch, Lt. General, USMC (Ret.)* Website: http://www.epnaao.com/BIOS_files/.../Fitch-%20William%20H.pdf
24. Online article, TWA Pioneer Lawrence (Larry) G. Fritz. Website: http://www.twaspirit.com/larry_fritz_.html
25. LIFE magazine, 11 November 1966, article by Keith Wheeler, *The Full Story of the 2.8 Seconds That Killed the XB70*, pg. 135
26. *Ibid*, pg. 140
27. E-mail correspondence with John and James Fritz, 22 December 2011–14 July 2012

28. Author interview with William Geiger, Los Angeles, CA, 28 May 2012
29. E-mail correspondence with Ann Glunt, 30 November 2011–12 December 2011
30. E-mail correspondence with Orville Johnson, 6 December 2011–9 June 2012
31. Shugarts, David A., article, *William P. Kelly, Jr., Test Pilot*. Website: http://testfly.org/williamkelly/index.html
32. E-mail correspondence with Barbara Kelly, 11 January 2012–3 February 2012
33. *Ibid.*
34. Shugarts, David A., article, *William P. Kelly, Jr., Test Pilot*. Website: http://testfly.org/williamkelly/index.html
35. E-mail correspondence with Brad McCann, 2 January 2012–27 February 2012
36. Biographical details previously supplied to author by Capt. J.R.C. Mitchell in e-mail messages, 4-7 February 2010
37. North Country Catholic (Ogdensburg, NY) newspaper, November 4, 1962, *Future Spaceman Starred for Catholic High*, pg. 5A
38. Wikipedia online encyclopedia, *Francis G. Neubeck*. Website: http://en.wikipedia.org/wiki/Francis_G._Neubeck
39. PBS Nova: *Astrospies: Secret Astronauts: Col. Francis G. Neubeck*. Website: http://www.pbs.org/wgbh/nova/astrospies/profiles.htm
40. Wikipedia online encyclopedia, *Francis G. Neubeck*. Website: http://en.wikipedia.org/wiki/Francis_G._Neubeck
41. E-mail correspondence with VADM William E. Ramsey, 17 November 2011–21 June 2012
42. Smith, Robert W., *The Robert W. Smith Autobiography, NF-104.com* (incomplete). Website: http://www.nf104.com
43. Biographical details previously supplied to author by Robin Solliday Heyne and Robert E. Solliday in e-mail message, 8 February 2011
44. E-mail correspondence with Al and Trip Uhalt, 19 December 2011–17 April 2012
45. E-mail correspondence with M/Gen. Ken Weir, 10 November 2011–28 May 2012
46 E-mail correspondence with Dick Wright, 3 December 2011–14 February 2012
47. Jones, Chris, *Too Far From Home: A Story of Life and Death in Space*, Vintage Books, London, UK, 2008, pg. 86
48. E-mail correspondence with Dick Wright, 3 December 2011–14 February 2012

4

The "Next Nine"

While many ambitious goals had been set for Project Mercury, from its inception there was always one objective above all others; to place Americans into space – and hopefully into orbit – and return them safely to Earth. Much as the orbital mission of Yuri Gagarin on 12 April 1961 would frustrate NASA, it also provided welcome reassurance that man could survive being launched into space.

The names of the Mercury astronauts had been announced to the world amid much fanfare in April 1959, and while there was considerable interest in those named in the second group three years later, very few people realized at the time that their accomplishments would create milestones in human space flight history. Included in their number was the first American to walk in space; two astronauts who not only set a space endurance record in Project Gemini but would be crewmembers on the first manned spacecraft to loop around the Moon; and the first person to walk on the Moon.

The nine chosen astronauts, referred to alternatively as the Class of '62, the Next Nine, the Nifty Nine, and the New Nine, would eventually total up 25 flights into space; they would be aboard all of the Gemini missions; six would fly to the Moon and three would walk on its surface.

But while there was triumph, there was also tragedy in their ranks. One would die in a horrifying launch pad fire during a rehearsal, and another in an airplane crash shortly before he was to fly his first space mission.

The new astronauts would have precious little time to adjust to their new life and all the personal and professional challenges that it presented. They would report to the Manned Spacecraft Center in Houston the following month to begin a two-year training course. Once that had been completed, they were expected to join the flight-experienced Mercury astronauts on two-man Gemini missions as NASA began the next stage of its thrust to the Moon.

NEIL A. ARMSTRONG

As one of the most famous, even immortal names associated with the 20th century, Neil Armstrong will always be remembered as the first person to set foot on another world.

Neil Alden Armstrong was born on a farm just out of Wapakoneta, Ohio on 5 August 1930. He gained his student pilot's license on his sixteenth birthday. A naval aviator from 1949 to 1952, he flew 78 combat missions with Fighter Squadron 51 (VF-51) during the Korean War. After graduating from Purdue University in January 1955 with a bachelor's in aeronautical engineering, he joined the National Advisory Committee on Aeronautics (NACA) as a test pilot at their High Speed Flight Station, where he flew a large number of experimental aircraft. He married Janet Shearon, whom he had met while both were studying at Purdue. They had three children, Eric, Karen and Mark, although Karen died of a malignant brain tumor in January 1962.

Armstrong next transferred to NACA's Flight Research Center at Edwards AFB, flying a number of high-performance jet aircraft, as well as participating in the U.S. Air Force's Man in Space Soonest (MISS) and X-20 Dyna-Soar programs. As a pilot on seven flights of the X-15 research rocket plane he flew the supersonic aircraft to 207,500 feet and achieved a top speed of just less than 4,000 mph. Selected in the second group of astronauts in 1962, he was named as command pilot for the three-day Gemini VIII mission along with pilot Dave Scott, performing the first docking of two vehicles in space on 16 March 1966 using an Agena target vehicle. However, a jammed thruster on the Gemini craft sent the docked vehicles into a wild spin that placed the astronauts in extreme danger. They managed to control the spinning and undocked, but the mission had to be aborted. Armstrong and Scott then carried out an emergency re-entry and splashdown just eleven hours after being launched on their mission.

Neil A. Armstrong. (Photo: NASA)

Along with lunar module pilot Edwin ("Buzz") Aldrin and command module pilot Michael Collins, mission commander Neil Armstrong was launched to the Moon on 16 July 1969. Once in lunar orbit, Armstrong and Aldrin entered *Eagle* and guided it to a landing on the Sea of Tranquility. At 9:56 p.m. (Houston time) on Sunday, 20 July, Neil Armstrong stepped onto the lunar surface and uttered the immortal words, "That's one small step for [a] man; one giant leap for mankind." After just over two and a half hours collecting soil and rock samples and setting up experiments, he and Aldrin returned to *Eagle* for a link-up the following day with Michael Collins aboard *Columbia*. Apollo 11 ended with a successful splashdown in the Pacific and recovery by USS *Hornet* (CVS-12).

In July 1970 Armstrong was appointed NASA's Deputy Associate Administrator for Aeronautics in the Office of Advanced Research and Technology, Washington, D.C. After resigning from NASA in 1971 he became a professor of engineering at the University of Cincinnati, and in 1984 was named as chairman of the board of Computing Technologies for Aviation, Inc., in Charlottesville, Virginia. That year he also received a presidential appointment to the National Committee on Space, and following the loss of the shuttle *Challenger* in January 1986 was immediately named vice chairman of the committee formed to investigate the tragedy.

In 1994 he divorced his first wife after 38 years of marriage, and in June that year married Carol Knight. His authorized biography, *First Man: The Life of Neil A. Armstrong*, was written by James Hansen and published in 2005.

Neil Armstrong died on Saturday, 25 August 2012 from complications following heart bypass surgery. He was 82 years old.

FRANK F. BORMAN II, USAF

Among his achievements for NASA, Frank Borman will always be remembered as the commander of Apollo 8 in December 1968, the first crew to fly out to the Moon.

Born on 14 March 1928, Frank Frederick Borman II was named after his paternal grandfather. The only son of Marjorie and Edwin Borman, a garage owner of Gary, Indiana, he suffered chronic respiratory problems as a child. As a result, the family relocated to the drier climate of Tucson, Arizona when he was five years old. After gaining his early education at Sam Hughes Elementary School, Borman enrolled at Tucson High School. When he was fifteen he began taking flying lessons at nearby Gilpin Airport, doing miscellaneous after-school jobs to raise the $9 per hour needed for dual instruction. While at high school, where he was a champion quarterback and gifted scholar, he met his future wife, Susan Bugbee.

In 1946 Borman won an appointment to the U.S. Military Academy at West Point. In his senior year he was ranked in the top 2 per cent of his class, and on 6 June 1950 graduated sixth in his class of 670 with a bachelor of science degree. He was also awarded his expressed first preference for military service by being commissioned a 2nd lieutenant in the U.S. Air Force.

Borman's flight training began in 1951 at Williams AFB in Arizona, and he was awarded his wings in August. He was then assigned as a fighter pilot with the 44th

Frank F. Borman II. (Photo: NASA)

Fighter-Bomber Squadron at Clark Air Base in the Philippines and promoted to 1st lieutenant in March 1952. From 1953 to 1956 he was a pilot and instructor with a number of fighter squadrons based in the United States, and in February 1955 was promoted to the rank of captain. In 1957 he gained his master's from the California Institute of Technology and went on to teach thermodynamics and fluid mechanics at West Point. In August 1960 he was assigned to Class 60-C at the USAF Test Pilot School at Edwards AFB. After graduating he was accepted for the first Aerospace Research Pilot School course, and later became an instructor for subsequent courses and project officer for F-104 Starfighter rocket modification. Promoted to major in February 1961, he was still at ARPS when he applied for NASA's second group of astronauts and was one of the nine successful candidates.

His proficiency at NASA earned him a place on Gemini 3 with Gus Grissom following the medical grounding of Alan Shepard, but Grissom then insisted on having John Young as his pilot, so Borman became backup command pilot for Gemini IV. His first space flight was as command pilot of Gemini VII, launched on 4 December 1965 along with James Lovell. In addition to serving as a target for a rendezvous by Gemini VI-A, Borman and Lovell spent two weeks in space, thereby claiming the space endurance record from the Soviet Union and establishing that the human body could survive the space environment for long enough to fly an Apollo mission to the Moon and back.

Following the Apollo 1 launch pad fire in which three astronauts lost their lives, Borman was the astronauts' representative on the investigating board. His next space assignment came as commander of the Apollo 8 mission, launched to the Moon on 21 December 1968. After entering lunar orbit on Christmas Eve, the three-man crew famously read an extract from the *Book of Genesis* during a TV broadcast to Earth.

After serving as field director of NASA's Space Station Task Force, Col. Borman retired from the Air Force and NASA in July 1970 to become a director of Eastern Airlines. He was promoted to chairman of the airline's board on 15 December 1976; a position he held until his retirement in June 1986. His autobiography, *Countdown*, was written with Robert Serling and published in 1989.

CHARLES CONRAD, JR., USN

Charles Conrad Jr. was born in Philadelphia, Pennsylvania on 2 June 1930 to Frances De Rappelage (née Vinson) and Charles Conrad, a balloonist during the First World War. When his mother began to call him Peter, the moniker stuck. While attending Haverford School in Pennsylvania he began working as an apprentice mechanic and general hand at the nearby Paoli airfield in order to take flying lessons.

Conrad then went to Princeton University on a Navy scholarship, completing the regular NROTC program and graduating in 1953. While there he also met his future wife Jane DuBose. They would go on to produce four sons before divorcing in 1990. After graduating with his ROTC commission he applied for flight training to become a naval aviator. Ensign Conrad reported to NAS Pensacola, Florida for flight training on 1 July 1953, and was assigned to Fighter Squadron 43 (VF-43) the following year. In 1956 he transitioned to the Grumman F9F-8 Cougar and became carrier qualified. On 1 August 1958 he was admitted to the U.S. Navy Test Pilot School at the Naval Air Test Center (NATC) Patuxent River, Maryland, joining future astronauts Wally Schirra and Jim Lovell in Class 20. On graduating he stayed at NATC as a test pilot, flight instructor and performance engineer.

Although Conrad failed in his bid to become a Mercury astronaut, he was selected to Group 2 in 1962. One of the first of his group to be given an assignment, he flew the long-duration Gemini V mission with Mercury astronaut Gordon Cooper. He was promoted to lieutenant commander in April 1963, the first of three rapid promotions in rank: commander in September 1965 and captain just ten days later. In September 1966 Conrad returned to orbit as the commander of Gemini XI, along with his good friend Dick Gordon. They docked with an Agena and used its rocket engine to set an altitude record of 850 statute miles.

Three years later, Conrad was launched to the Moon as commander of Apollo 12, along with naval aviators Dick Gordon and Alan Bean. After guiding *Intrepid* to a safe landing on the Ocean of Storms, Conrad and Bean conducted two moonwalks; the second to the nearby unmanned probe Surveyor 3, which had landed on the Moon two years earlier.

Had it not been canceled, Conrad might have commanded Apollo 20 as the last in

Charles ("Pete") Conrad, Jr. (Photo: NASA)

the series. Instead, he spent 28 days aboard the Skylab space station in the summer of 1973 with Paul Weitz and Joe Kerwin after having rescued it from damage suffered during launch.

Conrad retired from NASA and the Navy in 1973 and took on private sector work, principally as a Vice President of the Douglas Aircraft Company with responsibility for the reusable *Delta Clipper* launch vehicle. In 1988 Pete and Jane were divorced, and two years later he married attractive divorcée Nancy Crane from Denver. That year, sadly, he also lost his 29-year-old youngest son Christopher to bone cancer. On 8 July 1999 Conrad was injured in a motorcycle accident outside of Ojai, California and died in hospital. The irrepressible Pete Conrad was buried at Arlington National Cemetery with full military honors.

JAMES A. LOVELL, JR., USN

James Lovell became one of the best known of the Apollo astronauts, particularly as commander of Apollo 13, one of the greatest stories of modern-day survival against the odds.

James Arthur Lovell, Jr. was born in Cleveland, Ohio on 25 March 1928. He was only five years old when his father died in an automobile accident. For a time he and his mother lived in Indiana before moving to Milwaukee, Wisconsin in 1938. He got his higher education at Juneau High School. During these years he became

James A. Lovell, Jr. (Photo: NASA)

fascinated with rocket propulsion, experimented with crude rockets in his back garden, and read all that he could about astronomy, physics and engineering. While in high school he met his future wife, Marilyn Gerlach.

In September 1946 he entered the University of Wisconsin under the V-5 naval pilot training program, and after two years of study moved on to flight training. Two months later, in March 1948, he gained an appointment to the U.S. Naval Academy as a member of the Class of '52. Four hours after graduating with his bachelor of science degree, he and Marilyn were married.

Over the next six years Lovell's career as a naval aviator and instructor included advanced training in the hazardous business of night flying off aircraft carriers. In 1958, now with two of an eventual four children, he attended the Navy's Test Pilot School at Patuxent River as a student in Class 20, along with future astronauts Wally Schirra and Pete Conrad. He graduated top of his class, with Schirra and Conrad tied for second place. He spent the next three years at the Maryland base as a test pilot, serving as a program manager with F4JH Weapons Systems Evaluation. During this time, he also underwent testing as a Mercury astronaut finalist but was ruled out by an incorrectly diagnosed liver complaint. He applied a second time in

1962 and this time swept through the medical tests and other examinations. On 17 September 1962 NASA announced his appointment as a Group 2 astronaut.

His first flight assignment came as backup pilot for Gemini IV. Then, along with Frank Borman, he lifted off on 4 December 1965 for the Gemini VII mission which spent a fortnight in orbit. They served as a rendezvous target for Gemini VI-A, and the two craft flew in close formation for several hours. On 11 November 1966 Lovell commanded Gemini XII, the final mission in the program, with future moonwalker Buzz Aldrin. In 1968 he flew with Bill Anders and Frank Borman on Apollo 8, and they became the first humans to fly around the Moon. But it is for his fourth and final space flight, Apollo 13, in April 1970, that Lovell is best remembered. By conserving every possible watt of electricity and other supplies, the near-frozen astronauts made it back to a safe splashdown after six days of high drama.

Lovell resigned from the space agency and the Navy on 1 March 1973 in order to go into private industry. In the early 1990s he and Jeffrey Kluger wrote *Lost Moon*, which told the dramatic story of the aborted mission of Apollo 13. Released in 1994, the book was later filmed by Ron Howard as the movie *Apollo 13* with Tom Hanks playing the principal role.

Capt. James Lovell, the first person to fly into space four times, remains one of only twenty-four humans to have traveled out to the Moon; the first of only three to have flown there twice; and the only one of the returnees to do so without making a landing.

JAMES A. MCDIVITT, USAF

James Alton McDivitt was born 10 June 1929 in Chicago, Illinois, and grew up in Kalamazoo, Michigan. After Central High School he attended Michigan's Jackson Junior College from 1948 to 1950 before entering the U.S. Air Force as an aviation cadet in January 1951. He gained his pilot wings and commission as a 2nd lieutenant in May 1952 at Williams AFB, Arizona, and six months later completed his combat crew training.

Assigned to the 35th Bombardment Squadron, Lt. McDivitt was shipped to Korea, where he flew 145 combat missions in F-80 Shooting Star and F-86 Sabre aircraft. Returning to the United States in September 1953, he served as a pilot and Assistant Operations Officer with the 19th Fighter-Interceptor Squadron at Dow AFB, Maine. In November 1954 he entered advanced flying school at Tyndall AFB, Florida and in July 1955 was transferred to McGuire AFB, New Jersey, where he served as a pilot, Operations Officer, and later as Flight Commander with 332 FIS. He returned to the University of Michigan in June 1957 under the auspices of the Air Force Institute of Technology (AFIT) program, and received his bachelor of science degree two years later, graduating first in his class.

In June 1959 McDivitt entered Class 59-C at the U.S. Air Force Test Pilot School, Edwards AFB. Following graduation on 22 April 1960 he remained at the Air Force Flight Test Center as a test pilot. He completed the Aerospace Research Pilot School as a member of its first class and joined the center's Manned Spacecraft Operations

James A. McDivitt (Photo: USAF)

Branch in July 1962. Capt. McDivitt applied for NASA's second astronaut group and his selection was announced in September 1962.

Three years later he made his first space flight as command pilot for Gemini IV, a four-day mission launched on 3 June 1965. The highlight was the first spacewalk by an American astronaut, conducted by fellow crewmember and close friend from Test Pilot School, Capt. Edward H. White II. McDivitt's second flight was as commander of Apollo 9, launched on 3 March 1969. With David Scott and Rusty Schweickart, he tested the entire suite of Apollo flight hardware in Earth orbiting as a prelude to the first lunar landing later that year. In June 1969 he resigned as an astronaut to become manager for Lunar Landing Operations with responsibility for planning the lunar landing missions that would follow the historic first landing, and also for redesigning the two spacecraft to extend their lunar exploration capability. That September he became manager of the Apollo Spacecraft Program with responsibility for the entire program, a job that he held until his retirement from the space agency.

Promoted to the rank of brigadier general effective 17 February 1972, McDivitt retired from both the Air Force and NASA in June 1972 to pursue a career in private industry, most notably as Senior Vice President for Government Operations at Rockwell International.

ELLIOT M. SEE, JR.

Elliot McKay See, Jr., a civilian test pilot and flight test engineer, was born in Dallas, Texas on 23 July 1927. He attended elementary and high school in Dallas's Highland Park School District and joined the Reserve Officer Training Corps. See gained his private pilot's license while attending the University of Texas, and in 1945 received an appointment to the U.S. Merchant Marine Academy at Long Island, New York. On graduating in 1949 with a bachelor of science degree in marine engineering, he successfully applied for a position with his father's former employee, the General Electric Company (GE), and became a flight test engineer and later experimental test pilot with the firm in Ohio.

Except for active service as a naval aviator from 1953 to 1956, he remained with GE until 1962. He was project pilot of J70-8 jet engine development for the F4H Phantom II aircraft at Edwards AFB, California. Later, he conducted power plant test flights on the J-47, J-73, J-70 and CJ805. He performed flight tests of these engines in numerous aircraft, including the F-86 Sabre, XF4D Skyray, F-104 Starfighter, and T-38 Talon. In 1962 he earned a master's degree in engineering from the University of California. He married the former Marilyn Jane Denahy from Ohio, and they had two daughters, Sally and Caroline, and son, David.

Elliot M. See, Jr. with his son David. (Photo: NASA)

Selected by NASA as a Group 2 astronaut in 1962, See served as backup pilot for the Gemini V mission prior to being named commander of Gemini IX in September 1965, together with pilot Charles Bassett II. On 28 February 1966 they were killed when their T-38 struck a building during a missed approach in atrocious weather at Lambert Field in St. Louis, Missouri, just three months before they were to fly into space. Their places on that flight were taken by the backup crew, Tom Stafford and Gene Cernan.

THOMAS P. STAFFORD, USAF

Thomas Patten Stafford was born in Weatherford, Oklahoma on 17 September 1930, and would be the only child of Thomas Sabert Stafford, a dentist, and the former Mary Ellen Patten. He distinguished himself at school with straight A's and by being a formidable football player. On graduating from Weatherford High School in 1948 he served briefly with the 158th Field Artillery of the Oklahoma National Guard, prior to entering the U.S. Naval Academy.

After graduating in 1952 ranked 50th in his class of 783 he elected to join the U.S Air Force and was commissioned a 2nd lieutenant. After flight training, he received his wings in September 1953. Following interceptor training at Tyndall AFB, Florida he was assigned to the 54th Fighter-Interceptor Squadron at Ellsworth AFB, South Dakota. In December 1955 his next assignment took him to Hahn Air Base in West Germany, where he was Flight Leader and Flight Test Maintenance Officer with the 496th FIS, flying the F-86D Sabre jet.

Thomas P. Stafford. (Photo: NASA)

In August 1958 Stafford enrolled at the Air Force Test Pilot School at Edwards AFB, California, graduating in April 1959 as the recipient of Class 58-C's A. B. Honts Award as the outstanding graduate in his flying and academic achievements. He remained at Edwards until September 1962 as a project pilot and instructor with the Aerospace Research Pilot School. An inch too tall to qualify for selection as a Mercury astronaut, Capt. Stafford applied again for the second intake which relaxed the height limit by that one vital inch. Just in case, he decided to prepare for a career in administration, applying for admission to Harvard University's Graduate School of Business. In September 1962, three days after being admitted to the course, he was notified that his astronaut application had been successful.

His first assignment was backup pilot for the first manned Gemini mission, then he became pilot to mission commander Wally Schirra on Gemini VI. After a series of launch delays the two men were fired into orbit on 15 December 1965, achieving an historic rendezvous in space with Gemini VII, which was conducting a long-duration mission. Stafford's second flight was command pilot of Gemini IX-A, in the wake of the deaths of the original crew of See and Bassett in an airplane crash. Launched on 3 June 1966, the mission featured three types of rendezvous and a spacewalk by pilot Gene Cernan. Three years later, on 18 May 1969, Stafford was launched on his third mission as the commander of Apollo 10, together with Cernan and John Young, for a rehearsal in which the lunar module *Snoopy* closed to within 50,000 feet of the lunar surface.

With the rank of general, Stafford's final mission was to command the American spacecraft of the Apollo-Soyuz Test Project in July 1975, with Mercury astronaut Deke Slayton and rookie Vance Brand. They docked with a Soviet Soyuz spacecraft manned by Alexei Leonov and Valery Kubasov, greeting each other in the docking tunnel and sharing joint commemorative duties with the cosmonauts over a period of two days. The Apollo spacecraft splashed down on 24 July, concluding Stafford's astronaut career with over 507 hours in space on his four missions. He resigned from NASA in order to return to the Air Force. In 1979 he went into private business as an aerospace advisor.

EDWARD H. WHITE II, USAF

In 1965 he became a space legend by performing America's first spacewalk, but less than two years later Lt. Col. Ed White died under the most tragic of circumstances on the launch pad during a rehearsal in the run up to his second mission.

Edward Higgins White II was born in San Antonio, Texas on 14 September 1930. After attending Western High School in Washington, D.C., he reported to the U.S. Military Academy at West Point on 15 July 1948. Following his graduation in June 1952 with a bachelor of science degree, White joined the Air Force and obtained his initial pilot training at Bartow AFB, Florida and jet training at James Connally AFB, Texas. During this time he married Patricia Finegan.

After gaining his wings in 1953, Lt. White was sent to Luke AFB, Arizona for Fighter Gunnery School prior to joining the 22nd Day-Fighter Squadron, based at

Edward H. White II. (Photo: USAF)

Bitburg in West Germany, as an F-86 Sabre and F-100 Super Sabre fighter pilot. He also completed the Air Force Survival School in Bad Tolz, West Germany. By the time the Whites returned home they had two children, a son named Edward III and a daughter, Bonnie.

In September 1957 White entered the School of Engineering at the University of Michigan, where three of his classmates were future astronauts – Ted Freeman, Jim Irwin, and a man who was to become a very close friend, Jim McDivitt. In 1959 he received his master's in aeronautical engineering from the university and was sent to the Air Force Test Pilot School at Edwards AFB as a member of Class 59-C. Once again McDivitt was a classmate. After graduating, his last assignment before joining NASA was as an experimental test pilot with the Aeronautical Systems Division at Wright-Patterson AFB, Ohio.

The four-day flight of Gemini IV in June 1965 commanded by McDivitt achieved vital tests of the new spacecraft's capabilities, but it will always be remembered for Capt. Ed White floating freely in space at the end of an umbilical, propelling himself with a small gas gun.

In March 1966 White received his second space flight assignment, as senior pilot on the first manned Apollo mission, which was then scheduled for the first months of 1967. On 27 January of that year he was sealed inside the Apollo 1 spacecraft for a routine "plugs out" test along with mission commander Gus Grissom and pilot Roger Chaffee, when an electrical spark caused a flash fire in the pure oxygen environment. Within seconds, all three men were asphyxiated by toxic fumes. When the hatch was finally opened by the pad crew, the astronauts were pronounced dead.

On 31 January 1967, Gus Grissom and Roger Chaffee were buried at Arlington National Cemetery, while at the same time Ed White was laid to rest at his beloved West Point.

JOHN W. YOUNG, USN

John Watts Young was born in San Francisco, California on 24 September 1930 but grew up in Orlando, Florida, where he attended high school. On graduating from the Georgia Institute of Technology with a bachelor's in aeronautical engineering (with highest honors) and having served as an ROTC student, he entered the U.S. Navy in June 1952. After a year aboard the destroyer USS *Laws* (DD-558) he was reassigned to aviation training.

From 1955 to 1959 Young served with Fighter Squadron 103, then attended the U.S. Navy Test Pilot School as a student in Class 23. After graduation he remained

John W. Young. (Photo: NASA)

at the Naval Air Test Center for three years. As a test pilot and program manager, he was attached to test projects to evaluate the F8D and F4B fighter weapons systems for the Navy. He was also Maintenance Officer of All-Weather Fighter Squadron 143 at NAS Miramar, California. Early in 1962 he set time-to-climb records for the 3,000 meter and 25,000 meter events in Project High Jump.

Young was the first member of NASA's second astronaut group to be assigned to a space mission when he was appointed in April 1964 as pilot of Gemini 3, the first flight in the two-man series, under the command of Mercury astronaut Virgil ("Gus") Grissom. The three-orbit flight on 23 March 1965 was a successful check-out of the spacecraft. In July 1966 Young was command pilot for the Gemini X mission, along with Michael Collins, and they rendezvoused with one Agena target vehicle and used its engine to reach another one, in the process setting an altitude record of 475 statute miles.

In 1968 Young served as backup command module pilot for Apollo 7, which flew the mission meant for the lost Apollo 1 crew. In May 1969 he flew as the command module pilot on Apollo 10, which paved the way for the first lunar landing. In April 1972 he walked on the Moon as commander of Apollo 16, along with Charles Duke, having landed *Orion* in highland terrain near the crater Descartes.

Capt. Young's next assignment was to command the inaugural flight of the space shuttle. On 12 April 1981 (precisely 20 years after cosmonaut Yuri Gagarin made the first human space flight) Young and pilot Bob Crippen soared into space on the two-day STS-1 mission. In October 1983 he commanded STS-9 on a ten-day flight with a then-record crew of six, including scientists, to operate the European-built Spacelab module on its maiden flight. Prior to the loss of shuttle *Challenger* in January 1986, Young had been named to fly a historic seventh space mission in September of that year to deploy the Hubble Space Telescope. But the loss of *Challenger* and her crew prompted Young, who was chief of the Astronaut Office, to openly question NASA's safety standards and procedures. When the flight was later rescheduled, he had been replaced as mission commander and given the job of Special Assistant to the Director of the Johnson Space Center for Engineering, Operations and Safety.

John Young retired from NASA on 31 December 2004, aged 74. Throughout his 42 years with the agency he remained an active astronaut, still eligible to command space shuttle missions.

REFERENCES

The biographical material is derived from the author's own files, Michael Cassutt's *Who's Who in Space: The International Space Year Edition*, Macmillan Publishing Company, New York, NY, 1993, and Douglas Hawthorne's *Men and Women of Space*, Univelt, Inc., San Diego, CA, 1992.

5

Settling in

Project Gemini, for which the latest cadre of astronauts would commence training, had been given the green light to proceed in early December 1961. Twelve missions were manifested, and it was planned that ten of these would carry a crew of two astronaut pilots. The first flight of a Gemini spacecraft, mated to a Titan II launch vehicle, was planned as an unmanned orbital test of the spacecraft, which would be intentionally disintegrated upon re-entry into the atmosphere. The objective of the second unmanned flight would be to test the spacecraft's heat shield.

As a bridge between the Mercury and Apollo programs, Gemini introduced a number of crucial goals to be accomplished over the course of those missions. Most importantly, they included the need to demonstrate that astronauts could live and function effectively during Earth orbital flights which lasted at least eight days; the need to perform the rendezvous and docking maneuvers so essential to the later lunar landing missions, and maneuver docked vehicles using the propulsion system of the target vehicle; the need for astronauts to perform simple tasks during extravehicular activity (EVA), or spacewalking; and the need to perfect techniques of atmospheric re-entry and landing at a pre-selected location. The ambitious schedule called for the manned Gemini flights to be launched at two-month intervals. It was a bold plan, but NASA knew that the two groups of astronauts would integrate quickly and rise to the challenge.

BEYOND PROJECT MERCURY

As each successive Mercury mission was undertaken, so those astronauts became involved in development and training for Project Gemini, although Alan Shepard did lobby hard at the highest levels to have a final, long-duration Mercury mission which would have been designated MA-10. As he told NASA Administrator James Webb, there was a spare Mercury spacecraft, a spare rocket, and he was ready to fly. When Webb said that he felt Project Mercury had achieved all its goals and the agency was ready to move on, Shepard boldly asked if he could raise the question with President Kennedy at an upcoming meeting. Webb gave his blessing to this, but when Shepard outlined his plan at the White House, Kennedy's response was, "What does Mr. Webb think about it?" Knowing he had to give an honest answer, Shepard said Webb did not want to do it, whereupon the president retorted, "Well, I think I'll have to go along with Mr. Webb."[1] With that, Shepard's hopes of a second – this time orbital – Mercury mission were over. There was some later consolation for Shepard, however, when he was given command of the first manned Gemini mission, but to his chagrin a chronic medical problem would mandate his removal from that flight, which was handed to fellow Mercury veteran Virgil ("Gus") Grissom.

In October 1962 the nine newly recruited astronauts reported to Houston to begin their two-year training for Gemini. First was a two-week orientation period, during which they were familiarized with NASA's policies and organization and brought up to speed on the Mercury program – with the six-orbit flight that month of astronaut Wally Schirra, there was only one manned flight remaining. The new men were told that they would be rostered on a four-day training week, with a fifth day set aside for work-related travel or public relations duties. After the orientation period the trainees were joined by members of the Mercury astronaut group for a three-month classroom course covering such basic subjects as aerodynamics, astronomy, space

Elliot See and Neil Armstrong. (Photo: NASA)

medicine, rocket propulsion, and the physics associated with launch, orbital flight and re-entry. Significantly, about one-third of this classroom time was assigned to navigation and guidance techniques. These studies were interspersed with briefings on technical and other aspects associated with Project Gemini and, to some extent, Apollo.

When the time came to report to NASA in Houston, Elliot See was at Edwards AFB. Along with Neil Armstrong, he made the trip down to Texas in an automobile. Asked to recall this trip, Armstrong confessed that his memory was vague. "I don't remember whether the car was his or mine," he told the author. "If it were his, it would have been a convertible; he loved convertibles. He was enjoyable company; we enjoyed being together and wondering what our future would be."[2]

As Gus Grissom was already working full-time on the Gemini program when the nine new astronauts showed up on 1 October, Deke Slayton assigned him to act in a temporary supervisory capacity. "Pretty soon the nine new guys were out buying lots and building homes, most of them a little east of the new Manned Spacecraft Center site, in El Lago," Slayton recalled in his autobiography.[3]

SEALING DEALS

When the new astronauts turned up in Houston, the Manned Spacecraft Center, some 25 miles southeast of the city, was still under construction. Work on the sprawling complex had begun six months earlier in April 1962, with a planned completion date of September 1963. In the interim, most of NASA's offices were in rented buildings in the city itself.

Fortunately, the space agency had set up a transition office that proved extremely helpful in getting new MSC personnel settled into the area and into housing, among them the new astronauts. As they wanted to live near where they would be training, most of the men decided to rent apartments or houses while they chose building sites and designs for homes. Tom and Faye Stafford temporarily moved into an apartment on the Gulf Freeway near Ellington AFB, not far from those rented by Neil and Jan Armstrong, and John and Barbara Young.

Typically, Elliot See was the last of the new astronauts to select a plot of land for the new home he and Marilyn would share with their three children. Nothing that he pursued was ever marked by extremes, doing things in his own unhurried and methodical way, and this extended to carefully choosing a lot at Timber Cove and personally designing their home. Frank and Susan Borman were also busy looking around, and finally settled upon a building. "I signed my first home-building contract for a purchase price of $26,500," he recalled.[4]

On Sunday, 14 October, just as the new astronauts and their families were in the early throes of adjusting to life in Houston, the U.S. military suddenly went on high alert. Flying high above a pre-determined area on the island of Cuba, a U-2 spy plane had obtained reconnaissance pictures of newly constructed missile bunkers and some suspicious-looking hardware on the ground. After analyzing the photographs, the CIA's intelligence officers positively identified the shapes as medium range ballistic missiles capable of reaching the American mainland. Once this had been presented to President Kennedy and his military advisors it prompted a nuclear stand-off between the United States and the Soviet Union, which held the world in suspense. For a time, the Cold War adversaries stood on the brink of a nuclear war. Fortunately, the crisis was resolved a fortnight later by compromises on both sides, and the Soviets agreed to remove their offensive weapons from Cuba.

During the tense days of the "Cuban missile crisis", security had been beefed up at NASA, and the military officers among the new astronauts wondered whether they might be recalled to active duty. But with the welcome news that a solution had been successfully negotiated, the men and their families were able to relax and concentrate once again on their own domestic issues.

Financial help for the new astronauts' houses eventuated through an agreement that would be struck between their group and Field Enterprises. It was an expansion of a deal that existed between *Life* magazine and the original seven astronauts. Back then, through Washington lawyer Leo DeOrsey, the Mercury astronauts had offered to the highest bidder the magazine and book rights to their personal stories. DeOrsey, who accepted no fee for organizing this deal, had convinced the astronauts that an exclusive contract would shield them against unwanted and intrusive media interest in their flights, training, and even their families. Initially opposed to the idea, NASA granted approval after John Glenn explained its benefits to President Kennedy. Glenn maintained it would prevent media pressure during their training, and there would be far more control over what was being written about them. Kennedy saw the sense in this, and urged James Webb to allow the deal to proceed. Once NASA had relented, DeOrsey signed a contract with *Life* magazine. For the next three years, each of the astronauts benefited from the half-million-dollar deal to the tune of $24,000 per year.

With the selection of the second group of astronauts, a fresh contract was needed. This time Field Enterprises World Book Science Service joined the deal and would contribute $10,000 per man, supplementing the $6,250 from Time-Life. As with the original contract, the rights covered *Life* magazine articles, books, and television and movies. The publishing contract was worth half a million dollars over the four-year period, now to be shared equally by the sixteen astronauts, and included an option for another four years. This time it amounted to $16,250 annually per man for four years. Most of them were still on regular military salaries, and whilst the deal did not make them financially well off, it significantly eased the financial burden by enabling them to construct homes near the space center. As well, there was the added inducement of a $100,000 life insurance policy for each astronaut, which was provided without cost by both publishers. Because of their new occupation, insurance contracts would have otherwise been subjected to extremely high premiums.

ACADEMIC STUDIES AND SPECIALTY ASSIGNMENTS

The initial phase of training for the new astronauts would also involve the Mercury astronauts, and was principally academic. Over four months, starting in late

NASA's newest astronauts on a familiarization tour of the launch pads. From left: McDivitt, Young, See, Borman, White, Stafford, Lovell, Conrad and Armstrong. (Photo: NASA)

October, the two groups combined for a classroom program of basic science studies, covering such topics as rocket propulsion, flight and orbital mechanics, astronomy, computers, space physics, medical aspects of space flight, communications, guidance and space navigation, environmental control systems, meteorology and star recognition. These classes lasted around six hours per day, two days per week, and all sixteen astronauts were required to attend – including Gordon Cooper and his backup Alan Shepard, who were then in training for the final Mercury mission. There was also individual familiarization with the Gemini spacecraft, the Titan booster, the Atlas booster, and the Agena docking target. In addition, they made scheduled visits to contractors such as McDonnell Douglas, Martin, Aerojet, and Lockheed. This academic phase of their training would conclude on 6 February 1963.

Earlier, on 26 January 1963, Robert Gilruth, Director of the Manned Spacecraft Center, announced assignments in various areas of specialization for NASA's sixteen astronauts. These assignments were designed to ensure ongoing pilot input into the design and development of spacecraft and flight control systems, and to provide part of the broad training which the pilots would undergo. Deke Slayton, as Coordinator for Astronaut Activities, maintained overall supervision of these and other astronaut duties. Additional to any new duties that may be added, the six Mercury astronauts were already engaged in assignments as follows:

- Gordon Cooper: MA-9 pilot
- Alan Shepard: MA-9 backup pilot
- Gus Grissom: Project Gemini
- John Glenn: Project Apollo
- Scott Carpenter: lunar excursion training
- Wally Schirra: specializing in overall operations and training.

The new astronauts took the remaining areas of responsibility, designed to provide for pilot input "across the board" to assure thorough awareness of pilot requirements in the major manned space projects as well as operations and training. The areas of specialty were assigned as follows:

- *Trainers and simulators*
  Neil Armstrong was responsible for monitoring the development, design and use of trainers and simulators, including new training requirements that were not associated with specific mission simulators.
- *Boosters*
  Frank Borman was to concentrate on the booster design and development programs, especially abort systems and the development of abort-preventing procedures for mission success.
- *Cockpit layout and systems integration*
  Pete Conrad would specialize in cockpit layouts, instrument displays and pilot controls to ensure system displays were appropriately integrated into the cockpit panels.
- *Recovery systems*
  Jim Lovell was responsible for monitoring the design and development of all recovery systems, such as paraglider and parachute of a returning capsule, including resolving operational problems in the re-entry and recovery part of the mission, and the landing system of the lunar excursion module.
- *Guidance and navigation*
  Jim McDivitt would specialize in the design and development of guidance and navigation systems and aids for operational requirements.
- *Electrical, sequential, and mission planning*
  Elliot See would monitor the design and development of the electrical and sequential systems, and be responsible for aiding in coordination for mission planning.
- *Communications, instrumentation and range integration*
  Tom Stafford was responsible for monitoring the design and development of communications and instrumentation systems, ensuring that onboard

systems were compatible with pilot needs and integrated with the IMCC (Integrated Mission Control System), GOSS (Ground Operational Support System) and other communication links.

- *Flight control systems*
  Ed White monitored the design and development of flight control systems and related equipment.
- *Environmental control systems, personal and survival equipment*
  John Young monitored the design and development of environmental control systems, survival gear, pressure suits, couches, and personal equipment.[5]

DEVELOPMENT TRAINING

In these capacities, the astronauts were required to attend all major meetings and conferences on the design and mockup reviews, attend staff meetings, and have the authority to request input from specialists in other areas. Gilruth explained that the specialty assignments given to individual astronauts might change somewhat during the training program as required, but this would not imply crew selection for future manned space missions; specific crews would be selected prior to each flight.

With the academic part of their training coming to an end in early February 1963, the astronauts pursued different areas of tutorial development, including a series of seminars attended on space science and technology from 11 February to 9 May. Soon they would start specific mission and survival training, but in mid-May there was the very welcome and instructional distraction of Gordon Cooper's MA-9

The New Nine astronauts and Deke Slayton seem highly amused during a press conference at Cape Canaveral on 12 May 1963. (Photo: Associated Press)

mission as the final flight in the Mercury program. They monitored the flight, which lifted off from Launch Complex 14 at Cape Canaveral, as Cooper guided his spacecraft *Faith 7* on a 22-orbit flight which, despite several technical malfunctions, splashed down safely in the Pacific just four miles from the prime recovery vessel, USS *Kearsarge* (CV-33), making it the most accurate manned landing for NASA to that time.

In June of that year, the Soviet Union launched the first woman into space when Valentina Tereshkova, a former factory worker and eager parachutist with almost no flying experience, made her first and only space flight aboard Vostok 6. It was a tandem mission with Valery Bykovsky on Vostok 5.

Also in 1963, President Kennedy tentatively proposed that the U.S. and the Soviet Union jointly undertake an expedition to the Moon, but Congress ruled out spending money on such a venture. Later in the year, on 22 November, Kennedy was killed by an assassin's bullets as his motorcade drove through Dallas, Texas.

SWELLING THE RANKS

The year 1964 would pass without a single manned U.S. space mission, as NASA geared up for the first Gemini mission in the first quarter of 1965. Not surprisingly, being aware of America's plans for two-man space missions, the Soviet Union flew a trio of cosmonauts aboard their first Voskhod spacecraft in a risk-filled, single-day mission. In order to squeeze three men into a capsule designed for a single occupant, Soviet space officials ordered that they not wear pressure suits and not have ejection seats. Although the cosmonauts could only carry out some very basic scientific and medical tests, the mission had massive propaganda value. But the safety of the crew had been severely compromised – in particular, they would have been doomed if the launch vehicle had failed. Indeed, seven years later a three-man crew lost their lives by not having protective suits when their capsule depressurized as they returned to Earth.

For NASA's new astronauts, almost every day of 1963 and 1964 was filled with intense training for their upcoming space missions. This included jungle, desert and sea survival training and water egress exercises, weightless simulation in specially modified aircraft, parachute jumping techniques (although they did not perform any actual training jumps from aircraft), parasailing, centrifuge exercises, rendezvous and docking simulations, research investigations, pressure suit indoctrination, and a host of skills associated with space flight, space systems, and space survival. Classroom courses were taught by MSC and NASA personnel, as well as experts from various educational institutions and industry.

SKILLS FOR THE FUTURE

Towards the end of January 1963 the nine new astronauts went to Flagstaff, Arizona along with geologist Dr. Eugene Shoemaker on a two-day field trip to inspect Meteor

Crater and such volcanic features as lava flows. They also viewed the cratered face of the Moon through a powerful telescope at the nearby Lowell Observatory, and put questions to astronomers. This was part of a training plan that Shoemaker had put together with NASA to teach them the fundamentals of geology and selenology – the latter being the geology of the Moon. To this purpose, Shoemaker, whose own hopes of becoming an astronaut one day in order to visit the Moon had been ruled out by a medical problem, had assembled a resident staff of U.S. Geological Survey (USGS) geologists at MSC to supply knowledge through "know-how" courses set up as part of the astronauts' training. Since it was conceivable that one of the astronauts in this class would be the first to walk on the Moon (which indeed was the case) they were given instruction in how to "see" features through a geologist's eyes, identify and collect samples, and evaluate which items should be brought back to Earth for expert analysis by scientists in laboratories.

Responsibility for the astronauts' training activities rested with MSC's Raymond G. Zedekar, who, under the chief of Flight Crew Operations, Warren North, arranged the academic sessions, field trips, and operational training activities. "The classroom courses will be completed in early February," Zedekar said at the time, "and shortly thereafter, beginning with Dr. Homer Newell, who is in charge of NASA's Space Science program, as first speaker, we will begin a series of science seminars."[6] These seminars, on a one-week schedule, introduced the nine astronauts to leading science personalities such as Dr. James Van Allen, discoverer of the globe-girdling radiation belts through which Apollo crews would have to pass on their voyages to the Moon. Designed to keep the aerospace pilots abreast of American scientific space projects and technology, the seminars were primarily intended to mesh technical engineering thinking into scientific gear.

"We hope to highly educate the aerospace pilot-engineers to do scientific tasks, to be able to communicate with Ph.D. scientists in every field," Warren North added in support of Zedekar's comments. "The pilots were selected for their potential to keep up with and to become scientists. We hope to augment their technical backgrounds with broad scientific knowledge."[7]

The astronauts were to be exposed to scientific details of deep space and satellite programs, as well as the X-15, the Air Force's X-20 Dyna-Soar program, the status of nuclear and ion rocket engines, and large liquid and solid rocket developments. Scheduled around the science seminars were briefings on spacecraft and launch vehicle design and development, and trips to contractor sites.

As space flight historian David J. Shayler once noted, it was also a time when they had to face up to being a nation's latest heroes. "Many of the new astronauts found that one of the most difficult assignments was in learning how to deal with the press and meet the public, and in preparing their families for the attention of the world's media as a new celebrity. For many astronauts, this was by far the most difficult part of space flight training. Another aspect not covered in detail during the early years of the program, and from which many suffered (if only temporarily), was in preparing to adjust to life after a space flight, both for themselves and their families. In some cases this was to prove much more difficult than preparing to make a flight into space, or the mission itself."[8]

The dynamic phase of operational training included proficiency flights in Air Force aircraft assigned to NASA, allowing them to fly T-33 Shooting Star and F-102 Delta Dagger jets out of Ellington AFB.

Although the new astronaut pilots had experienced weightless conditions in the cockpits of supersonic jet aircraft during test-flight experiences, these were always under the tight constraint of their seat harness. On 20 May they were able to take this experience to new levels in weightless training aboard a modified Boeing KC-135 at Wright-Patterson AFB, Ohio. With the support of the 6750th Aerospace Medical Research Laboratory, each of the nine astronauts took part in two flights involving twenty zero-g parabolas, each of which provided unrestrained weightlessness for 20 to 30 seconds before the force of gravity once again took over. To accomplish this, the inner fuselage of the aircraft had been stripped of all but the essential fittings and equipment and then padded out for the safety of the occupants as they floated about while gaining valuable experience of untethered weightlessness.

The zero-g training was followed in late May at MSC by a two-week program of Gemini systems briefings by personnel from McDonnell Douglas, which was making the spacecraft, after which there was a more practical hands-on training program on static trainers which demonstrated cockpit layouts and how the spacecraft operated. Later on, a full-scale Gemini trainer (also called a mission simulator) would produce flight conditions programmed by computers capable of replicating the spacecraft in every detail and enable the pilots to practice actual flight plans.

THREE DAYS IN THE JUNGLE

For four days beginning on 3 June 1963, all sixteen of NASA's astronauts received detailed instruction in jungle survival at the U.S. Air Force Caribbean Air Command Tropic Survival School at Albrook Air Force Base, Canal Zone, followed by field training in the local jungle. This school was chosen because it was the only one of its kind run by a U.S. agency. This training was deemed necessary by NASA officials when they realized that the longer Gemini missions would require the spacecraft to travel over a greater area of the globe, raising the possibility, albeit remote, of a crew making an emergency landing in a tropical area. This not only marked the first time that astronauts received tropical survival training, but also the first time that all seven Mercury astronauts and nine new astronauts took a training program together.

Four MSC personnel also participated in the training: Dr. George B. Smith of the Aeromedical Section, Ray Zedekar and Bud Ream of the Flight Crew Operations Division, and James Barnett of the Life Systems Division.

The course of instruction was given by H. Morgan Smith, Director of the Tropic Survival School, and some of his staff. It included classroom instruction on a variety of subjects, including identification of poisonous tropical plants, their location, safety precautions and rudimentary first aid; the identification of animals, reptiles and birds, their habits, location, likelihood of attack, palatability, and safety precautions. The astronauts were familiarized with the customs and food of local

Tropical survival training at Albrook AFB near the Panama Canal in March 1963. From left: unknown trainer, Neil Armstrong, John Glenn, Gordon Cooper and Pete Conrad. (Photo: NASA)

indigenous people, the correct way to approach and communicate with them, and how to enlist their aid.

Following their "classroom" instruction program, the group spent three days and two nights in the jungle. During this phase of their training they were paired off into two-man teams and sent into the jungle with sufficient distance between the teams that they could not communicate with each other, even by calling out. Each instructor was assigned to supervise two of these two-man teams and monitor their activities by radio. Primitive field activities included the establishment of a suitable campsite and using local flora and other materials to erect protective shelters. The teams gathered dry wood for campfires, purified water for drinking, hunted using only items from their spacecraft, gathered edible plants, and constructed a river-worthy two-man raft from vines and branches. Each man was also required to construct an animal trap and build a usable hammock.

Following their arrival in the jungle, and in order to make the training as realistic and beneficial as possible, the group was dressed only in boots and long underwear – the underwear was the same as they would wear beneath their pressure suits, with the suits being discarded after egressing from the spacecraft.

Desert survival training near Stead AFB, Nevada. Front row, from left: Borman, Lovell, Young, Conrad, McDivitt and White. Back row: Raymond Zedekar (Astronaut Training Office), Stafford, Slayton, Armstrong and See. (Photo: NASA)

Strict conservation principles were adhered to by the personnel participating in the exercise, with jungle vegetation only cut to serve a specific purpose, while maximum use was made of any creatures killed during the three days of the program.

On Thursday afternoon, 6 June, the group returned to Albrook AFB, where the training program continued with a critique and debriefing session at the headquarters of the Tropic Survival School.[9]

Having survived the rigors of the Panamanian jungle, the next survival exercise for the new astronauts took place at Stead AFB, Nevada, where from 5 August they undertook a five-day desert survival course oriented towards Gemini missions. The course was divided into three separate phases; the first comprised one and a half days of academic presentations on characteristics of the world's desert areas and survival techniques. The second phase was a one-day field demonstration on the use and care of survival equipment and the application of the spacecraft's parachute in the making of protective clothing, shelters, and signals. For immediate protection from the fierce heat of the sun they were shown how to fashion a loosely fitting robe and headgear from parachute silk, and to make a makeshift tent using the left-over material and the oar of their life raft. The third phase involved two days of remote site training, with two-man teams being taken out and left alone in the desert at Carson Sink, Nevada, where the daytime temperature could reach 43°C, to apply what they had learned from the academic and demonstration phases of the program. Each astronaut could carry only four litres of water to last two days, but were sustained by food items from their spacecraft emergency packs, containing fudge, corn flakes, chocolate bars, fruit cake and soup. They spent their days learning how to conserve body water and their nights learning the secrets of desert travel in darkness.

WATER LANDINGS AND SURVIVAL

Starting in mid-September 1963, all sixteen astronauts took part in the first segment of a three-phase program lasting several weeks. Overall, this program demonstrated safe parachute landings on land and water, and water survival techniques in the event of a Gemini launch aborting at low altitude (under 60,000 feet) in which case the two crewmen would eject to descend by their personal parachute.

In the first part of this exercise an astronaut was suspended beneath a parachute canopy that was already inflated and raised to an altitude of 400 feet while traveling at 30 mph at the end of a 600-foot tow rope pulled by a pickup truck at Ellington AFB. He was then cut loose for a free descent to the ground. For these land-based exercises the astronauts wore flight suits, crash helmets and jump boots.

Following this, only the nine newer astronauts were sent to the U.S. Naval School of Preflight in Pensacola, Florida for the second phase of the program, which was a one-day water survival course. It began with training in an enclosed water tank for techniques of underwater egress from a spacecraft, pressure suit flotation, boarding a life raft, parachute extraction, disentanglement from a chute, and helicopter pick-up.

Frank Borman undertaking parachute landing exercises. (Photo: NASA)

Tom Stafford taking part in parasailing exercises over Galveston Bay, Texas. (Photos: NASA)

The underwater egress used a Dilbert Dunker, which resembled an aircraft cockpit. With the trainee strapped aboard, it slid down a track to strike the water at moderate speeds. On contact with the water, the device flipped over in a way designed to upset the pilot's equilibrium and simulate experiences that he might encounter in a sinking spacecraft. In a parachute drag escape, the key action was rapid extraction from the harness.[10]

The final phase of this water training was in Galveston Bay. In a similar fashion to the earlier parachute landing exercises, the astronauts, now in Gemini-style pressure suits, were towed by a power boat until they had attained an altitude of 400 feet. For this (as with the earlier ground landings) they were suspended beneath a new type of ring-sail parachute called a para-commander, which had a glide capability and was used by many parachute jumping clubs. Once the tow line was released, the astronaut had to guide himself to a safe landing in the water.

MORE ASTRONAUTS IN TRAINING

There was one other major event for the NASA astronaut corps late in 1963, when (as planned) their numbers suddenly swelled with the selection of the latest cadre of fourteen astronauts to support space efforts in the immediate and projected future. As with their predecessors, many of this group would go on to achieve fame as America faced up to its slain president's bold commitment to send one or more astronauts to the Moon and returning them safely to the Earth before the decade was out.

They, too, would be well up to this enormous challenge.

REFERENCES

1. French, Francis and Colin Burgess, *Into That Silent Sea: Trailblazers of the Space Era, 1961-1965*, University of Nebraska Press, Lincoln, NE, 2007, pg. 285
2. Burgess, Colin and Kate Doolan, with Bert Vis, *Fallen Astronauts: Heroes Who Died Reaching for the Moon*, University of Nebraska Press, Lincoln, NE, 2003, pg. 45
3. Slayton, Donald K., and Michael Cassutt, *Deke! U.S. Manned Space from Mercury to the Shuttle*, Forge Books, New York, NY, 1994, pg. 123
4. Borman, Frank with Robert J. Serling, *Countdown: An Autobiography*, Silver Arrow Books/ William Morrow, New York, NY, 1988, pg. 97
5. NASA *Space News Roundup*, issue July 8, 1964, *Astronauts Are Given Specific Assignments*, MSC, Houston, TX, pg. 1
6. NASA News Release 63-10, *Training of Astronaut-Candidates*, MSC, Houston, TX, 26 January 1963
7. *Ibid*
8. Shayler, David J., *Gemini: Steps to the Moon*, Praxis Publishing Ltd., Chichester, U.K., 2001, pg. 88
9. NASA *Space News Roundup*, issue June 12, 1963, *Three Days in Panama Jungle Teach Astronauts to Survive*, MSC, Houston, TX, pg. 8
10. NASA *Space New Roundup*, issue October 2, 1963, *Astronauts Complete Part of Gemini Water Survival Phase at Navy School*, MSC, Houston, TX, pg. 2

Part Two

6

The boy from Barren Run

In so many ways, John Yamnicky was a giant of a man; not only in size and personality, but with an accomplished life to match. A resourceful and decorated Navy test pilot who had survived amphibious assault landings in Korea and aerial combat missions over Vietnam, he was once described as a magnificent man devoted to his faith, his family and his country. Gentle by nature but exceptionally strong, his daughter Jennifer revealed that even at the age of 71 he loved swimming and weight-lifting, his arms were rock hard, and he could still perform one-armed push-ups.

John David Yamnicky, USN. (Photo: USN courtesy Yamnicky family)

In the fall of 1963, Yamnicky might easily have appended the words "NASA Astronaut" to his name. By 1 July of that year, the space agency had received 271 applications for their third group of astronauts: 200 from civilian hopefuls (including two women) and 71 from military personnel. When that number was winnowed down to just 34 for medical, psychological and stress testing, his name remained penciled in as a finalist for the ten to twenty projected places in the third astronaut team.

HUMBLE BEGINNINGS

The family name Yamnicky has Slovakian origins. Although once spelt Jamnicky with a capital J, the first letter is pronounced with a softened Y. When his forebears reached the United States the spelling was changed to reflect this arrangement.

John Yamnicky came from a humble background, having been born in the upstairs front bedroom of the Yesko farmhouse on 8 June 1930 in a rural area of Western Pennsylvania known as Barren Run. He was delivered by widowed Marie Yesko, his maternal grandmother, who served as a midwife when needed to families and friends in the area.

Barren Run was dotted with farms back then, located close by the coal-mining borough of Smithton, on the east side of the Youghiogheny River. Smithton is the hometown of acclaimed Hollywood actress Shirley Jones, whose grandfather, the Welsh immigrant William B. "Stoney" Jones, founded the Jones Brewing Company in Smithton in 1907 after winning the brewery in a poker game.

Geographically, Barren Run is also located some 25 miles southeast of the city of Pittsburgh. John Yamnicky's relatives told him that back in the early 20th century, chapters of the dreaded Ku Klux Klan were quite active in the Barren Run area. His mother Mary's younger sister Sophie told him how she and her mother would often hide under the kitchen table whenever large bands of klansmen wearing masks and conical hats gathered on neighboring farms to symbolically burn their large wooden crosses. The two women feared for their lives, as they were Catholics and usually home by themselves.

John's father, John Joseph Yamnicky, had worked in the nearby coal mines until he became seriously ill with a lung disease and was forced to take on night work in a candy factory. Of necessity, this brought about a family move to the nearby town of McKeesport. At first they lived in the small rented part of a house in the town, where his mother Mary soon found work as a domestic cleaner in order to supplement the family income. She would also take young John to Mass every morning.[1]

"My uncle John spent most of his growing-up years in McKeesport, Pennsylvania, then a thriving city," nephew Craig Rutter told the author in backgrounding John Yamnicky's early life in the area. "U.S. Steel was the main employer. McKeesport, located in the same county as Pittsburgh, has been in decline since the 1980s."[2] He also revealed that one of his uncle's earliest jobs as a young teenager was assisting a Menzie Dairy home delivery driver.

For his early education, Yamnicky attended Holy Trinity Church School in the

Slovak parish of McKeesport, then Shaw Avenue Grade School. Former classmate Tom Macko once reflected on the rudimentary street football games in which he and Yamnicky participated. "They used to block off Seventh Avenue in front of the school during lunchtime [and] we'd play touch football. We balled up our lunch bags and wrapped them in rubber bands. That was our football. Johnny loved to play football. He dominated those games," Macko added. "He was in seventh grade, but he was taller than the rest of us. He was the best player. I was an eighth-grader and I always had to cover him. I didn't have much luck."[3]

The following year, Yamnicky would meet a student from South Park School who soon became one of his closest friends. As George Miller recalls, "John and I met 68 years ago in the late summer of 1944 when we were starting in the ninth grade, and we both answered the call for freshman football players. While John and I were both athletic, neither of us had ever played on an organized team before. We struck up a conversation while standing in line to get our practice gear. It didn't take long for us to discover that we had something in common – neither of us knew any of the other people there. We also both liked to laugh and to see the humorous side of things. When we picked up our game uniforms we looked for two numbers close together: I got 38 and he got 39.

"After a couple of days of conditioning, coach Bill Clees said we were going to arbitrarily form some teams. He said, 'Yamnicky; you stand here, this is the right end position.' Then he said, 'Miller; you stand on his left. You will be the right tackle.' John and I worked so well together as a strong blocking team that the coaches kept us in those positions all during high school." Today, Miller still remembers the strong bond between them. "We were inseparable," he mused. "Always together."[4]

In the spring of that year of 1944, Yamnicky joined the track and field team and took part in the 440 and 220 yard events, as well as the relay team. He was good

The South Park football team circa 1944. George Miller (#38) and Yamnicky (#39) are both in the second row. (Photo courtesy Yamnicky family)

enough to place in the Western Pennsylvania Interscholastic Athletic League. He and George Miller would walk home each evening by way of Miller's house, where they shot hoops until dinner time. In fact they enjoyed this so much that they formed a team and persuaded Miller's preacher, the Reverend Norman Kieffer, to coach the team in a church league. They got off to such a good start, winning several of their first games, that they were invited to join the City League. "At this point we needed a sponsor to buy us uniforms," Miller stated, "so by chance the local Salvation Army volunteered to sponsor and buy us uniforms with their name on them. This worked well for the church league and the City League."

After an enjoyable season of basketball it was back to the high school football field, where Yamnicky developed into a football powerhouse. Under the guidance of head coach J. Harold ("Duke") Weigle he became a standout right end receiver for the McKeesport Tigers. In his senior year the Tigers not only finished the regular season with an 11-1 record and won the state championship, but then played against Miami in the Scholastic Orange Bowl on Christmas Night, 1947, unfortunately going down in front of 25,000 fans. An all-round, talented athlete and letterman, he still played basketball and set records as a member of the school's championship 800-yard track relay team – while also shining as a scholar.[5]

To maintain his physical fitness on the football field during high school summer breaks, Yamnicky worked hard in the steel mills shoveling coal into hot furnaces. Russ Goetz, a retired American Baseball League umpire who played basketball at McKeesport, regaled Yamnicky as an outstanding athlete. "Johnny was tall and very fast. He was smart, and a very good football player [and] he also was a standout in track. He was one of the best sprinters ever to come out of McKeesport High."[6]

Perhaps another influence in Yamnicky's early years was the fact that pioneering female aviator Helen Richey was born in McKeesport in 1909. In addition to being the first female commercial airline pilot and the first woman sworn in to fly air mail, she was one of the first female flight instructors, and a friend and fellow competitor with legendary aviatrix Amelia Earhart.

A MAN OF ANNAPOLIS

By the time he graduated high school in 1948, Yamnicky's outstanding academic and athletic record meant that he was eagerly sought after by a number of colleges and universities, as explained by George Miller. "John and I had several offers from small Pennsylvania colleges, and it appeared that we would probably get offers from University of Miami, as did several of our team mates. We did not hold U of M very high academically, nor the Pennsylvania colleges we had heard from. We also were concerned that if we got a football-ending injury, we would lose our scholarship and neither of our families could afford to keep us in school. Meanwhile, John and I had been approached by Mr. Arthur Landis from U.S. Steel and Mr. Dinty Moore, President of Exide Batteries, offering us the opportunity to compete for entry into the U.S. Naval Academy. So, all considered, we decided to try for the academy without a clue as to what that entailed.

"We found we had about three weeks to get ready for the entrance exams. We didn't have a clue what the exams would consist of but we expected math, science, as well as English and history."

To support his application, Yamnicky had even received a letter from Philadelphia Congressman Frank Buchanan, a former teacher at McKeesport High (and, it turned out, his friend George's future father-in-law).

"The first week in February we hopped on the streetcar to Pittsburgh to be there for the exams," Miller continued. "This was the routine for the next three days. I never did very well on testing but John, who was an outstanding student, always did. He was first to get his letter advising him he had passed and would be admitted with the class of 1952. A couple of days later I got my letter which said I had passed all the tests except trigonometry, where I needed a 2.5 and had gotten 2.4. We were both disappointed and John said he would turn it down and we could go to Miami. I said no, that we'd decided to go to the academy and I would retake the tests and pass the next year, which did in fact happen."

When pressed, Miller remembers one episode more than any other in his enduring friendship with John Yamnicky. "One spring day while we were still in high school, John told me he wouldn't be going to school the next morning. He seemed kind of secretive about it. When I asked him why, he said he was having his name changed. I had no idea why he would do that. I couldn't wait to stop at his house on the way home from school to find out who he was now. His mother, a true saint, welcomed me into the kitchen, poured some milk and cut some cake for us. She was always smiling, but that day she seemed even brighter. So I said, 'Okay John, what's your new name?' His mother couldn't wait and said, 'Tell him John.' He said something like, 'Well, I didn't really change it. I just added a name. I am now John David Yamnicky.' I was puzzled, and said 'David is *my* middle name.' John looked at me and said quietly, 'I know.' Wow, that really got to me. We already felt like brothers and that sealed the deal. When I told my mother she cried because she treated John like a son."[7]

Yamnicky's enrolment at the USNA would begin a 53-year career dedicated to that service. While at the Naval Academy he and Miller saw each other occasionally, but it was difficult to maintain a close friendship while in different classes and living on opposite sides of Bancroft Hall. However, they were able to spend Christmases together at home in McKeesport. After graduation they chose different services, and slowly drifted apart. George Miller shared his friend's interest in flying and went on not only to become a general in the Air Force, but later to head the U.S. Olympic Committee. Their paths would converge many years later in Vietnam under the most unusual of circumstances. The last time the two men met was at a 50th anniversary high school class reunion.

John Yamnicky was just 16 years old when his father died on 31 December 1946. Early in 1948, his older sister Mary, a registered nurse, learned that her brother had gained entrance to the academy and selflessly postponed her wedding to fiancé Don Rutter in order to support their mother while John was away. She and Rutter were finally married at the Holy Trinity Church in McKeesport on 14 June 1952, a week after John's graduation, whereupon he took over the principal support role for their mother.

At Annapolis, Yamnicky maintained both his academic and athletic prowess. In a newspaper article in June 1950 it was reported that Midshipman John Yamnicky was home on leave and maintaining a 3.3 Grade Point Average (GPA) at the academy, as well as being a football letterman. As a lineman, he was credited with being a key member of the academy team when it pulled off a stunning 14-2 defeat of the Army in the 1950 Classic, ending the Cadets' 28-game-winning streak in what is now seen as one of college football's greatest upsets. Also in that year's academy team was quarterback William Porter Lawrence, who later became a commander of the U.S. Third Fleet in the Pacific, a decorated Vietnam War fighter pilot and POW, the Chief of Navy Personnel at the Pentagon and later Superintendent of the Naval Academy, eventually retiring from the service in 1986 with the rank of vice admiral. Lawrence was also a finalist in NASA's first astronaut selection process, but missed out owing to a minor heart complaint. Some years later, his Navy captain daughter Wendy was selected as an astronaut and flew on four space shuttle missions. Admiral Lawrence was once asked to comment on his former teammate and fellow fighter pilot. "John was a very tough, solid football player and a tremendous fighter pilot, but he was also very pleasant, friendly and outgoing .... You wanted him on your team when the chips were down."[8]

LEARNING TO FLY

Graduating from Annapolis with 738 fellow class members and a commission as an ensign on 6 June 1952, Yamnicky was given a very high assignment selection and shipped out to Korea, where he served his required "black shoe" term as a gunnery officer on USS *Calvert* (APA-32), a *Crescent City*-class attack transport. During this time *Calvert* ferried troops to and from the West Coast and Korea, and Yamnicky was an assault wave commander supporting the U.S. Marines. In September 1953 his shipboard service came to an end in Puson, Korea with orders to attend flight school at NAS Pensacola, Florida.

After completing his basic and advanced training and receiving his Wings of Gold as a naval aviator on 26 January 1955, Lt. (jg) Yamnicky reported to Utility Squadron 10 (VU-10) based at Guantanamo Bay, Cuba, which was mostly engaged in towing targets for anti-aircraft gunnery practice, radar exercises, providing rescue missions for ships or aircraft in distress, and delivering emergency supplies to the base. While there, he flew a variety of aircraft and was assigned electronics, legal and education duties.

Monte Cassels from Tampa, Florida was a plane captain in the squadron and flew Douglas JD-1 Invaders with Yamnicky. "We were both kids at that time and we both enjoyed our time flying," he recalled. "One day when we were waiting clearance to take off, he took out his shark knife and started scratching his initials in the yoke. I fussed at him for that and he joked it off. Because he was large in stature, we joked about greasing him down so we could get him into the cockpit of the small fighters. He broke the parking brake pedal on my plane and it took several hours to repair it. After that I watched and cautioned him when he used it. He was truly a gentleman and treated me like an equal even though I was an enlisted man. He was one of the best officers I served with."[9]

After finishing his tour with VU-10 in April 1957, Lt. Yamnicky was assigned as Administrative and Maintenance Officer of Attack Squadron 46 ("Clansmen") flying the F9F-8 Cougar at Cecil Field, Florida. After transitioning to the A4D2 Skyhawk, VA-46 deployed on USS *Franklin D. Roosevelt* (CV-42) and USS *Intrepid* (CV-11).

His next assignment in December 1959 took him to NAS Jacksonville, Florida as a Replacement Air Group instructor pilot and Squadron Standardization Officer with VA-44 ("Hornets"), which was then under the leadership of Cmdr. Damon Cooper, and where he would remain until October 1960. He was involved in the squadron's transition from the AD-4 propeller-driven Skyraider to the A4D2N Skyhawk ground-attack jet. It was during this time that his mother, Mary Elizabeth Yesko Yamnicky, passed away at just 51 years of age. However, one happy event did eventuate from his posting, since it was there that he also met his future wife, Janet ("Jann") Wilson, a nurse at Jacksonville Naval Hospital. Following what she described as a whirlwind romance they were married on 23 December 1959, and would eventually be blessed with four children; John David, Jr., Lorraine Elise, Mark Stirling and Jennifer Lynn Yamnicky.

A SCHOOL FOR TEST PILOTS

In September 1960 the Commanding Officer of VA-44, Cmdr. Arthur Detweiler, proudly presented Yamnicky with orders to attend the U.S. Navy Test Pilot School at Patuxent River, Maryland the following month. When he reported in on a chilly day in late October he was formally greeted by TPS Director Leo Krupp. His class had sixteen members in all: ten Navy, two Marines, one Army, one Canadian Navy, and two civilian engineers.

Dick Liljestrand was one of those civilian engineers. "I was a member of the academic portion of the class. I was loaned to the School by contractor Ling-Tempo-Vought, Inc. to teach aerodynamics. Each contractor on loan (for approximately 18 months) goes through the academic portion to get a 'feel' for the course material. Personally, I had a good familiarity with the program because I'd previously been on active duty at the Test Center as a flight test engineer." Whilst he only had routine classroom contact with Yamnicky during the course, in later years they met up in far more relaxed circumstances for class reunions. Liljestrand noted that Yamnicky and future NASA astronaut C.C. Williams got on well together. "C.C. and John were close until C.C.'s untimely death in a crash of one of NASA's [T-38A] aircraft."[10]

Although they were given weekend leave (which some used as extra study time), for the remainder of the week Yamnicky and his fellow students had to work hard, attending full-day classes on the base, returning home to their wives and families (if married), grabbing a quick meal, and then hitting the books and writing reports until exhaustion finally forced them to bed at one or two o'clock in the morning.

As recalled by Elizabeth Carroll Foster, the wife of U.S. Army class member John

Class 28, U.S. Navy Test Pilot School, Patuxent River. From left: Robert Brace, unknown TPS staff member, David Glunt, Lee Bausch, Gary Van Slyke, Joseph Sosnkowski, Harlan Cheuvront, TPS staff 'Sky King', Leo Krupp (TPS Director), John Winkowski, unknown TPS staff member, Clifton C. Williams, John Kirkland, John Yamnicky, James Tyson, Russell Farley, Lloyd Jones and John Foster. (Photo: USN courtesy N. Lee Bausch)

Foster (herself a prolific writer and the award-winning author of the memoir *Follow Me: The Life and Adventures of a Military Family*), "TPS was so rigorous we had few opportunities to socialize." As a result the wives – obviously calling themselves the 'TPS Widows' – formed a social group and became members of the larger Navy Officers' Wives Club, with relaxing teas and luncheons to attend.[11]

Fellow class member Lee Bausch has a particularly vivid recollection of an evening during NTPS training which he says had a lot to do with Yamnicky. "I met him in the fall of 1960 at the Naval Air Test Center, Patuxent River, Maryland. John was in our U.S. Navy Test Pilot School class. It was Class 28. Back in 1960-61 I was a bachelor. John was married and with family. For the most part, our time together was at the school. As you would expect our routines did not ordinarily coincide.

"During the midpoint of our test pilot training calendar it was an established custom to hold an evening together with the preceding class. It was the terminus for that class. The occasion was held at the Officers' Club with ample spirits. When it came time to introduce our class members to the assembly, the stage curtains were opened to reveal a Volkswagen Bug. The driver's door opened and the name of the member 'disembarking' was announced. This was followed by second member then a third. And so it followed until all sixteen of us were assembled on stage. The vehicle had been cleverly arranged so that the entire class could pass through it with our movements into the car hidden from view. The effect was that somehow the entire class arrived in the Bug. I mention this very emphatic and well received class introduction because, as memory serves me, it was conceived and staged by John."[12]

Like Yamnicky, Lee Bausch would later apply to NASA for astronaut selection.

Unfortunately, at 6 foot 2½ inches his height precluded him from progressing to the interview stage.

In June 1961, as recalled by Elizabeth Foster, Class 28 finally completed their grueling studies at NTPS and attended a graduation ceremony presided over by Vice Admiral Paul H. Ramsey, Director of the Navy Air Test Center. It was a chance for the graduates to let their hair down and relax after nine months of solid work. They put on amusing performances, including one by a group of four top-hatted graduates (John Yamnicky, C.C. Williams, John Foster and one other) who had painted eyes, noses and mouths on their bared midriffs and proceeded to belly-dance while at the same time whistling the Colonel Bogey March. Some wives also took part in the fun, with Jann Yamnicky reading out a self-penned poem on what it was like for them to be TPS 'widows'.

After the graduation ceremony the former students were sent on a week-long tour of aircraft manufacturing plants across the nation, including Grumman in Bridgeport, Connecticut, Boeing in Seattle, Washington, and Ling-Tempo-Vought in Dallas, Texas. On their return, service representatives handed the men their eagerly awaited orders and told them to take two very welcome weeks of leave.

DANGEROUS WORK

Lt. Yamnicky's orders assigned him to remain at the Naval Air Test Center, where he served until October 1963 as a Flight Test Project Officer in the Carrier Suitability Branch. One of his first tasks was to determine the minimum acceptable airspeed for the A4 Skyhawk following a catapult launch from an aircraft carrier.

As his friend and later co-worker Dennis Plautz recollected, "He worked with ground-based engineers to see what those minimum speeds should be, but confirmed the final calculations himself. The stories of shipboard catapult launches at slower and slower end-speed until the point of sinking below the carrier deck after launch were frightful." This process involved computing the end-speed of a catapult with a safety margin of perhaps 20 knots excess end-speed. Yamnicky would fly an A4 to the carrier with an empty external fuel tank fitted. Water was pumped into the tank prior to take-off, and once the aircraft was airborne the water was dumped in order to reduce the airplane to its maximum landing weight, normally thousands of pounds less than the maximum take-off weight. The first launch would be with the normal excess end-speed. The next launch would be two knots slower, and so on.

As Dennis Plautz recalls, Yamnicky once told him about these risky tests, saying "It was just me and my aircraft – in the end it is your ass on the line. On my last launch, when my aircraft sank below the flight deck [nominally 60 feet above the water], the Air Boss screamed, 'Eject! Eject!' Instead, I stayed in my aircraft because I had done the math and I *knew* the aircraft would fly. When I was safely airborne a new voice was heard on the radio: 'This is the captain. You are done on my ship. Take your aircraft home and don't come back!'"[13]

Jennifer Yamnicky can remember her father telling her about another incident which could have ended in complete disaster. He was on a test flight when one of his

In 1962, Capt. Yamnicky (foreground) joined other Patuxent River Naval Air Test Center pilots in weightless training aboard a modified KC-135 aircraft. (Photo: USN courtesy Yamnicky family)

jet engines malfunctioned and abruptly cut out. When he turned to head home the other engine suddenly flamed out and he was in deep trouble.

"His superiors told him to ditch it," Jennifer stated, "but he thought he could make it. He rode it in and hit the end of the runway so hard it took his landing gear off. So now the aircraft is on its belly sliding down the runway. It slid off the runway and went through a volleyball game some Marines were playing. He said Marines were jumping everywhere." Eventually the aircraft smashed at speed into several large 55-gallon drums and came to a shuddering halt, but not before some of the drums that had become airborne in the impact came crashing back down on the top of his jet. "If it hadn't been for his canopy, he would have gotten squashed," Jennifer reflected. He later told his wife Jann that he knew how much that airplane cost and he didn't want to lose it by ejecting without at least trying to land.[14]

COMBAT DEPLOYMENTS

In November 1963, having missed out on astronaut selection due to his height and weight, and now with the rank of lieutenant commander, Yamnicky was ordered to VA-125 ("Rough Riders"), the Replacement Air Group at NAS Lemoore,

Officers of VA-146 on the flight deck of USS *Constellation* in August 1964. John Yamnicky is fifth from left in the back row. (Photo: USN courtesy Yamnicky family)

California, for refresher training in A4 Skyhawks. In March 1964 he was assigned to VA-146 ("Blue Diamonds") as Operations Officer, and between June 1964 and June 1966 he served two deployments to Southeast Asia. During its first deployment aboard USS *Constellation* (CV-64) the squadron participated in supporting photo reconnaissance missions over Laos and South Vietnam during the first days of retaliatory air strikes by U.S. forces on North Vietnam after the Gulf of Tonkin incident. Initially, their A4C aircraft provided tanker and rocket-armed escort support for the reconnaissance missions, but they later flew night sorties and conducted retaliatory air strikes against North Vietnamese targets, marking the first use of the A4 in combat.

The squadron's first tour finished in February 1965, and in December of that year it was redeployed to WestPac, this time on USS *Ranger* (CVA-61).

Hugh Magee flew on this tour, and he remembers John Yamnicky well. "I had the unique honor to serve as John's Assistant Operations Officer in VA-146. We made a combat deployment together on USS *Ranger* in 1966. John was a great leader, a true warrior, and an American patriot."[15]

It was while flying in the Vietnam conflict that Yamnicky unexpectedly fell back in contact with his closest friend from high school, George Miller, who had joined the Air Force after graduating from Annapolis. On one combat sortie Yamnicky had attracted the unexpected attention of a surface-to-air missile, which hotly pursued his aircraft around the skies. Eventually, he managed to maneuver free and the defeated missile fell away. However, he realized he was now quite low on fuel and was unable to make it back to his carrier. Instinctively, he crossed into Laos where the U.S. Air Force had a base. He radioed the base, identified himself and asked if he could land to refuel. To his annoyance, the response came back in the negative. In frustration he demanded to speak to the commanding officer. Amazingly, this turned out to be none other than George Miller. Once Miller realized who he was speaking to, he granted permission to land and refuel, and the old friends took the unexpected opportunity to catch up on each other's news. After they had shared a quick dinner, Yamnicky took off in his fully fueled aircraft and returned to his ship.

While serving in WestPac, Lt. Cmdr. Yamnicky was awarded the Distinguished Flying Cross, eleven Air Medals, the Navy Commendation Medal and two Navy Unit Commendation Medals for strikes against the enemy in Vietnam.

FAMILY COMES FIRST

John Yamnicky did not serve out the full WestPac tour, because in August 1966 he returned to the mainland having received orders to attend the Naval Warfare (Senior) Course at the U.S. Naval War College in Newport, Rhode Island, where he and his family set up their new home. He concurrently studied for a master's in international affairs at George Washington University in Washington, D.C.

As Dennis Plautz observed, it was a busy work period for his friend, with precious little time to spend with his family. "For many people, the course of study at the War College was enough. But for others, John included, it was an opportunity to get a master's degree at the same time. The family didn't see too much of him, as he was always studying. He wrote two dissertations, one for each course of instruction. He would write the pages out longhand and, periodically, bring them up to Jann to be typed."[16]

In July 1967, having graduated from the Naval War College and gained his degree from George Washington University, Cmdr. Yamnicky reported to VA-44 at NAS Cecil Field, Florida for refresher training in the A4 Skyhawk. He and Jann then took their three young children back to Jacksonville, Florida, where he had been assigned as Executive Officer of VA-172 ("Blue Bolts"). A second daughter, Jennifer, was born just a week before he joined his squadron aboard USS *Franklin D. Roosevelt* (CV-42), deployed with the Sixth Fleet in the Mediterranean. In November 1968 he was made the squadron's Commanding Officer, and later led a second Mediterranean deployment, this time aboard USS *Shangri-La* (CVA-38).

On the first deployment he was due to spend Christmas in Cannes on the French Riviera serving as Senior Shore Patrol Officer throughout the holiday season, which was not an enviable way to spend Christmas. In a letter home on 17 December 1967 he expressed his regret at being away from his family at that time. "I really hated to leave on this cruise," he wrote. "Each parting gets more and more difficult and there seems to be no end in sight. It's rough on the group back at home."

Following his later deployment aboard the *Shangri-La* he was offered command of a carrier air wing, but politely declined because he felt he had spent too little time with his family in recent years and wanted to be more of a father to his children.

On 23 August 1966 John Yamnicky was promoted to the rank of commander during a ceremony in the Naval War College's Mahan Library. From left: Chuck Anderle, Toxey Califf, Joe Jackson, Admiral Tom Hayward, John Yamnicky and Pak Pakradooni. (Photo: USN courtesy Yamnicky family)

COMMANDANT AT PATUXENT RIVER

After being promoted to captain, Yamnicky returned to NAS Patuxent River to serve on the staff of Rear Admiral Roy Isaman, Commander, Naval Air Technical Center. Its responsibilities included approving all aircraft test and evaluation projects, setting priorities, and maintaining an overview of operations at the test pilot school. Several months later, on 18 May 1972, the Director of the Test Pilot School, Capt. James L. Gammil, was killed while flying one of the school's X-26A gliders when it went out of control and crashed onto a road. Admiral Isaman appointed John Yamnicky in his place.

Dennis Plautz says that Yamnicky introduced some very welcome changes to the school. "He felt his greatest achievement was his part in restructuring the curriculum. The school used to run three classes a year, with each class being eight months with no breaks ... John took the position that although the school was important, this was shore duty, and the students needed to have time to spend with their families, too. He prevailed – the culmination of an effort started in 1958

Capt. John D. Yamnicky, USN. (Photo: USN courtesy Yamnicky family)

resulted in a change to two classes a year, each class running for twelve months."[17] Yet again, John Yamnicky was demonstrating that while obligations such as the school were vitally important, so too was quality family time.

Former Navy Lt. Cmdr. Harry Errington was the Aircraft Maintenance Officer at Test Pilot School at the time, and became a lifelong friend of the Yamnickys. "John was the perfect person for the job and the situation," he reflected. "Everyone had the highest respect for Capt. Gammil and it took a special person to replace him. Capt. Yamnicky was great; he immediately enjoyed the respect of the instructors and spent many hours on the hangar deck with my CPOs and the enlisted troops who were maintaining our school's aircraft. You have to be the best to be the commander of the Navy Test Pilot School, and John was the best. He ranks among the top Americans I know because of his lifestyle, his values, the way he cared about people, and his obvious pride for the Navy and for his country. He served as our director until he was relieved by Capt. Carl "Tex" Birdwell on 26 September 1972."[18]

In 1973, following the Paris Peace Accords, North Vietnam finally released 591 American prisoners of war during a long-overdue exercise known as Operation Homecoming. Many of them were airmen shot down over North Vietnam or Laos. John Yamnicky was selected to be part of a medical control group whose members would study the long-term medical and psychological effects of their imprisonment. This entailed traveling to Pensacola once a year for three days of physical and mental examinations.

TAKING LEAVE OF THE U.S. NAVY

Yamnicky's final assignment for the Navy was in the Pentagon, where he worked on vital security programs in the office of Secretary of Defense Frank A. Shrontz. In September 1976 Shrontz completed a mandatory evaluation of Yamnicky, saying he was one of the finest officers with whom he'd been associated in the Department of Defense. "He is absolutely outstanding," Shrontz wrote (in part) in his summary, "a real self-starter of rare competence and dedication, with exceptional common sense – a true professional of great value to the Navy and DoD." At the end of his report, Shrontz added, "I am convinced that Captain Yamnicky is qualified in every respect for flag rank *now*, and I urge that he be promoted to Rear Admiral and given a major Navy command immediately."[19] Unfortunately, his recommendation was not heeded. Three years later, on 30 June 1979, John Yamnicky retired from the Navy. While in service, he had flown 43 different models of aircraft and made over 900 landings on aircraft carriers.

Following his retirement after 27 years of active service with the Navy, plus four in training at the Naval Academy, Yamnicky continued to serve his country, this time in private industry. Three days after retiring, he took up a position with VEDA (later Veridian) Engineering, a Virginia-based military contractor, becoming senior advisor to the Under Secretary of Defense, Research and Engineering, for major weapons system development, procurement and production matters. Among other projects, he worked on the development of the F-14 and F/A-18 fighter jets, and air-to-air missile programs which included the Phoenix, Sidewinder, Sparrow, and the Advanced Air-to-Air Missile. In the evening, however, he would leave his office in the Pentagon and head home to a place of greenery, horses, and an active family life.

FINAL DAYS

In the mid-1970s John and Jann Yamnicky bought a 10-acre horse farm in Waldorf, in southern Maryland, which they lived on and worked steadily to improve. He often said that his best times were spent away from the Pentagon, riding one of his tractors. According to Jann, he spent most weekends on those tractors, mowing, dragging the fields, and checking the fence lines. "He loved his tractors – he had two – and really enjoyed taking care of the farm on weekends or evenings. He loved seeing the improvements and changes, something not always obvious in his work with the Navy and government."[20]

Their son John David agreed. "He loved being out there. His nature was not to stand around. He was always out in the fields, always working on something."[21]

One serious problem did cause an amount of grief for John Yamnicky around this time, with the onset of a rare eye disease that turned the vision in his left eye into a prism, distorting the vision in his other eye. His doctor advised him that there was no treatment for this problem, and he ought to consider wearing a patch over his left eye in order to protect the vision in his good eye. He tried commercially manufactured patches, but could not find any he felt comfortable wearing. One day he happened to discuss the problem with his local shoe repair man. Intrigued, the repairer offered to make a custom patch, and Yamnicky liked the result so much that he ordered several. Sometime later, he and Jann were on vacation in Thailand when they were ushered into an umbrella factory where women hand-painted designs on items such as hats and shirts for tourists. Jann had an elephant painted freehand on her shirt, and was pleased with the result. The artist then suggested painting something on John's patch. They liked the idea, and came up with a golden dragon, which they loved. "From that day on," Jann recalls, "it was his 'dress' patch." He wore it everywhere and it soon became synonymous with him, featuring in many of his later photographs.[22]

Lee Bausch recalls that his last meeting with his TPS friend John Yamnicky was at the Tailhook 2001 reunion, just before the tragic events of 11 September 2001.

"It was at the symposium/seminar/annual gathering of carrier pilots traditionally scheduled in early September. The event encompassed the entire second level of John Ascuagha's Nugget hotel in Sparks, Nevada. I was there for five days, and it was on the fourth night our paths crossed. With well over three thousand persons in attendance, it was a matter of pure chance that we met and spent the entire evening together. This involved strolling through the displays of a hundred-plus contractors. The booths ran the gamut of highly technical hardware presentations including fully operational flight simulators, works of art, books and flamboyant T-shirt

John Yamnicky with granddaughter Devon, wearing his distinctive eye patch with a golden dragon. (Photo courtesy Yamnicky family)

collections. There were strategically placed cocktail bars throughout, and the drinks ran the full range from beer to gin slings. Unofficial bars sponsored in the more elaborate and well-funded contractor booths offered Mai Tais in bountiful quantities. These were served in complimentary glasses each emblazoned with the contractor's name. A sumptuous buffet was offered for several hours of the evening. Seating for this was both in the main ballroom and in a number of areas among the booths.

"John was intercepted many, many times by friends, squadron mates and other acquaintances as we strolled about. He had a story to relate, an incident to recall, or a joke to tell on each intercept after the handshake, shoulder grasp or slap on the back he received. He was well known in our profession and quite obviously well liked. He displayed a casual but warm-hearted attentiveness to both men and women during conversation exchanges. It was rarely I can remember the smile absent from his face.

"Personally I feel fortunate to have known John and to have worked with him on projects at the Test Pilot School. And the memory of our last several hours together will not cloud."[23]

Within days of the Tailhook reunion, John Yamnicky found himself caught up in

Jann and John Yamnicky, San Diego, 2000. (Photo courtesy Judy Bausch)

one of history's most notorious days, suffering with others an overwhelmingly cruel fate that would not only completely devastate his family and friends, but also cause unimaginable horror and profound anxiety across America and around the world.

REFERENCES

1. Yamnicky family background material submitted to author by Dolores Sebastian, 10 and 17 January 2012.
2. Mail correspondence with Craig Rutter, 7–21 December 2011

3. *Pittsburgh Post-Gazette* newspaper, article, *Pentagon Crash Claims Ex-Tiger Yamnicky*, by Norm Vargo, Friday, 14 September 2001
4. E-mail correspondence with George D. Miller, 1–5 May 2012
5. *Ibid*
6. *Pittsburgh Post-Gazette* newspaper, article, *Pentagon Crash Claims Ex-Tiger Yamnicky*, by Norn Vargo, Friday, 14 September 2001
7. E-mail correspondence with George D. Miller, 1–5 May 2012
8. *The Daily News* (McKeesport, PA) newspaper, article, *Local Victims of 9/11 Have Not Been Forgotten*, Saturday, 10 September 2011, author unknown
9. Monte Cassels entry, Guest Book, *Legacy.com, John D. Yamnicky, Sr.*. Website: http://www.legacy.com/guestbook/guestbook.aspx?n = john-yamnicky&pid-91836
10. E-mail correspondence with Dick Liljestrand, 18–20 January 2012
11. Foster, Elizabeth Carroll, *Follow Me: The Life and Adventures of a Military Family*, iUniverse Books, Bloomington, IN, 2010
12. E-mail correspondence with N. Lee and Judy Bausch, 8 December 2011–20 May 2012
13. E-mail correspondence with Dennis Plautz, 23 January 2012
14. E-mail correspondence with Jennifer Yamnicky, 30 October 2011–2 February 2012
15. Hugh McGee entry, Guest Book, *Legacy.com, John D. Yamnicky, Sr.*. Website: http://www.legacy.com/guestbook/guestbook.aspx?n = john-yamnicky&pid-91836
16. E-mail correspondence with Dennis Plautz, 23 January 2012
17. *Ibid*
18. E-mail correspondence with Harry Errington, 12–29 February 2012
19. Shrontz, Frank A., *Report on Fitness of Officers: Captain John David Yamnicky*, Office of Secretary of Defense, Washington, D.C., 2 November 1972
20. E-mail correspondence with Jann Yamnicky, 5 November 2011–19 April 2012
21. E-mail correspondence with John D. Yamnicky, Jr., 4 February 2012
22. E-mail correspondence with Jann Yamnicky, 5 November 2011–19 April 2012
23. E-mail correspondence with N. Lee and Judy Bausch, 8 December 2011–20 May 2012

7

Answering the call

In announcing the Group 3 selection process NASA again maintained a preference for test pilots, but this time made a judicious allowance for other applicants pressing for inclusion with only 1,000 hours of high-performance jet flight. Overall, the space agency was encouraging applications from candidates who not only had the ability, experience and attitude to operate in high-stress environments, but equally possessed engineering expertise in highly technical development programs.

SETTING PARAMETERS

On 15 May 1963, NASA announced that it would be recruiting "nine to twelve" new astronauts. The deadline was 1 July, and applications were open to both military and civilian volunteers. However, military services, which pre-screened their pilots, were granted a fortnight's extension to pass on to NASA the names of their recommended applicants. With minor exceptions, the selection criteria were similar to those applied to the second group the previous year. In order to qualify, an applicant had to:

- Be a U.S. citizen.
- Have been born after 30 June 1929, so that they did not reach their 34th birthday until after 30 June 1963.
- Be six feet or less in height.
- Have earned a degree in engineering or physical sciences.
- Have 1,000 hours of jet pilot time, or have attained experimental flight test status through the armed forces, NASA, or the aircraft industry.
- Be recommended by their present organization.

NASA also took the opportunity to announce that discussions would be held with representatives of the nation's scientific community on how to go about recruiting scientists as astronauts for lunar missions – a process that would lead to the selection of six scientists for Group 4 on 28 June 1965.

Compared with the 1962 selection criteria, the maximum allowable age for the third group was reduced by a year in order to ensure a broad age spread to the pilot pool. Certification as a test pilot, while still preferred, was no longer mandatory, and there was increased emphasis given to academic credentials.[1]

Having eliminated the requirement for a test pilot school certificate, NASA began soliciting recommendations from relevant organizations such as industrial aerospace firms, the Air National Guard, various reserve organizations, the military services, the Society of Experimental Test Pilots, the FAA, the Airline Pilots' Association, and the three NASA field centers which used pilots in their work – the Langley Research Center in Hampton, Virginia, the Ames Research Center at Moffett Field, California, and the Flight Research Center at Edwards AFB, California. NASA also contacted organizations involved in aviation development and test programs, including Boeing of Seattle, Washington, General Dynamics of New York, Douglas Aircraft of Santa Monica, California, Lockheed Aircraft of Burbank, California, General Electric of New York, Republic Aviation of Farmingdale, New York, Ling-Temco-Vought of Dallas, Texas, McDonnell Aircraft of St. Louis, Missouri, Westinghouse Electric of Pittsburgh, Pennsylvania, Grumman of Bethpage, New York, and Northrop Space Laboratories of Hawthorne, California.[2]

A BLACK ASTRONAUT?

Even as NASA began the process that would choose the third group of astronauts, political pressure was being exerted at the highest levels for the space agency to select an African-American pilot. For some time, President John F. Kennedy had wanted the minority electorate to regard him as doing something positive on the issue of equality in the military. On 24 June 1962 he appointed an advisory committee to study equal opportunity policies in the military, charging its members with ensuring that "any remaining vestiges of discrimination in the armed forces on the basis of race, creed or national origin" were removed. One of the initiatives he pressed for was for a black serviceman to be inducted into the high-profile astronaut corps.

At the specific behest of the president, the Department of Defense was contacted to determine whether the Air Force had any suitable candidates, but even though the records were thoroughly scoured the response coming back to the White House was apologetic. No black Air Force officers had the required amount of flying time or the requisite academic background, let alone meeting other stringent requirements for consideration. President Kennedy did not like being denied his initiative. In response the Air Force was essentially instructed to locate a suitable black candidate and have him enrolled in the next Aerospace Research Pilot School course at Edwards AFB. Once the airman had passed the course, and even without the necessary flight hours, background, experience and academic qualifications, pressure would then be exerted on NASA to include the officer in its next astronaut group. Once again the Air Force searched through its records and, to the relief of the researchers, finally came across something that might fit the bill – a hope-filled application from a serving black Air Force officer requesting test pilot and astronaut training. The name on the application was 28-year-old Capt. Edward Joseph Dwight, Jr., USAF.

Growing up poor on a small farm, the son of a Negro League baseball player, and

Capt. Edward J. Dwight, Jr., USAF. (Photo: USAF)

graduating from Kansas City Junior College in August 1953, Dwight had set his eyes early on the sky and joined the Air Force. Having obtained his pilot's wings in 1955, Lt. Dwight was assigned to Williams AFB, Arizona to begin jet aircraft training. During his two-and-a-half years there he also attended night classes at Arizona State University, where he earned his degree in aeronautical engineering in 1957. After a brief tour of duty in Japan flying B-57 jet bombers, Dwight was reassigned to Travis AFB, California. In 1962, knowing he had little chance of being accepted owing to inherent racism in the services, he nevertheless submitted an application for test pilot and astronaut training, citing his 2,000 hours of flight experience and engineering degree.

Relieved to have found a suitable candidate, the Air Force seized upon Dwight's application and advised the president, who then wrote a personal letter to the young airman, strongly urging him to apply for ARPS training. Capt. Dwight was enrolled

Ten members of Ed Dwight's ARPS Class IV during an aircraft factory tour. Front row from left: William Campbell, Mike Adams, Dwight, James McIntyre and Frank Frazier. Back row: Robert Parsons, Dave Scott, Russell Scott, Lachlan Macleay and Alex Rupp. (Photo: USAF courtesy Frank Frazier)

in the next ARPS class on the direct orders of the Air Force Chief of Staff, General Curtis LeMay, much to the outrage of the school's commander at that time, Col. Charles ("Chuck") Yeager. Since it had been intended to have only eight students in this particular class, and Dwight's average qualifications placed him well down the suitability list of candidates, a grudging compromise was reached to accept him: he could join the class only if all those who were better qualified were also included. This resulted in Class IV being doubled in size to sixteen.

Once his training began, Dwight tried his hardest to maintain a positive outlook, but it was difficult since everyone knew he was there only because of White House intervention, and his time there was one of frustration compounded by prejudice and isolation. Yeager was vehemently opposed to Dwight's attendance, and let him know it at every opportunity. At one point Yeager even said, "Why in hell would a colored guy want to go into space anyway? As far as I'm concerned, there'll never be one to do it. And if it was left to me, you guys wouldn't even get a chance to wear an Air Force uniform." Yeager would later state that although Dwight wasn't a bad pilot, he wasn't exceptionally talented either.

In desperation, unable to take the situation, Dwight lodged a complaint with the White House. Department of Defense investigators arrived at Edwards to assess the problems and apply further pressure on the school. They confronted an irate Yeager, reminding him and his flight instructors in no uncertain terms that Dwight was to graduate. Period.

However, things would change dramatically a month before the class was due to graduate. On 22 November 1963 President Kennedy was felled by an assassin's bullet. The following Monday, Dwight unexpectedly received orders assigning him with immediate effect to Germany as a liaison officer for what was essentially a non-existent German test pilot school and space program. To his disbelief he – and he alone – had been dropped from the astronaut training program. Dwight contacted the White House in despair, but was told that nothing could be done: with Kennedy gone the door had been firmly closed, even though the president's brother Robert Kennedy tried unsuccessfully to have NASA accept Dwight for astronaut training.

Edward Dwight was never officially informed that he was no longer in the USAF space program, but *Ebony* magazine, in reporting on his sudden transfer, said he had been shelved from the program due to his letter to the White House claiming "racial pressure" during his training at the aerospace school.

Dwight refused to accept the politically motivated assignment to Germany, in the process initiating a lengthy and heated dispute with the Air Force. His life now in turmoil with the ongoing publicity and outrage following his premature transfer from the aerospace school, he eventually resigned his commission in 1966. After tentative ventures into private business he turned to one of his childhood loves – sculpture. In the mid-1970s he earned a master of fine arts degree from the University of Denver, Colorado and went on to become recognized as a noted sculptor. Today, his work is prominently displayed all across the United States.[3]

WINDOWS OF OPPORTUNITY

According to Edwin ("Buzz") Aldrin, "At the time I started my doctorate [of science] degree, I also applied to NASA to be a member of the second group of astronauts, asking that the requirement of attendance at test flight school be waived. The waiver was denied .... Nevertheless, I believed when my thesis ['Line of Sight Guidance Techniques for Manned Orbital Rendezvous'] was well underway my chances would improve." He tried again for the third group the following year. "Early in the spring, just a few months after I had made my application, my hunch proved true. NASA officially announced that candidates need not have attended test flight school. Not only that, but I sensed a trend toward pilot-academicians."[4]

This was the announcement which Russell ("Rusty") Schweickart had also been hoping for. "Just before the opening for applications was announced, we found out through the grapevine that the requirement for test pilot was going to be dropped, and all of a sudden that opened up the possibility now, not after I somehow managed to become a test pilot."

As he filled out his application, Schweickart only had one minor problem – he didn't quite have enough jet hours. "I can't remember exactly the first step. But the prerequisites were listed, and now I met them all, although I must admit ... I sort of cheated slightly, in that one of the prerequisites in place of it, you had to be either a test pilot or have 1,000 hours of high-performance jet experience. Well, at the time that I actually filled out the application, I had about 995 hours or something like that, so I said 1,002 hours and sent it off, then I said goodbye to my wife and went to the airport and flew seven more hours so that by the time they opened the letter I was an honest person again."[5]

The selection committee for Group 3 consisted of astronauts Deke Slayton, John Glenn, Wally Schirra and Alan Shepard, and the chief of Flight Crew Operations Warren North. A total of 720 applications were received; 492 from recommended military personnel and 228 by civilian hopefuls. After initial screening, 490 were still eligible for selection, and after further screening 136 were referred to the selection panel. They in turn narrowed the final interview number to 34.

First, however, those hoping to make the astronaut group were asked to undergo an intense, week-long set of medical and psychological tests at Brooks AFB in San Antonio, Texas. One applicant was Capt. David R. Scott, USAF. "Luckily we were not subjected to the indignities endured by the Original Seven," he observed. "When they were recruited the survival of human beings in space was such a big unknown that their physical tests included having every bodily orifice probed and checked. By the time we went through the selection process, at least they knew the human body could survive space flight."[6]

Group 3 Candidates' Arrival at the Aerospace Medical Center, San Antonio, Texas, July–August 1963

(Weekends excluded)	
July 31	Gordon, Ramsey
August 1	Adams, Bell, Freeman
August 2	Irwin, Rupp, Scott
August 5	Anders, Eisele, Furlong
August 6	Cernan, Davis, Shumaker
August 7	Chaffee, Guild, Vanden-Heuvel
August 8	Brand, Ebbert, Evans
August 9	Cornell, Cunningham, Smith
August 12	Phillips, Williams, Yamnicky
August 13	Aldrin, Cochran, Schweickart
August 14	Bean, Kirkpatrick, Swigert
August 15	Bassett, Collins

Navy RADM (then Lt.) Bob Shumaker recalls that for a while during medical and psychological screening at Brooks AFB he was rooming with fellow aspirant Buzz Aldrin. "One night he told me he didn't think he'd be selected. Yeah, sure ... he was Corps Commander at West Point, had shot down MiGs over Korea, and he wrote his doctorate on orbital mechanics.

RADM Robert H. Shumaker, USN. (Photo: USN)

"One day a psychiatrist interviewed us, showing us pictures of several individuals in various action scenarios. We were supposed to make up a story that included the events leading up to this scene, what the people in the scene were experiencing in the way of feelings, and wrap it up with a logical conclusion. As a bachelor, I was at a disadvantage in not having told stories to children. One scenario showed a sort of weak-willed guy standing over a bed on which lay a woman partially under the sheets, but with exposed breasts. The guy was partially turned away and his arm was bent at the elbow, covering his eyes. My story was that Joe Smith was a married salesman on a convention; he'd met a call girl and had just finished having sex with her, and he was experiencing regret. I thought it was a reasonable story for me to concoct. I later asked Buzz what story he came up with for that scene. 'That was *my* story!' he said. 'You try that again, buster, and I'll squirt you in the other eye!'"[7]

Aldrin recalls that the interviewing Air Force psychiatrist was Dr. Don Flinn. "We went through the whole battery of tests, numbers, read backs, puzzles, Rorschach inkblots, symbols, and everything else they could think of to test our mettle. That was in July 1963, and I passed the tests easily. I was rock-solid stable."[8]

Another of Shumaker's roommates was Walt Cunningham, who wrote in his autobiography that Shumaker, the young naval lieutenant, "was your typical All-American boy with a good flying background, and I conceded him better odds than myself of being selected."[9]

After extensive medical tests, further scrutiny of all the applicants' qualifications along with their medical, psychological evaluation and interview results narrowed the number still under consideration. These men would be invited to Houston for a series of panel interviews and final examinations.

Cunningham later recalled that the tests at Brooks AFB involved some 45 hours of activity. He considered his fellow candidates, from varying backgrounds, to be, "with rare exception, excellent in health, from good to excellent in physical condition, well above average in intelligence, and with good flying credentials. We shared a strong feeling of self-identity, self-confidence, and awareness of where we were going and how we would get there. Some might call it ego. The competitive instinct surfaced quickly."[10]

Following the Brooks tests, Bob Shumaker was upbeat, reckoning he had a better than even chance. "But at the last moment the interviewing physician noticed from an X-ray that my lymph nodes were larger than normal. That ended my space career. It was particularly embarrassing because I had just returned from a six-month cruise on an aircraft carrier with, of course, no women on board. My problem was due to mononucleosis, the so-called 'kissing disease'. So how does one explain *that* to his friends?"[11]

Shumaker was one of four finalists to be disqualified after a final review of their medical results, leaving just thirty to face the final interviews and then the anxious wait for the phone to ring – hopefully with good news from Houston.

Navy Lt. Bill Ramsey had been a dual applicant for the second and third groups, but unfortunately missed selection this time due to a minor hypervascular condition. He would, however, go on to many notable achievements, including command of the nuclear-powered aircraft carrier USS *Dwight D. Eisenhower* (CVN-69), eventually retiring from the Navy with the rank of rear admiral.

TRYING OUT AGAIN

Having missed out on selection for NASA's Group 2, there was mixed news in store for Capt. Ken Weir, USMC, in relation to the Aerospace Research Pilot School at Edwards AFB.

"Headquarters succeeded in obtaining a Marine ARPS billet and they then went through a selection process of their own that narrowed the choice to Bill Geiger and me. The next thing I knew was that the commander of the Marine Air Detachment called both of us into his office at the same time and told us who had been selected to go to ARPS. I always thought that was a classless display of insensitive leadership by informing my long-time friend Bill Geiger what the decision was with me present in the room.

"I flew to USAF Brooks Aeromedical Facility to go through all of the enhanced physical, psychiatric and psychological exams all over again. Same demanding drill, treadmill and all. But at least I didn't have the Houston interview to look forward to.

"I attended the ARPS in Class IV with Lt. Walt Smith, who had been in the Navy Test Pilot School class right after mine. It was an incredible experience. I had been upgraded to a full member of the Society of Experimental Test Pilots with a gold 'X' lapel pin and was promoted to the rank of major shortly after the class began, which meant I replaced Air Force Capt. Jim Irwin as the class leader. Marine promotions were running ahead of the Navy and Air Force. I was in tall cotton.

"Then NASA unexpectedly announced the third astronaut selection in the fall of 1963 and I was the class expert as far as the NASA selection process went, being the only one who'd participated in the second astronaut selection. Everyone in the class who met the height limit applied for the NASA astronaut openings and filled out all of the necessary paper work. I had kept copies of everything possible and sending in the requested information was easy. Besides, NASA already had most of everything about me they needed."

Weir's classmates kept pestering him to find out what was going on. One day Col. Chuck Yeager's secretary entered the class room and announced for everyone to hear that John Glenn was on the phone and wanted to talk to Ken Weir. In fact both John Glenn and Dr. Lamb were on the phone and told Weir they had been going over all of his records and they were very concerned about his sinus issues. Weir responded that he did not have any sinus issues and had experienced explosive decompressions at least two dozen or so times in controlled environments without any problems at all. They said they were relieved to hear that, thanked Weir and hung up. But like his colleagues, Weir found it hard to focus on the academic work while wondering what was in store from the NASA astronaut selection panel.

"One day after class, those of us applying for the third astronaut selection stayed over to discuss the process a little. I told the candidates to watch the mail carefully in the next week or so. If they received a large NASA franked manila envelope then they'd made the second cut and were headed towards additional evaluations. If they got a small NASA franked white envelope they were eliminated, end of the exercise. The next Friday I received a small NASA franked white envelope. It simply said 'Thank you for your interest in the NASA manned space program.'"

Three of Weir's ARPS classmates, Dave Scott, Jim Irwin and Ted Freeman, were eventually selected as NASA astronauts in the third group. Dave Scott and Jim Irwin would go on to walk on the Moon together as members of the Apollo 15 crew. Ted Freeman was killed during a T-38 training flight following a bird strike before he had the opportunity to fly into space.

Although Ken Weir's dreams of becoming an astronaut had been finally shattered, he retains many memories – fond and otherwise – of that period in his life.

"Many years later at a formal Marine Corps function in Irvine, California to honor Bob Hope's lifetime contributions to the Marine Corps we were all decked out in our dress uniforms. I was in my major general's full dress as were others. Alan Shepard attended in his rear admiral's evening outfit.

"Alan was prancing around, table-hopping and pressing flesh with everyone who wanted to be seen talking to America's first man in space. He came over to our table and I stood up and said, 'Good evening, Admiral.' I then said, 'You won't remember me, but years ago I spent several minutes being interviewed by you at the NASA Manned Spacecraft Center.' He interrupted me, and said, 'Ken Weir! I remember you very well.' We stood there looking at each other for what felt like an eternity without saying anything else and then he turned and walked away. Last I ever saw of him.

"When Bill Fitch, Bill Ramsey and Bill Geiger and I reminisce of our astronaut selection disappointments, we frequently remind each other about how much more extensive and exciting the flying of airplanes was after becoming NASA rejects.

"Brigadier General Chuck Yeager, the first to break the sound barrier, once said at Pancho Barnes' Happy Bottom Riding Club at Edwards that he never wanted to ride around in a tin can, sitting in a seat that some monkey had previously crapped in. It wasn't quite like that, but colorful it was, and we were proud to have participated to the extent we did."[12]

MAKING THE GRADE

Two days after completing their one-week-long physical at Brooks AFB, the thirty remaining finalists received orders to travel to Houston for a further evaluation and final interview by the selection panel early in September.

Although the selection process had been widely reported, NASA was still anxious to keep a lid on things, and the remaining candidates were asked to fly to Houston in civilian clothing and not to inform anyone, other than their wives, just where they were headed. On arrival, they were picked up individually by drivers and transported to the Rice Hotel, where once again the whole charade of checking in under the alias of hotel manager Max Peck was re-enacted. As before, many of them came across one other in the hotel bar. Then they presented themselves at MSC for a further test of their knowledge and personal interviews with the selection panel.

"They hit us with another blizzard of paperwork," observed Gene Cernan, "which now included questions about space travel and orbital mechanics. Although I didn't know much about these subjects, I wrote essay-length answers in long-hand to every query. If they were looking at how a candidate handled the unknown, I must have rated pretty high, for I didn't know much at all."[13]

Two of those selected as finalists had missed the final cut for Group 2, and been encouraged to try again. With this added experience, Mike Collins and Dick Gordon swept through the process and were eventually named in the fourteen-man Group 3 recruitment. Collins later recalled that he had been really "psyched up" for the final interview hurdle. "Oddly enough, the interview itself seemed a lot easier this time, despite the importance I placed on it. For one thing, I was not among strangers any longer, and for another, the ARPS training had provided me with a good deal of information I hadn't had the year before. Also, I had a better idea of what they might ask, and had studied up for their questions. Even Deke Slayton and Warren North seemed to have mellowed a bit."[14]

As Bill Anders explained to the author, one question at the final interview caused him some concern. "Initially I believe they had several thousand applicants, of which I think they invited about three hundred, and then cut us down to 60 or 100. Pretty soon we figured out that every time we'd go for a physical or review they would cut the number down by half. When we were finally down to about twice the number of actual selectees needed, we had to individually meet a review board of astronauts and space center executives, plus the chief flight surgeon, Dr. Chuck Berry. Deke Slayton was on the panel, as was Al Shepard and Wally Schirra, and I believe Warren North was there as well.

"When it came my turn I went into a long room and sat down in front of this table of judges who were thumbing through my records; a thick stack of medical reports and other papers from various tests that we'd had. At one stage in the interview the chief flight surgeon said, 'Well everything looks good here, and we're basically satisfied – except for this concussion you had a few years back.' Well, for me, time stopped, because to my knowledge I had never suffered or reported a concussion. Straight away I wondered what was going on here; were they somehow testing me, to see if I would fess up to a concussion I'd never had? Was Al Shepard deliberately throwing me something of a red herring to see how I might react? Maybe they had the medical reports mixed up and they had somebody else's information; someone who had already washed out? In those microseconds before responding, my mother's words came back to me; she had told me, 'As you're going through life you should never lie ... but you don't always have to blurt out the truth.' Wise words indeed, and so I thought that while I couldn't lie, I needed a response. So with hardly any pause – nothing like what it's taken to explain this – I said, 'Well, I've never had a problem with a concussion.' Which I think is a true statement. However, it avoided being entrapped in case they did have my records and someone else's mixed up. They all looked at me, and Berry just said, 'Fine.'"

"Now that many years have gone by, I have asked a friend of mine, now a NASA flight surgeon for the astronaut group, to go back and take look at my pre-acceptance medicals as well as the more recent medicals I now get from NASA, to see if things like maybe my blood type had somehow changed in that time; just to see if I actually had been cross-examined at the personal interview based on the medical records of someone else who perhaps got dumped. It would be interesting to find out."[15]

Once the final interviews had taken place, the candidates enjoyed a semi-relaxing cocktail evening where they mingled with other pilots they knew and met several of the Mercury astronauts. They were still on edge and would remain so throughout that tense period.

Buzz Aldrin was initially confident that he would make the cut, but then grew nervous. "During the week-long process, as much to ease the strain as to see how we performed socially, there were various social activities. At the official cocktail party for the group of us, I was the only candidate to be accompanied by his wife ... none of the other guys lived in Houston and their wives were not invited.

"During the week, a number of the men I had known at various points confided that they thought, of all of us, I had it wired the best. I was the most obvious choice, or certainly one of the more obvious choices.

"Nevertheless, when the week ended I was a bundle of nerves. We were told that only thirteen to fifteen of us would be selected and that we'd be informed within the next three weeks."[16]

Once the finalists had all departed MSC, Slayton and the rest of his panel began their final selection process, as he explained in his memoir, *Deke!*

"I'd already developed a point system that we used in making the final evaluations on astronaut candidates. There were three parts: academic, pilot performance, and character/motivation, ten points for each part, with thirty being the highest possible score. Some of it was cut and dried: you got points for a certain amount of flying time and for education. Some of it, by design, was subjective and based on face-to-face interviews."[17]

Eventually, once the list of the fourteen successful candidates had been assembled and agreed upon, the pleasant duty of calling them fell to Slayton. The finalists all knew this, but also understood that in the past it had been left up to Warren North to break the news to those who had fallen short. While some recall being phoned by North, others such as Vance Brand say it was Jack Cairl of NASA's personnel office who broke the disappointing news to them.

In his book *Last Man on the Moon*, Gene Cernan described the elation he felt when he received the hoped-for call from Slayton inviting him to come to Houston and become an astronaut. But at the same time he also felt sorry for one of his good friends and classmates at the Naval Postgraduate School in Monterey, California, Ron Evans, who was also in the running. Evans received a call at the same time, and to Cernan the elation that he personally felt was "a bittersweet experience, because I soon saw the face of a dejected Ron Evans, who had received the dreaded call from Warren North. He missed the cut and my heart bled for him. No fortune-teller could have predicted that in not too many years Ron and I would be together in a rocket ship flying to the Moon."[18]

Four of the finalists who missed the cut would later be selected as members of NASA's Group 5 in 1966: Vance Brand, Ron Evans, Jim Irwin and Jack Swigert.

None of the remaining fifteen finalists would ever be selected by NASA, although Mike Adams did fly an X-15 to an altitude just in excess of 50 miles, thereby earning his astronaut wings from the Air Force, albeit awarded posthumously because he was killed returning from that same record flight when his aircraft went into a fatal spin and he failed to eject before it broke up over the desert.

MICHAEL J. ADAMS, USAF

Mike Adams qualified as an Air Force astronaut at ARPS and was awarded astronaut wings for flying the X-15 rocket plane, but never achieved his ambition of becoming a NASA astronaut and flying faster, higher and farther than ever before. Sadly, death got in his way.

Michael James Adams was born in Sacramento, California on 5 May 1930, the son of Michael L. Adams and the former Georgia Domingos, and the older of two brothers. He took his grade school education at Donner Elementary, Stanford Junior High and Sacramento High prior to completing two years of study – specializing in forestry – at Sacramento City College, where he was also a varsity javelin thrower and baseball outfielder. In his late teens he achieved great mechanical satisfaction in rebuilding the engine of his 1937 Chevrolet. Just as he was completing his second year at college the conflict in Korea escalated into a full-scale war and he promptly enlisted in the Air Force, entering service on 22 November 1950. Having completed basic flight training at Lackland AFB, Texas he served with the 3501st Pilot Training as a Link trainer instructor until his selection as an aviation cadet. Then he underwent primary flight training at Spence AFB, Georgia. On 25 October 1952 he graduated from pilot training at Webb AFB, Texas and was commissioned in the Air Force as a 2nd lieutenant. After gunnery school instruction at Nellis AFB, Nevada flying F-80s and F-86s, Adams was deployed to Korea in April 1953. He initially served with the 80th Fighter-Bomber Squadron, based at Suwon Air Base (K-13), flying the F-86. The following year, as a 1st lieutenant, he was flight commander for the 618th FBS. Overall, he flew 49 combat missions and earned an Air Medal.

After returning from Korea in February 1954, Lt. Adams completed another three years of service, first as a flight commander in fighter squadrons in the United States and then a six-month rotational duty at Chaumont Air Base in France. In January 1955, during his 30-month tour with the 813th FBS at England AFB, Louisiana, and now with the rank of major, he married Freida Beard. Soon thereafter, he left on a six-month temporary duty tour in Germany.

In 1956 Adams entered the University of Oklahoma as part of an Air Force career development program for promising officers. During this time Freida gave birth to Michael, Jr., their first son. He earned his aeronautical engineering degree in 1958 and went on to complete graduate work in astronautics at the Massachusetts Institute of Technology. He also welcomed second son Brent into the world. After completing his studies, Adams was assigned to Chanute AFB, Illinois as an instructor for the Maintenance Officer Course, and while based there the family was completed with the birth of daughter Liese. It was during this time that he was selected to attend the USAF Experimental Test Pilot School at Edwards AFB, California as a member of Class 62-C. He graduated as the outstanding pilot and scholar in his class, and was awarded the prestigious A. B. Honts Trophy. He was subsequently selected to attend the Aerospace Research Pilot School, also at Edwards, which was then commanded by Col. Chuck Yeager.

One memorable day, while he and front-seat fellow student and good friend Dave Scott were practicing a low lift-to-drag ratio landing in one of the F-104 Starfighters operated by ARPS, they came perilously close to being killed after their jet engine unexpectedly lost most of its thrust over the landing strip and the aircraft descended rapidly. Knowing they would hit hard, in a split second both pilots fortuitously made opposing decisions which saved their lives. Adams chose to eject on impact, safely blasting from the crippled aircraft moments before the airplane's jet engine slammed into the rear cockpit. Scott, meanwhile, elected to stay with the F-104, which proved to be the right decision, since his ejection seat was damaged during the first, violent impact, and if he had pulled the ejection handle he would have been killed.

In 1963 Adams and Scott both applied for NASA's third astronaut intake, but the temporary back injury Adams had suffered as a consequence of his F-104 ejection seems to have caused him to miss out on medical grounds. His disappointment was somewhat tempered in October by the news that Scott was one of the fourteen new astronauts.

Michael J. Adams, USAF. (Photo: USAF)

Mike Adams with the X-15 rocket plane. (Photo: NASA)

Maj. Adams graduated with honors from ARPS in December 1963. He then began conducting stability and control test flights in the Northrop F-5A, and later served as the Air Force project pilot on the Cornell Aeronautical Laboratory's variable stability T-33 program at Buffalo, New York. He was next involved in working on the Lunar Landing Research Vehicle (LLRV), an ungainly contraption nicknamed "The Flying Bedstead", specifically constructed to simulate lunar landings. His assignment to this program began on 13 January 1964 and the first of the LLRV test vehicles arrived at Edwards three months later.

Adams now began to consider a new direction that might one day take him into space. On 12 November 1965 he was named as an astronaut candidate for the Air Force's Manned Orbiting Laboratory (MOL) program. This was to be an all-military version of NASA's Gemini program, operated by the Air Force, but it soon began to run into serious funding problems. Eight months after entering the program, and realizing that there was an ever-diminishing chance of actually making a MOL space flight, Adams requested a transfer to the X-15 rocket plane program, where he felt he might be more effective. His request was granted on 14 July 1966, and he became the twelfth and final pilot to enter that program.

He successfully complete six flights in the winged craft, but on his seventh flight, designated 3-65-97, everything went wrong. The 191st flight in the X-15 program occurred on 15 November 1967, with the rocket plane being released at an altitude of 40,000 feet above Delamar Lake, Nevada from beneath the wing of a modified B-52 carrier aircraft. After falling free, Adams ignited the rocket engine and was propelled upwards. On ascent, however, an electrical problem in an onboard experiment caused an arc to flash across the exposed experiment terminal, creating a coronal discharge that affected the operation of several critical systems, including the adaptive control, as the X-15 passed through 100,000 feet. He then began to receive persistent and disorientating signals indicating he had computer instrumentation problems. As the X-15 neared its peak altitude he was concentrating on trying to center a needle that he thought was indicating pitch, but which was actually indicating yaw, causing the aircraft's nose to yaw sideways relative to the direction of flight. Shortly thereafter, the aircraft reached 266,000 feet (50.38 miles). This was a little higher than planned, but that fraction of a mile of altitude would qualify him for his Air Force astronaut wings under the service's guidelines. At that extreme altitude no air was passing over the wings, and with the nose still pointed upwards he could not see the ground. As a result he did not notice the anomalous yaw, and facing ongoing computer problems he decided to switch to manual control.

Still with no visual indication that he was on the wrong heading, Adams check-fired his controls jets, causing the X-15 to yaw even more to the right. This deviation increased dramatically as the aircraft started to descend until – still unaware of any directional problem – he was actually tail first. Then, when he reached 225,000 feet and was traveling at around Mach 5, Adams suddenly reported to the ground that he was caught in a hypersonic spin. As he struggled to regain control, it is believed he was rendered unconscious. Without any human intervention, the superb aerodynamic design of the X-15 caused the aircraft to stabilize itself. But at around 120,000 feet, Adams seems to have regained consciousness and mistakenly resumed his struggle to regain attitude control by feeding back into the control system. This quickly resulted in a pilot-induced oscillation in pitch which, combined with an incremental build-up of atmospheric resistance as the X-15 hurtled towards the ground, caused the vehicle to buck wildly, with ever-increasing energy. It was later reasoned that if Adams had simply taken his hands off the controls the X-15 may have stabilized itself yet again. Unfortunately, Adams was probably incapacitated by now, as he did not initiate an ejection from the crippled aircraft. At an altitude of nearly 62,000 feet the airframe could no longer sustain the violent g-forces and began to fail. The fuselage bent, and moments later the aircraft began to disintegrate. Large pieces of wreckage eventually fell over a 12 mile area of hilly desert terrain. Adams' body was recovered from the wreckage of the forward fuselage.

Two months on, the results of the crash investigation were released, concluding that X-15 No.3 had broken apart due to excessive g-loads "induced by severe pitch oscillations".

Maj. Adams' funeral was held at Edwards on 18 November 1967. He left behind his widow Freida, sons Michael Jr. and Brent, and daughter Liese. Two months later, on 16 January 1968, Freida was invited to Barksdale AFB, where she was presented with her late husband's silver astronaut wings for having exceeded the altitude that the Air Force defined as the boundary of space.[19]

TOMMY I. BELL, JR., USAF

The McDonnell Douglas F4C Phantom II, a long-range supersonic jet interceptor and fighter-bomber, gave invaluable service during the Vietnam War at the hands of many superbly talented pilots. One of those was Capt. (later B/Gen.) Tommy Bell. In just ten months he flew 243 combat missions in the F4C, his calmness and tenacity almost resulting in his becoming one of NASA's pioneering Apollo astronauts.

Tommy Ila Bell, Jr. was born in Jacksonville, Texas on 4 December 1930, one of three children for Tommy Sr. and the former Francis Alice Stewart. Sadly, both his parents died at a comparatively young age; he was only 13 years old when he lost his father. Obtaining his high school education at the School of the Ozarks in Hollister, Missouri, he graduated in 1948 and began a course in aeronautical engineering at the University of Illinois. But in March 1952 he enlisted in the Air Force as an aviation cadet. He received his pilot wings and commission as a 2nd lieutenant in March 1953 at Laredo AFB, Texas. Over the next four years he served as a fighter pilot with the 71st Fighter-Interceptor Squadron, initially based at the Greater Pittsburgh Airport, Pennsylvania and from August 1955 at Selfridge Field Michigan, flying the F-86D Sabre.

Around this time he met his future wife Lillian Villiger. They were married and would eventually have three children, Carolyn, Leslie and Anthony.

In June 1957 Lt. Bell was assigned to the Air Force Institute of Technology at Wright-Patterson AFB, Ohio, later returning to the University of Illinois to complete his degree in aeronautical engineering. Immediately following his graduation in May 1959 he was selected for Class 59-C of the Experimental Test Pilot School at the Air Force Flight Test Center, Edwards AFB, California. Two of

Lillian and Tommy Bell on their wedding day. (Photo Courtesy Carolyn Bell Phillips)

Tommy Bell (left) with good friend and fellow Air Force pilot, Ronnie McGuire. (Photo courtesy Carolyn Bell Phillips)

his fellow students were USAF Captains James McDivitt and Edward White II, who would later be selected by NASA as Group 2 astronauts and go on to fly the Gemini IV mission together. After graduating on 22 April 1960, Capt. Bell was ordered to Wright-Patterson AFB, initially as an experimental test pilot and later as chief of the Aero-Mechanical Branch of Flight Test Engineering. He returned to the Test Pilot School in May 1963 as an instructor in the Aircraft Stability and Control Division, specializing in the control of aircraft spins. It was during this time that Bell was one of 34 finalists for the third astronaut group, unfortunately without success, and returned to his test work at Edwards AFB. He was in a dangerous occupation, as his daughter Carolyn reflected. "During that period – around 1964 to 1965 – nine pilots lost their lives in accidents at Edwards. It was hard at times; I remember we lost some neighbors and my dad lost many friends due to the hazardous nature of their work."

In 1966, Bell graduated from the University of Southern California, Los Angeles with his master's degree in mechanical engineering. In June of that year he was sent to MacDill AFB, Florida for combat crew training in the F4, and after volunteering for active duty in Vietnam was transferred to Cam Ranh Air Base, from where the 12th Tactical Fighter Wing conducted air-to-ground missions, including close air support and interdiction, as well as air-to-air missions such as rescue protection and combat air patrols. It was from this base that Bell flew his 257 combat missions in the F4C between November 1966 and September 1967. In the latter half of this combat tour, he served as Operations Officer for the 391st Tactical Fighter Squadron.

Bell's next assignment came in October 1967, when he was ordered to Air Force Headquarters for ground duties in the Directorate of Development and Acquisition. In 1972 he graduated from George Washington University in Washington, D.C. with a master's in management. That same year he graduated from the Industrial College of the Armed Forces in Washington, D.C. He was then assigned to the Aeronautical Systems Division at Wright-Patterson AFB and served two years as System Program Director for the F4/RF4 Phantom aircraft. In August 1974 he was reassigned within the Aeronautical Systems Division as Commander of the 4950th Test Wing, which operated and maintained numerous testbed aircraft. In June 1977 he returned to Air Force Headquarters as Assistant Director, Tactical Systems Test and Evaluation, in the Office of the Director of Defense Research and Engineering. He was promoted to the rank of brigadier general on 17 July 1977. Three months later he became Special Assistant for the General Dynamics F-16 Fighting Falcon. In October 1979, he was appointed Director of Development and Production in the Office of the Deputy Chief of Staff, Research, Development and Acquisition, Air Force Headquarters.

B/Gen. Bell retired from the Air Force on 1 September 1980, and went to work for Northrop Grumman on a top-secret development program which led to the B-2 Spirit stealth bomber. He died on 14 February 2011 as a result of a pulmonary

B/Gen. Tommy I. Bell, USAF. (Photo courtesy Carolyn Bell Phillips)

embolism at the Providence Little Company of Mary Medical Center in Torrance, California. He was 80 years of age. His many decorations and awards included the Legion of Merit, Distinguished Flying Cross, Bronze Star Medal, Meritorious Service Medal, Air Medal with 14 Oak Leaf Clusters and NASA's Distinguished Service Medal.[20]

JOHN K. COCHRAN, USMC

John Kenneth Cochran was born on 23 July 1931 to Frank Miller Cochran and Jesse Hughes Cochran at the Jewish Hospital in St. Louis, Missouri. The family resided in Kirkwood, Missouri, where Frank was employed as a coal salesman and his wife as a school teacher.

Adventuresome like his mother, John attended local schools and graduated from Kirkwood High School in June of 1949. He was subsequently awarded a NROTC scholarship to attend Northwestern University in Evanston, Illinois, from which he graduated on 4 June 1954 with a degree in industrial engineering and a commission as a 2nd lieutenant. After basic training at the Marine Corps station in Quantico, Virginia he attended Naval Flight School at Pensacola, Florida and Corpus Christi, Texas. He was promoted to 1st lieutenant on 4 December 1955, and got his wings as a naval aviator on 1 June 1956. Fifteen days later he married Patricia Murphy of St. Louis, Missouri. The newlyweds then moved to Cherry Point, North Carolina where Cochran had been assigned to VMF-114 ("Death Dealers"), an all-weather fighter squadron that was transitioning from the F9F-8 Cougar to the F4D-1 Skyray. It was deployed to Guantanamo Bay Naval Base and to Puerto Rico, conducted a number of short deployments, and concluded with a Mediterranean cruise on USS *Franklin D. Roosevelt* (CV-42) as part of Carrier Air Wing 1.

On 1 January 1959, Cochran was promoted to the rank of captain. The following year he was sent to Patuxent River, Maryland, to attend the Navy Test Pilot School as a member of Class 26, graduating on 21 October 1960. One of his fellow students, Lt. Alan Bean, went on to become a NASA Group 3 astronaut. Cochran remained at the Naval Test Center as a project officer for the Weapons System Test Branch, and during this period he experienced a dramatic flameout approach in a Vought F8U-2NE Crusader.

In May 1962, Cochran, along with two other officers (and later NASA astronaut aspirants), Bill Geiger and Ken Weir, submitted a proposal to set a round-the-world speed record. Two months later a memo arrived from the Commandant of the Marine Corps denying the request owing to "lack of resources, especially GV-1 [Lockheed Hercules] aircraft, and the requirements to realize the earliest possible introduction of the F4H [Phantom] into the fleet squadrons militate against such a flight at this time. Additionally, the staging of personnel and equipment necessary to support such a mission are considered excessive to the direct benefits to be derived." Nevertheless, the three pilots were commended for their proposal. On 11 October 1962 Cochran became an associate member of the Society of Experimental Test Pilots.

Capt. John K. Cochran, USMC. (Photo courtesy Kenneth Cochran)

Then, in the fall of 1962, Cochran was transferred to El Toro, California where he reported for duty to Marine Attack Squadron VMF(AW)-513 ("Flying Nightmares"). During his time there, the squadron transitioned from the F4D-1 Skyray to the F4B-1 Phantom II, training at the base through October 1964. Redesignated as VFMA-513, the squadron was shipped to Atsugi in Japan, and in June 1965 a further deployment took it to Da Nang, Vietnam for five months of combat duty. For a time, it was the only Marine jet fighter squadron involved in that conflict. In August 1965 it provided support to the 7th Marine Regiment in Operation Starlite, the first major American ground operation of the Vietnam War. The deployment ended in October 1965, and on its return home VFMA-513 was reformed at NAS Cherry Point, North Carolina.

On 20 June 1963 Capt. Cochran applied for NASA's third astronaut intake, but did not progress beyond the final 34 because of an ophthalmology examination. The medical officer reported that "the cataract changes are of a type which we expect to progress. It is our opinion that there is a good probability that, if Captain Cochran were selected as a space pilot now, he would have to be eliminated from the program within the next few years." He was disappointed, but there was some consolation in April 1966 when his best friend, Capt. Gerald ("Jerry") Carr, was accepted into the fifth astronaut group. Seven years later, Carr commanded the third and final mission to the Skylab space station mission and spent a record-breaking 84 days in orbit.

After being promoted to the rank of major in September 1965, Cochran reported in November to the Marine Corps Landing Force Development Center at Quantico, Virginia for a brief assignment prior to commencing graduate school. While there he spent four months writing a white paper detailing the Marine Corps' involvement in the American space program. He graduated from Rensselaer Polytechnic Institute in Troy, New York with a master's in management on 9 June 1967. Following this, he served as Management Services Officer to the Naval Air Rework Facility, Cherry Point. In January 1969 he embarked on his second deployment to Vietnam, this time with VMFA-323 ("Death Rattlers"), which operated F4B-1s out of Chu Lai, South Vietnam. When that squadron left Vietnam, he was given command of VMFA-122 ("Werewolves") which promptly transferred from MCAS Iwakuni in Japan to Chu Lai. Amongst other medals for his activities during this deployment he was awarded the Legion of Merit and the Distinguished Flying Cross.

Back home again and promoted to lieutenant colonel on 1 August 1970, Cochran returned to the Naval Air Rework Facility, Cherry Point, where he served in various management positions. He was then assigned to MCAS Beaufort, South Carolina, to take command of Headquarters and Maintenance Squadron 31, and later command of VMFA-333 (known variously as the "Fighting Shamrocks" and the "Triple Treys") flying Phantom II F4Js. In June 1972 the squadron deployed aboard USS *America* (CVA-66) to conduct combat missions with a dozen F4Js that had previously flown over Vietnam, mostly in the close air support role. This time, however, the "Triple Treys" were engaged in armed reconnaissance. On 11 September 1972 the squadron made its only air-to-air kill when Maj. Lee Lasseter, along with his radar intercept officer Capt. John Cummings, shot down a MiG-21 near Hanoi.

On 18 December, VMFA-333 was ordered to commence instrument-controlled bombing through cloud cover over North Vietnam. Five days later Lt. Col. Cochran and his backseater Maj. Henry ("Hank") Carr were flying Phantom AJ 201 (call sign "Shamrock 201") near the island of Danh Do La, east of the Haiphong Channel inlet on what should have been a routine sortie, escorting a Vigilante of RVAH-6 flown by Cmdr. James Thompson. They had just made a turn and were flying about 480 knots when their aircraft took a direct hit in the belly tank from ground-based 85-mm anti-aircraft fire. The crippled aircraft pitched straight down and both men were forced to eject into the Gulf of Tonkin. Even as they struggled to stay afloat in the water about 400 yards offshore they came under ground fire. Taking shelter behind their raft was, as they put it later, "like hiding behind tissue paper when

The F4J Phantom that was shot down over Vietnam, causing Cochran and Carr (presumably the crew seen in this photo) to eject. (Photo: USMC)

The superbly crafted and finely detailed model of Phantom AJ-201 bearing the names of Cochran and Carr below the cockpit. Compare it with the photograph of the actual aircraft above. (Photo and permission for use courtesy Hobbymaster, Hong Kong)

they're shooting cannons at you". Thanks to a quick response by a rescue helicopter from USS *Horne* (CG-30), Cochran and Carr were rescued. Back on board *America*, it was discovered that Carr had suffered second degree burns on his face and neck, while Cochran had sustained an injury from the ejection shot deemed serious enough that he had to be relieved of his command by Maj. Lasseter, the squadron's Operations Officer. According to the medical report, "The patient ejected from an

F4J aircraft in December 1972 and sustained a compression fracture of L-1, and a cervical cord contusion of C-5, 6 and 7." He was evacuated to Parris Island Hospital, South Carolina, and later awarded the Purple Heart and Bronze Star Medal for meritorious service with the squadron. The squadron was also presented with the Robert M. Hanson Award for its outstanding performance.

Owing to his injury, Cochran was not permitted to fly ejection seat aircraft for an extended period, and was assigned to the MAG-31 staff at Beaufort, South Carolina. Offered the opportunity to attend the War College in Rhode Island, he opted to retire in February 1973 after 20 years of service. He then worked in various management positions for U.S. Furniture Industries, Inc., High Point, North Carolina.

Lt. Col. John Cochran died on 22 September 2008 in High Point from a ruptured abdominal aortic aneurysm, leaving behind his wife Patricia, their three children – Kathleen Cochran Clayton, Marie Cochran and Kenneth Cochran – as well as four grandchildren.

Following his death, Marie became aware of die-cast metal model F4 jets from VMFA-333 for sale. "Interestingly," she points out, "the models bear our father's name and that of his radar intercept officer, Maj. Hank Carr."[21]

DARRELL E. CORNELL

Darrell Eugene Cornell, who would become Northrop's chief test pilot in the early 1980s, was born in Moorhead, Minnesota on 19 August 1932 – but he lived much of his early life in Fargo, North Dakota. After leaving Fargo High School in June 1950 he attended North Dakota State University until June 1953, where he met his future wife, Beverly Jean Mclean.

Beginning in January 1954, he participated in fighter pilot training with the Air Force, as well as combat crew training in the F-86E Sabre from July to August 1955. He then returned to North Dakota State University, graduating in June 1956 with a bachelor's in architectural engineering. Over the next five years he was involved in Air National Guard activities in direct support of Air Defense Command. In parallel, in February 1961 he resumed his university studies and graduated the following year with a master's in mechanical engineering.

In 1962 Cornell affiliated with the Northrop Corporation as an engineering test pilot for the all-weather interceptor F-89 Scorpion, involved in the development and operational employment of a high-altitude missile target. The next year he applied to NASA for the third astronaut group, reaching the final 34 but not being selected. In 1965 he moved to Northrop's Aircraft Division as Chief Production Acceptance Pilot at the Flight Test Unit at Palmdale, California, and participated in the F-5 flight test program. In January 1979 he was appointed lead test pilot for the RF-5E Tigereye, a single-seat tactical fighter/reconnaissance aircraft first flown in January 1982, funded and developed entirely as a low-cost fighter by the company for the export market. In July 1980 he was appointed Chief Test Pilot and Manager of Flight Operations.[22]

The Northrop F-20 Tigershark firing a Maverick air-to-ground missile. (Photo: USAF)

Cornell was in France in 1983, flying Northrop's F-20 in a display during an air show at Le Bourget airport, Paris, and earned the praise of NASA test pilot Fitzhugh Fulton, who noted that Cornell "did a super job. I've always had a lot of respect for his professional abilities."[23] Northrop had only built three of the $150 million high-performance aircraft, also known as the Tigershark, for testing and demonstration. Its extensive flight test program at Edwards AFB was described at the time as extremely successful. While at Edwards, Cornell was featured in an article on test pilots at the air base for the April 1984 issue of *Life* magazine. A two-page color spread showed a gray-haired Cornell lounging in a leather wing chair placed, rather surrealistically, in the desert. Six months later he was killed doing the very job featured in that article.

In September 1984, Cornell conducted another highly praised display of the F-20 at the air show at Farnborough in England. Four weeks later, on 10 October 1984, the 52-year-old Cornell had reached the final venue on this exhausting tour of nineteen countries, and he and the pair of F-20 aircraft were due to return to California two days later. The final display was for military officials of the South Korean Air Force (ROKAF) at the Suwon Air Base in Korea. That fateful day, as usual, he disdained to wear a restricting G-suit.

As part of his normal tactical demonstration, Cornell flew a series of climbing rolls with flaps and gear extended, thus creating considerable aerodynamic drag. He had just completed a simulated strafing run at an altitude of around 300 feet when he pulled the F-20 up to an altitude of 1,200-1,800 feet and began a 360-degree aileron roll, again extending the flaps and landing gear. This was a regular part of the display of the aircraft's capabilities, which usually ended with the F-20 about 1,000 feet near the downwind end of the runway, from which a base leg was entered for landing. But on this occasion the roll was not completed and the aircraft continued

Northrop test pilot Darrell E. Cornell. (Photo: Northrop Corporation)

inverted in an arc, stalling with insufficient altitude to effect a recovery. The aircraft plowed into rice paddies near the runway, throwing Cornell's body clear on impact. On hearing of Cornell's death, the shocked company chairman, Thomas V. Jones, said the test pilot community had lost "the best of the best".[24]

A three-month investigation into the cause of the accident, conducted largely by Northrop officials, declined to make public its report, claiming it to be "proprietary". Air Force advisors who participated in the investigation later said "the investigation was thorough" and "reflects logically derived conclusions". Northrop found that the aircraft and all its systems functioned properly and, in the absence of any evidence of airframe failure, attributed the crash to pilot error. This left open the question as to whether Cornell, who was not wearing a G-suit, had suffered a black out as a result of the high forces imposed by his final maneuver.[25]

Seven months later, on 14 May 1985, a second F-20 crashed at Goose Bay, Labrador, Canada, killing Northrop test pilot David Barnes while practicing for the forthcoming air show at Le Bourget. The F-20 program was canceled in November 1986. The sole surviving aircraft in the series was retired from service and donated to the California Science Center in Los Angeles, where it is on display today.[26]

A member of the Society of Experimental Test Pilots since May 1983, Darrell Cornell had accumulated over 6,700 hours of flight time, including 3,300 in jets. He was survived by his wife, Beverly; son Jeffrey; daughters Gail, Linda and Deborah; his mother Edith; brother Duane; sister Karen; and granddaughter Michelle.

REGINALD R. DAVIS, USAF

Reginald R. ("Rod") Davis was born on 9 March 1933 in Klamath Falls, Oregon, the son of Rex Q. Davis and the former Emma Mariah Grimes, and later an older brother to Patrick. His father would prove a strong influence in Rod's life, having received his Army Air Corps wings in 1929 after completing pilot school while attending high school at Hill Military Academy in Portland, Oregon. When World War II broke out, even though he was comfortable in civilian life and had a family, Rex volunteered for service. On being informed that he was too old to fly, he went into the Armored Branch of the Army. As Rod Davis explained, his father's life provided him with his own motivation to attend the U.S. Military Academy at West Point and become an aviator. "In addition, due to my natural interest in airplanes, from the time I was a small boy I always wanted to be a fighter pilot."

Davis attended Roosevelt Elementary, Fremont Junior High, and Klamath Union High School, with temporary attendance in both Louisville, Kentucky and Leesville, Louisiana when in fourth grade during the war while his father was stationed at Fort Knox, Kentucky and Camp Polk in Louisiana. He graduated third in his high school class and lettered in three sports. He then attended the USMA, playing football in his third and second class years, and as a first classman he was a battalion commander in the cadet chain of command. He graduated in the top ten per cent of the class with a bachelor of science degree and was commissioned a 2nd lieutenant in the USAF in June 1955. "I married the former Mary Mauch of Tulelake, California on graduation day," he added.

Davis subsequently attended Primary Flight Training at Marianna, Florida, where he flew the Piper PA-18 Super Cub and T-6 Texan aircraft, followed by Basic Flight Training at Greenville, Mississippi flying the T-28 Trojan and T-33 Shooting Star. After receiving his wings he received Combat Crew Training in the F-89 Scorpion at Moody AFB in Valdosta, Georgia. While there, he was promoted to 1st lieutenant in December 1956.

His next posting was to the 460th FIS at Portland International Airport, Oregon where he flew the F-89, F-102 Delta Dagger and T-33 from 1957 to 1959. "When Sputnik was launched by the USSR, I was accepted to attend graduate school at the Air Force Institute of Technology, Wright-Patterson AFB, Dayton, Ohio. While there I was promoted to captain in December 1959 and my son Scott was born December 1960." He graduated in 1961 with a master's in astronautics, meanwhile maintaining his currency in the T-33. He was assigned to the Warfare Systems School at Maxwell AFB, Alabama, where he was an instructor in space dynamics and nuclear weapons courses for staff officers. His daughter Laura was born in April 1962.

In 1963, while at Maxwell AFB, Davis undertook the one-week-long physical as a candidate for NASA's third astronaut group but was disqualified "for a problem that was surgically correctable but was not disqualifying as an Air Force pilot." He chose to stay as a fighter pilot in the USAF, without surgery. Then, as an additional duty at Maxwell, he worked as an instructor pilot in the T-33. His time there was completed as a student at the Air Command and Staff College, where he was further promoted to the rank of major in March 1964.

Capt. Reginald R. ("Rod") Davis, USAF, then teaching at the Warfare Systems School. (Photo courtesy R.R. Davis)

Perrin AFB, Texas proved to be the next assignment, renewing his currency in the F-102, and then it was to Goose Air Base in Labrador, Canada, where he was a Flight Commander and later Assistant Operations Officer in the 59th FIS, a squadron larger than most, with 39 assigned F-102s and several T-33s. In 1965 he participated as one of the four flying team members from the 59th FIS in the World Wide Gunnery and Weapons meet (now known as "William Tell") at Tyndall AFB, Florida which was sponsored by the Air Force. In May 1966, while at Goose Bay, his younger son Mark was born.

In 1966 Davis was posted to MacDill AFB, Florida for Combat Crew Training in the F4 Phantom. While there, he was promoted to lieutenant colonel in December 1966. Next he was assigned to the 33rd Tactical Fighter Wing at Eglin AFB, Florida. He was initially Operations Officer of the 25th TFS, and then the 4th TFS. He later took over command of the 4th TFS, one of the two squadrons that were the first to transition from the F4D to the F4E, which was the first model to have a 20-mm gun installed internally. In March 1969 the squadron relocated as a unit to Da Nang Air Base in the Republic of Vietnam. In August, he was reassigned as Assistant Division Chief, Current Plans Division of the Tactical Air Control Center, Headquarters 7th Air Force at Tan Son Nhat Air Base, where his primary duty was the scheduling and coordination of the daily air activity over Vietnam.

Maj. Davis participating in the "William Tell" gunnery meet at Tyndall AFB, Florida. (Photo courtesy R.R. Davis)

Returning home in 1970, Lt. Col. Davis served as Chief, International Relations Branch, Directorate of Plans, Deputy Chief of Staff/Operations and Plans at USAF Headquarters. This branch was responsible for developing and coordinating the Air Staff positions on prisoner-of-war issues and the Medal of Honor Awards that would ultimately be determined by the Joint Chiefs of Staff. In addition, it was involved in International Law of the Sea negotiations and the Strategic Arms Limitation Talks. In December 1970 he was promoted to full colonel. He returned to Southeast Asia in May 1972 and served with the 8th Tactical Fighter Wing, based at Ubon Royal Thai AFB, Thailand, initially as Deputy Commander for Operations, then as Vice Wing Commander. At that time it had six fighter squadrons, four permanently assigned and two on a rotational basis that came from the United States and were cycled every six months. The 8th TFW also had an AC-130 Spectre gunship squadron attached. There was a Forward Air Controller detachment of OV-10 Bronco aircraft on the base, but it was not operationally under 8th TFW control. During his two Southeast Asia tours, Col. Davis flew a total of 277 combat missions, more than 100 of which were over North Vietnam.

After his tour in Thailand in 1973, Davis was assigned as Director of Inspections, Inspector General, Headquarters Pacific Air Forces, Hickam AFB, Hawaii. In March 1976 he was given command of the 8th TFW, which had been moved to

Col. Davis, June 1974, then Director of Inspection, Pacific Air Force, Hickam AFB, Hawaii. (Photo courtesy R.R. Davis)

Kunsan AB, Korea. During this tour, the North Koreans murdered two U.S. Army soldiers in the Demilitarized Zone and the 8th went to full preparedness. Eventually the situation cooled and the wing resumed routine operations. In April 1977 he returned to the United States to take command of the 474th TFW at Nellis AFB, Nevada, in charge of transitioning the wing from F-111 Aardvark aircraft to F4s. In October 1978 he was sent to the Pentagon in the Directorate of Operations and Readiness, initially as Assistant Deputy Director for Readiness Development and in May 1979 as Assistant Deputy Director for Operations.

Col. Davis retired from the Air Force on 31 January 1983 with the aeronautical rating of Command Pilot. After taking a year off, in 1984 he enrolled at Willamette University in Salem, Oregon to study law, completing the course in December 1986 and becoming a member of the Bar in 1987. Meanwhile, he and Mary had divorced in 1985. He went into private practice in his hometown of Klamath Falls in the firm of high school classmate William P. Brandsness, focusing on business and contract law. He went to work for Klamath County in November 1990 as the County Counsel attorney for civil matters, and stayed in that position until 2005 when he retired. His replacement as County Counsel was also in the Air National Guard, and when he was called upon for one year of active duty in Iraq, Davis reinstated his lapsed license in order to fill in that position until his replacement returned.

In September 2000, Rod Davis married the former Kathie Hull and now has an extended family. "I have two sons, Scott and Mark, and daughter Laura; stepdaughter Kristy Yeadon and stepson John Hull; two granddaughters, three grandsons, a great-grandson, and three step-grandsons."

Amongst the decorations awarded to Col. Davis were two Legions of Merit, three Distinguished Flying Crosses, the Bronze Star, the Meritorious Service Medal, and a total of 21 Air Medals. He is a member of the Order of Daedalians, the Air Force Association, and the Red River Valley Fighter Pilots Association.[27]

DONALD G. EBBERT

Donald George Ebbert was born on 24 May 1930 in Kansas City, Missouri. He was the second son of Fredrick Henderson Ebbert and the former Kathro Marion Jones, whose American ancestry can be traced all the way back to 1620 and Miles Standish, a prominent passenger aboard the *Mayflower*. Like his older brother Richard, Donald spent much of his youth in the Boy Scouts. He also enjoyed spending his summers at Conneaut Lake Park in Pennsylvania, where he was a lifeguard and a pin-setter in the local bowling alley – in the days before automated pin-setters. He and several of his lifelong buddies also enjoyed pole vaulting.

He graduated from Paseo High School in Kansas City in 1947. After a short stint at a local junior college he was accepted to Purdue University, Indiana, graduating in June 1952 with a bachelor's in engineering. Having joined the ROTC as a student, he immediately entered the Navy and was commissioned in September, during his final midshipmen cruise. He was assigned to USS *Sicily* (CVE-118), which was operating off Korea. In May 1953 he was ordered to Pensacola, Florida for basic flight training. After initial training in an SNJ, he received instrument training in a Twin Beech and advanced training in the F6F Hellcat before finally transitioning to the T-33 Shooting Star for jet training. Around this time he also purchased his first private aircraft, an Aeronca Chief, for the princely sum of $400.

On completion of his flight training, Ebbert was assigned to NAS Miramar in San Diego, California, flying the FJ-3 Fury. It was there that he met his future first wife Elizabeth ("Betty") Wright at a party in a La Jolla beach house where he and several of his Navy buddies were living. They were married on 6 December 1956 at La Jolla Presbyterian Church.

Deciding that it was time to leave full-time service, Lt. Cmdr. Ebbert resigned and went to work for aircraft manufacturer Convair as a flight test engineer on the F-102 Delta Dagger and F-106 Delta Dart programs. However, he continued to serve with the Navy on weekends as a reserve officer in VF-771 based at NAS Los Alamitos in Long Beach, California, flying the F9F-6 Cougar. He and Betty celebrated the birth of their first child, Gregory, in October 1957, and their second son, Daniel, in December 1959. He then took a year off to gain his master's in engineering from San Diego State University.

Donald G. Ebbert during his service with the Navy. (Photo courtesy Greg Ebbert)

Although Ebbert did not apply for NASA's second group of astronauts in 1962 he knew some of those who did, and, suitably armed with the necessary qualifications, decided to try for the third astronaut group in 1963. He reached the final list of 34 but was not selected. In the meantime he continued to work for Convair, now involved in the development of the Atlas intercontinental ballistic missile, and flying the Douglas A4 Skyhawk with the Navy Reserve.

In 1966 he decided to get back into the airplane business and that October joined Lockheed in Palmdale, California to work on the Italian F-104 Starfighter program as a flight test engineer. While there he became good friends with Ken Weir, who had unsuccessfully applied for NASA's second astronaut group. On 23 January 1968, not long into Ebbert's career at Lockheed, USS *Pueblo* (AGER-2), a radio intelligence gathering ship, was captured by North Korean forces, creating a major incident. Four days later VA-776 Reserve Attack Squadron based at NAS Lemoore, California was recalled to active duty and he was assigned as its Executive Officer with the rank of commander. It was initially scheduled to deploy to Vietnam but was eventually stood down in September 1968. He was then given the choice of becoming a civilian once again or remaining in the Navy Reserve until his two-year call-up was completed. He accepted a billet at the Naval Air Test Center at Patuxent River,

Early Atlas rockets on the Convair production line at Plant 119, San Diego.

Maryland, where he enjoyed flying a variety of airplanes including the F8, A7, T-38, A4, OV-10, T-28, and gliders. He was so delighted with gliders that he started up a sailplane school at the base which attracted local high school students seeking flying lessons. According to Greg Ebbert, the best part of his father's time at Pax River began when he was offered a two-year extension with a place at the Navy Test Pilot School. He rounded out his final two years of call-up duty at NTPS, graduating as a member of Class 54 on 13 February 1970. It was during this time, in October 1969, that his third child, Cindy, was born. "He always complained that he wasn't sure if she was a good deal or not," son Greg laughingly recalled, "as she had cost him a whole five dollars to be delivered at the Navy hospital on base."

Ebbert returned to Lockheed in Palmdale, and in early 1972 was assigned as an engineering test pilot on the S3A Viking. As this program wound down in 1978, he became an engineering test pilot on the Lockheed L-1011 airliner program. As Greg Ebbert recalls, "It was during this period of time that his interest in sailplanes really took off and he flew many flights including one in an old rented Schweizer 1-26 that he took out one day and caught a mountain wave off the Tehachapi Mountains and ascended all the way to 26,000 feet. He had a small oxygen bottle to take breaths from and said it was very, very cold."

The L-1011 program was canceled in 1983, but Ebbert wasn't ready to retire at just 53 years of age. Fortunately, Lockheed had just purchased a Gulfstream III (G3) executive jet and was in the process of setting up a flight department to operate the aircraft. He and several other L-1011 pilots were given the opportunity to fly the G3 and a pair of Kingair turboprops out of Burbank, California. Over the years, he flew the G3 all over the world, ferrying Lockheed executives on business trips. He retired in December 1989 and joined Betty in a home he had earlier purchased in San Diego Country Estates outside of Ramona, California.

In retirement he enjoyed traveling in his motor home, sailing (with many trips to the British Virgin Islands), playing golf, snow skiing, flying gliders, camping, hiking, and just enjoying life. In July 1995 Betty, his wife of nearly 39 years, passed away from cancer. Not long after, he met his second wife, Patricia Jackson, and they were married in June 2000. The couple enjoyed eight years together, traveling extensively and taking many ocean cruises, including one with all his children to Alaska in July 2007. Sadly, Patricia succumbed to cancer in December 2008. He passed away on 19 November 2011 at the relatively young age of 81, considering that his father lived to be 94 and his mother 105.

Acting as spokesperson for the family, son Greg says they miss the man who had so much influence in their lives. "Beside the fact that our father did so many amazing things in his life, all three of his children are grateful to him for developing in us a love of the outdoors and being active." Greg currently lives in Palmdale, California and is planning on retiring this year from Lockheed after a 32-year career working on aircraft such as the U-2, SR-71, F-117, F-22 and numerous Skunk Works programs. Younger son Dan lives in Castle Rock, Colorado, where he owns his own business. Daughter Cindy and her husband Dave reside in Silverthorne, Colorado, where she works for the Forest Service.[28]

GEORGE M. FURLONG, JR., USN

The no-nonsense movie star and renowned patriot John Wayne was once quoted as saying, "America is Babe Ruth pointing into the stands before he hits a home run, and Merle Haggard singing 'Okie from Muskogee'." The city of Muskogee lies in Eastern Oklahoma, at the confluence of the Arkansas, Verdigris and Grand Rivers. It was in this formerly raucous cow town and historic rail city that another modern-day patriot came into the world.

George Morgan Furlong, Jr., the son of George Sr. and Anna Moore Furlong, was born in Muskogee on 23 November 1931. A few days after his seventeenth birthday, still in high school, he joined the Naval Air Reserve V-6 program. After graduating in June 1932 he entered active duty with the Navy, serving as a PBY-5A aircrewman (flight engineer) for over a year prior to being sent to the Naval Training Center in San Diego, California and then NAS Alameda, California. From December 1951 to June 1952 he attended Naval Academy Preparatory School at Bainbridge, Maryland, where he was an instructor in the "speedup program" for students arriving late in the school year before Naval Academy examinations. Then,

George Furlong in his flight gear. (Photo courtesy RADM George Furlong)

as a fledgling midshipman, he passed through the portals of the U.S. Naval Academy at Annapolis for four years of study and instruction.

Following his graduation with honors on 1 June 1956 he undertook three months' temporary duty at NAS Pensacola, Florida prior to flight training there in September. Next was advanced instruction at nearby Whiting and Barin Fields and NAS Corpus Christi, Texas. After receiving his wings as a naval aviator on 30 July 1957 he was assigned to VF-121 ("Pacemakers") at NAS Miramar, California, which operated the Grumman F-11 Tiger and North American FJ-3 Fury. In June 1958 Lt. (jg) Furlong was sent as a replacement pilot to VA-156 ("Iron Tigers"), serving on USS *Shangri-La* (CV-38) with Carrier Air Wing 11 on WestPac deployment. In January 1959 the squadron was redesignated VF-11 ("Sundowners") and he went on to complete two further WestPac deployments, this time on USS *Hancock* (CV-19).

At the completion of the squadron's second deployment in July 1961, Lt. Furlong entered U.S. Naval Postgraduate School in Monterey, California, graduating with his bachelor's in aeronautical engineering. His next assignment in June 1963 was a two-year duty as F4 (Phantom II) Project Officer at the U.S. Naval Weapons Evaluation Facility at Kirtland AFB, New Mexico. Shortly after commencing this duty he was selected by the Navy as an astronaut candidate for NASA's third group of astronauts, but was not chosen. He was not overly disappointed. "For the next

The Change of Command ceremony at NAS Miramar, California when Cmdr. Furlong took command of VF-142 in 1970, with his great friend and NASA astronaut Gene Cernan (right) in attendance. (Photo courtesy RADM G.M. Furlong)

two years I flew Mach 2-plus test flights in the Phantom II, evaluating a number of weapons at the outer edges of the F4 envelope," he recalled. "What a job; flying one of the fastest aircraft in the world."

In November 1965 Lt. Cmdr. Furlong was told that he had again been chosen as an astronaut finalist, this time in NASA's fifth group selection. Having been through the process once before, he had high expectations of being selected the second time around but once again he missed the final cut. He believes it may have been due to his height, at six foot one. When the names of the nineteen selected astronauts were announced on 4 April 1966, Furlong was on the first of five combat deployments to Vietnam. In June 1965 he was assigned as Aide and Flag Secretary, Commander, Carrier Division 9. In mid-1967, now promoted to the rank of commander, he joined VF-142 ("Ghostriders") flying Phantom F4B/J aircraft over Vietnam, initially as its Maintenance Officer, then Executive Officer and finally Commanding Officer (1970-1971). During this period VF-142 was named the outstanding F4 fighter squadron in the Navy.

From April 1967 to May 1970, VF-142 made three extended combat deployments to Vietnam on USS *Constellation* (CVA-64), during which time they transitioned from the F4B to the F4J. In September 1972 they were again deployed to Vietnam, this time on USS *Enterprise* (CVN-65). Altogether, Cmdr. Furlong flew 226 combat missions in the F4 during a total of five deployments to Vietnam and received the Navy League's John Paul Jones Award for Inspirational Leadership while in charge of VF-142.

His next assignment was Fighter Training Officer at COMNAVAIRPAC, but this tour was cut short by orders to attend a six-month course at the Navy Nuclear Power School at the Mare Island Naval Shipyard in Vallejo, California. After the academic phase of that program he returned to AIRPAC as Director of Fleet Introduction for the F-14 Tomcat, a job that involved a lot of traveling between San Diego, Washington, D.C. and Bethpage, New York, as the aircraft approached fleet introduction. In mid-1973 he became Commander, Carrier Air Wing 14, the first F-14 and S3 Viking air wing on USS *Enterprise* (CVN-65), a post he retained until relieved in Hong Kong in late 1974. During this period he maintained currency in numerous aircraft types and carrier qualification in the F-14, A6, EA6, A7, and S3, in the process becoming the first fleet aviator to day-and-night qualify in the Tomcat. "The fleet introduction of the F-14 was accident-free and on schedule," he noted with pride.

Between November 1974 and February 1976, Capt. Furlong commanded the Pearl Harbor-based fleet oiler USS *Ponchatoula* (AO-148) with an eight-month WestPac deployment and subsequent overhaul. He then took command of USS *Independence* (CV-62) in Norfolk, Virginia, seeing it through a seven-month Mediterranean cruise and remaining in command for the ship's conversion to F-14 capability in the yard in Portsmouth, New Hampshire, and the post-overhaul sea trials. In September 1978 he began a two-year tour as Chief of Staff, U.S. Sixth Fleet in Gaeta, Italy. Next was a Pentagon assignment. "[This] was my first and only in Washington, D.C. I served as OP-50W (Ops analysis) from September 1980 to May 1981. It is interesting to note that out of a 37-year career in the Navy, I spent less

George M. ("Skip") Furlong, Jr. as Commander, Fighter Airborne Early Warning Wing, U.S. Pacific Fleet at NAS Miramar in 1983. (Photo courtesy RADM G.M. Furlong, Jr.)

than nine months in D.C. and I was on the move worldwide for most of that time as a trouble-shooter for Naval Aviation."

Next, Capt. Furlong assumed the position of Commander, Fighter Airborne Early Warning Wing, Pacific Fleet, a command he described as "the most fun of my naval career. The command, back then, included my HQ base, Naval Air Station Miramar, my staff, Naval Air Facility El Centro, five carrier air wings including sixteen fighter squadrons and eight AEW squadrons, Top Gun, one experimental fighter squadron (VX-4), and one composite squadron split between Hawaii and Cubi Point [in the Philippines]. Naturally, I got to fly with all of them."

In August 1983 he was assigned the role of Deputy Chief of Naval Education and

Two distinguished guests at the National Flight Academy in 2010 were former Apollo astronauts Gene Cernan and Neil Armstrong, shown here flanking RADM Furlong. (Photo courtesy RADM G.M. Furlong, Jr.)

Training, headquartered at NAS Pensacola, Florida. Soon thereafter his wife Ryland ("Ry") began treatment for a serious illness, which prompted him to retire from the Navy on 1 January 1986. Fortunately she recovered from her illness. Over the next decade Furlong flew small aerobatic aircraft for recreation along with some fighter aviation friends. In 1990 he was inducted into the Arkansas Aviation Hall of Fame. Over his service career he had accumulated some 4,500 flight hours and 930 fixed carrier/ship landings, and his many awards include twelve Air Medals (strike/flight), two Legions of Merit, two Meritorious Service Medals, three Navy Commendation Medals with combat "V", and two Navy Achievement Medals.

Following his retirement from the Navy, RADM Furlong assumed the position of Executive Vice President for the Naval Aviation Museum Foundation, a not-for-profit organization that has built and supports the development of the National Naval Aviation Museum in Pensacola. He retired from that position in 1997, although he "re-enlisted" in 2001 to assist fund-raising for the Phase IV Museum expansion and the National Flight Academy. He and Ry have two sons, George and Bill (one of whom served his own Navy tour and went on to captain a Boeing 757 for FedEx) and three grandchildren.[29]

SAMUEL M. GUILD, JR., USAF

These days, Murt Guild (rhymes with "wild") thrills more to a stiff sea breeze and a straining spinnaker as he sails yachts off Pensacola, Florida, but he recalls his days as a test pilot and as a would-be astronaut with undisguised fondness and a twinkle in his eyes.

Samuel Murton ("Murt") Guild, Jr. was born on 3 January 1930 in Staunton, Virginia, which was the birthplace of U.S. President Woodrow Wilson. His mother, the former Beatrice Ogg, sadly died of tuberculosis when he was three years old and for the most part he was raised by his paternal grandparents after they were invited to move into the family home. In 1938 they all relocated to Clifton Forge, Virginia. His grandfather died in 1940, about the same time that his father was recalled to service and stationed at Bolling Field, Washington D.C. prior to being deployed to England. Murt and his grandmother moved to Delta, Pennsylvania where she had connections, and lived in a boarding house for the next two years. He later attended the Miami Military Academy, a military-style boarding school in Miami Shores, Florida.

Young Murt would often travel alone by train to visit his father at his apartment in Anacostia, Maryland, near Bolling Field. One vivid recollection dates back to 28 June 1942, when his father (recently remarried) was about to be posted overseas. On that day he and his father watched from a grandstand at Bolling Field as Gen. "Hap" Arnold decorated B/Gen. Jimmy Doolittle and the first "Tokyo Raiders". "It was one of those things one doesn't forget and [it] probably helped attract me to the newly organized Air Force."

In 1946 he was enrolled at what was then called Florida State College for Women, but would soon be renamed Florida State University because there were some 3,000 women and 300 men in attendance. His father, who fought in Europe during the Second World War, was eager for young Murt to go to a military college. In 1947 he had an appointment at the U.S. Coast Guard Academy, but also had backups at both the U.S. Military Academy at West Point and the U.S. Naval Academy at Annapolis. He recalls ending up at West Point by chance. "While traveling through Washington on the way to New London, my father decided to stop by the Pentagon to check on the status at West Point. It turned out that the two higher alternates had declined at the last moment, so I chose West Point the night before entry."

Guild entered West Point with Cadet Company B2 in the Class of '51 – the same class as future astronaut and moonwalker, Buzz Aldrin. "Sure I remember him," says Guild. "We were a relatively small and tight class. He was a very good athlete in addition to his academic talents." Guild did not participate in collegiate sports; his favorite extra-curricular activity was as a member of the Cadet Chapel Choir and the Glee Club.

On graduating 79th in general order of merit from a class of 475, Guild went into the Air Force and took his basic training at Hondo AFB, Texas flying the T-6 Texan. Among his fellow students were some twenty West Point graduates and forty Naval Academy graduates who had elected to join the Air Force, including another future astronaut, Jim Irwin. From there Guild was sent for advanced training to

Williams AFB, Arizona. He flew the Lockheed T-33 and then the F-80A/B Shooting Star, and got his pilot wings on 2 August 1951. "I was scored in the top three in the advanced training," he says, "but because of my diminutive height I got diverted from F-86s at Nellis AFB, Nevada to F-94s at Tyndall AFB in Florida."

He later joined the 319th FIS, the first all-weather fighter-interceptor squadron of the Air Force, which was then at Suwon Air Base (K-13) in Korea, and he flew night missions in F-94 Starfires between the Yalu and Chongchon rivers in North Korea. At the time of the armistice he had 39 combat missions to his credit. He later served with the 339th FIS at Chitose Air Base in Japan and transitioned to the F-86D Sabre all-weather fighter. On his return to the United States he was assigned to the 15th FIS at Davis-Monthan AFB, Arizona, where he was chosen to lead a flight of six F-86Ds through an atomic cloud in the Atomic Energy Commission's May 1957 Operation Plumbbob-Boltzmann test explosion near the Tonopah Test Range in Nevada.

In August 1958, Guild entered the USAF Test Pilot School Class 58-C at Edwards AFB, California, graduating on 24 April the following year along with future NASA astronauts Ed Givens and Tom Stafford, as well as Group 2 aspirant Al Uhalt. His next posting was to the Fighter Test Section of Test Operations, Air Proving Ground Center at Eglin AFB, Florida, but he was granted time off in the fall of 1962 to return to Florida State University. Sadly, his stepmother Dorothy (née Thiele) passed away in the spring of 1963 during his comprehensive exams studies, so he was not able to complete his degree course. Capt. Guild was back at Eglin when NASA announced in June 1963 that applications were being accepted for its third group of astronauts. He volunteered and was recommended by his superior officers,

USAF Test Pilot School Class 58-C and instructors. Murt Guild is fifth from left, front row. (Photo: USAF TPS, Edwards AFB)

A recent photo of Murt Guild. (Photo credit: Pensacola Yacht Club)

together with Capt. Robert Vanden-Heuvel, who was also attached to the fighter section. Both men made it to the 34 finalists but were not selected. In a third stint at Florida State University in the summer of 1965 Guild finally got his master's degree in mathematics – with a minor in physics.

After serving with the Air Force Advisory Group for the Republic of Vietnam, in 1969 he was appointed as Station Commanding Officer of the Grand Turk Auxiliary Airfield, Turks and Caicos Islands, which was situated in the Eastern Test Range and was heavily involved in supporting the Apollo 11, 12, 13 and 14 launches, "with at least [the] theoretical capability to terminate thrust should South African population centers be threatened".

In 1970 Guild was selected to attend the Armed Forces Staff College in Norfolk, Virginia, then assigned to the Air Force's Special Weapons Center at Kirtland AFB, New Mexico. His final service assignment was to the Armament Development Test Center, Eglin AFB in 1976.

Now retired from the Air Force, Lt. Col. Murt Guild lives in Pensacola, Florida.[30] Sailing is his true passion these days, but he works part-time at Ascend Performance Materials, the world's largest nylon production factory.

JAMES E. KIRKPATRICK, JR.

James Earl Kirkpatrick, Jr. (also known as "Jim" or "Kirk") was born in Manchester, Iowa on 12 November 1929. His interest in flying began at a very early age. Always curious to see how things worked, he would design all manner of paper airplanes. Looking for more of a challenge and the best design, he decided the ultimate test was to climb onto the roof and set them on fire before takeoff; an interesting experiment, and fun, until his mother caught him.

His father, Dr. James Earl Kirkpatrick, Sr., received his Ph.D in education in 1933 at the University of Iowa and became head of the Education Department. His mother, Mabel Kate Dorman, graduated from Iowa State Teachers College in Cedar Falls, Iowa, taught in several rural schools, and was secretary/assistant to Dr. Kirkpatrick. Young Jim attended grade and middle school and in 1942, prior to commencing high school, achieved a treasured ambition when he earned the rank of Eagle Scout. He then attended East High School in Sioux City for three years from 1944 to 1947, and also joined the Iowa Air National Guard. The family, including his younger brother, Richard Thomas, moved to Tulsa, Oklahoma in 1947 when his father took a job at Tulsa University as professor of education. In 1958, with both sons through college, his parents moved to Spearfish in the Black Hills of South Dakota, where his father was Dean of the Black Hills State Teachers College.

In 1948, Kirkpatrick finished his final year at Will Rogers High School in Tulsa, Oklahoma and won a competition which awarded him a full tuition scholarship to the Spartan College of Aeronautical Engineering in Tulsa. He also joined the Oklahoma ANG. He graduated from Spartan College in July 1950 with an associate's degree in aeronautical engineering. It was during these years he met Patsy "Pat" Jane Johnson, his future wife, at a Sunday School picnic where he offered her a ride in his yellow Ford Model-A. She was a choir member at the United Methodist Church which they both attended and where they were married in 1952.

Soon after graduation, Kirkpatrick was called up for active duty in the Oklahoma ANG during the Korean conflict, where he pursued his dream of flying. In 1951 he got primary pilot training at Goodfellow AFB, near San Angelo, Texas as a member of Class 52-E and had his first solo flight that year. Basic training followed at Vance AFB in Enid, Oklahoma, where he was rated number one for academics in his class and gained his wings. The Korean conflict ended while he was still in training, so he never had to engage in combat. He returned to Tulsa and attended the University of Tulsa from 1952 to 1954, graduating with a bachelor's in aeronautical engineering. Then he and his wife moved to Wichita and he joined the Kansas ANG. In June 1954 he joined the Military Division of the Cessna Aircraft Company as an aerodynamics engineer, preparing data for Phase I testing of the XT-37 jet trainer, including the calculations on the aircraft's performance and stability. In January 1955 he became a flight test engineer for the airplane, undertaking in-flight observation, data reduction, test planning, and report writing on its engine and systems. In the (more than) eight years that he spent on the T-37A, he received a string of promotions, advancing to Production Test Pilot and then to Engineering Test Pilot, flying aggravated spin tests and tip tank and armament development.

In 1958 Cessna sent Kirkpatrick to the USAF Experimental Test Pilot School at Edwards AFB to refine his skills in high-performance aircraft. He joined Class 58-B, flying and then reporting on the characteristics of a number of aircraft, including the F-86E/F Sabre, T-28A Trojan, B-57E Canberra, TF-102 Delta Dagger and the T-33A Shooting Star. One member of his ten-strong class was Ed Givens, who later

James Kirkpatrick at Edwards AFB, TPS Class 58-B. (Photo USAF courtesy Kirkpatrick family)

became an Air Force and NASA astronaut. On graduation day, 2 October 1958, Kirkpatrick was rated number two in his class. The following year he also attended the ANG Jet Instrument School, Class 59-S, graduating first in his class. Two years later, in 1961, he achieved his Senior Pilot rating and Instrument Green Card with the Air National Guard and retired as a captain in the Air Force Reserve. In his fourteen years with the ANG from 1948 to 1961 he flew F-51, F-80, F-86L and F-100 aircraft.

In December 1962, he resigned from Cessna and joined the Lear Jet Corporation in Wichita, Kansas as Engineering Project Test Pilot. During the next four years he advanced to Chief Test Pilot and then Chief of Engineering Flight Test, responsible for all flight testing of the Model 23/24 aircraft – development and certification, stick control system, aileron development, optimizing the vortex generator, and designing the autopilot.

In 1963 Kirkpatrick applied for NASA's third group of astronauts. His daughter, Terry Loewen, said her father once explained the reason for his not being a finalist. "I remember taking my parents to the Astronaut Hall of Fame at the Kennedy Space Center here in Florida about fifteen years ago. Dad told me some interesting stories about the qualifying process and all the testing that they had to pass in order to be considered. We had a good laugh when he explained about the centrifuge test. NASA was apparently looking for perfect physical specimens, and the wild G rides causing 'black outs' in that centrifuge seems to have sorted out a few of them. His own flying career had successfully prepared him for that test, but ultimately it was the reduced hearing in his left ear due to his jet flying career that caused him to be passed over." Kirkpatrick's roommate during the astronaut selection program was John L. ("Jack") Swigert, who later flew on the Apollo 13 mission.

Early in June 1964, Kirkpatrick escaped serious injury in the crash of Lear Jet's first experimental design, the Model 23, during FAA certification testing. On that day, FAA flight test certification pilot Donald Kuebler had the left-side pilot seat. Kirkpatrick was acting as co-pilot. Kuebler was evaluating the jet's performance on

Kirkpatrick (closest to camera) during Lear Jet Model 23 testing. (Photo courtesy Kirkpatrick family)

one engine, but after several successful runs he neglected to retract the wing spoilers following a landing. On the next take-off, the aircraft flew a short distance and then crash-landed in a muddy wheat field at the end of the runway where the heat of the engine and a jet fuel leak from the hard landing set the wheat field and plane on fire. The fire crews were not able to arrive in time to save the plane, but both pilots leapt out of the wreckage unhurt and sprinted to safety. Lear Jet Model 23 #2 became the new experimental plane, and is now on display in the Smithsonian National Air and Space Museum in Washington, D.C.

During this period Kirkpatrick also became an associate fellow with the Society of Experimental Test Pilots. In 1965 he served as Chairman of the Central U.S. Section of the organization, and in 1967 was elected Treasurer.

In July 1966, Kirkpatrick left Lear Jet to become Senior Engineering Test Pilot at the Beech Aircraft Corporation, also conveniently in Wichita, and the following year was promoted to Senior Project Pilot. His assignments included the Beechcraft 99 Airliner development and certification program, the Beech 60 Duke initial flight characteristics investigation, and the Beech King Air category II instrument landing

A little rodeo-style fun for Kirkpatrick during Lear Jet testing. (Photo courtesy Kirkpatrick family)

system approach testing. On being promoted in August 1968 to Chief of Engineering Flight Test he developed and flew on the King Air engine-icing program, which was a hazardous enterprise involving repeated penetrations into severe icing conditions, including contouring thunderstorms.

When Beech became a subsidiary of Raytheon in February 1980 it began to move in different directions. Soon after, Kirkpatrick was offered and accepted a position as Chief/Manager of Experimental Design with the Swearingen Aircraft Company in San Antonio, Texas, which was designing and developing a new jet aircraft. Founded in 1959 by aircraft designer Ed Swearingen, Jr., the company was created solely for the purpose of modifying twin-engine Queen Air business aircraft built in Wichita by Kirkpatrick's former employer, Beech. He and Pat moved from Wichita to the Texas Hill Country near San Antonio, where Pat still resides.

Swearingen's fortunes fluctuated as it worked to develop the Swearingen Metro commuter turbo-prop aircraft. The company was bought by Fairchild Aviation, Inc., and was known as Fairchild Swearingen until the founder's name was dropped from the title. In 1987 the company was restructured, and the San Antonio part was sold to Los Angeles-based GMY Investments. After all of the buy-outs and restructurings, Kirkpatrick decided it was time to retire from full-time work. However, he undertook engineering consultancy work on jet aircraft design for several years.

In 1984, while still at Fairchild Swearingen, Kirkpatrick became Vice President of Aeronautical Research and Management, Inc., which was a joint venture with James W. ("Wally") Leland to design, produce and patent the "Accutold" (an acronym for Accurate Take Off and Landing) safety warning module designed to visually inform a pilot of how his take-off was progressing, with sufficient time to abort if warranted. As Chief Engineer for the project, Kirkpatrick took Leland's concept, engineered and designed it, transformed it into a working prototype, physically built the first module using a Heath Kit, and adapted a Hewlett Packard calculator to prove the concept. He also built further prototypes as the project progressed. As Leland says, "Kirkpatrick was especially adept at taking a concept, developing it into a product, and building the prototype." Unfortunately the two men were ahead of their time with the product and the joint venture later dissolved.

Kirkpatrick was very family oriented, and included his wife and daughters in his favorite hobbies, including racing sports cars and sailing – he was an accomplished and competitive sailor who built and raced one-design monohull sailboats. While in Wichita, the family sailed on local lakes from Easter to Thanksgiving and attended regattas across the Midwest. He actively promoted one-design racing, and the local sailing club soon grew to include a preponderance of local pilots and engineers who enjoyed the camaraderie and challenge of sailboat racing. After moving to the Texas Hill Country the Kirkpatricks sailed on Canyon Lake nearby their home. In addition, he was able to fix virtually anything – be it a boat, an automobile, or an appliance in the home. Indeed, his children lovingly called him "MacGyver". He was interested in photography, and spent many hours experimenting with computers. At a time when personal computers were in their infancy, Kirkpatrick was always eager to see what he could do with both hardware and software. His grandchildren proudly called him a "computer genius". One of his ideas was even published in *PC Magazine*. He was elected president of the local computer club, where he enjoyed teaching and sharing his knowledge with others.

James Kirkpatrick passed away on 4 June 2005 at the age of 75, after a valiant battle with prostate cancer. He was survived by his wife Pat, who is an accomplished artist, and daughters Terry L. Loewen (married to Richard W. Loewen) of Orlando, Florida, and Patti S. Cole (married to Larry C. Cole) of Wichita, Kansas, and three grandchildren: Michael A. ("Andy"), James A. ("Alex"), and Patricia C. Cole.[31]

CHARLES L. PHILLIPS, USMC

Historically known as the "Buckeye State" and only the seventh most populated U.S. state, Ohio can nevertheless lay claim to the fact that three of the best-known NASA astronauts were born there – John Glenn, Neil Armstrong and James Lovell. One of those who tried to join that elite group of space travelers back in 1963 was a pilot in the U.S. Marine Corps from the city of Portsmouth, situated at the confluence of the Ohio and Scioto rivers in the southernmost tip of the state. Amongst other luminaries with roots in Portsmouth, it is celebrated as the home town of cowboy Roy Rogers.

Charles Laurence Phillips (later known as "Chet") was born in Portsmouth, Ohio on 13 January 1933. After gaining his secondary education at Holy Redeemer High School between 1946 and 1950, he enrolled for the Class of '54 at the U.S. Naval Academy. While there, he soloed for the first time in 1953. After graduating with his engineering degree, Phillips undertook his basic and advanced Navy flight training at Pensacola, Florida and Corpus Christi, Texas. With a commission and his wings, he became a Training Officer for VMF-311 ("Tomcats") at MCAS El Toro, California. During the conflict in Korea, the squadron had counted amongst its pilots baseball legend Ted Williams and future NASA astronaut John Glenn. Having returned home, many of the squadron veterans were reassigned to other units while new pilots were absorbed to familiarize themselves with the squadron's operations and to train using aircraft that had been crated, returned from Korea and then reassembled. By the end of 1955 these pilots were flying like the veterans they had replaced. After this, Capt. Phillips was assigned to Naval Air Advanced Training Command, NAS Chase Field, Beeville, Texas as an instructor on the F9F-8 Cougar jet fighter.

In February 1961, Phillips joined Capt. Ken Weir in Class 29 at the Navy's Test Pilot School at Patuxent River – the only two Marines in that class. On graduating in October 1961 he remained at the Naval Air Test Center as a project pilot for external weapons system tests. Over the next 21 months he evaluated the X4C practice bomb rack and Mark 44 missile adapter on the A1 Skyraider, A4 Skyhawk and FJ-4B in terms of airspeed, acceleration and separation limit using various practice bombs. He also tested the maximum flight limits and operating characteristics of a prototype spray tank. It was during this assignment that he unsuccessfully applied to NASA for consideration in the 1963 astronaut selection process.

His next assignment was as Aircraft Maintenance Officer for VMF-251, based at

Charles L. Phillips. (Photo: Society of Experimental Test Pilots)

MCAS Beaufort, South Carolina, where he took part in the squadron's transition to the F4B Phantom II. On rejoining VMF-311 in June 1965 as Aircraft Maintenance Officer he saw combat action flying A4Es out of Chu Lai, Republic of Vietnam. At the end of his tour, Phillips attended the U.S. Naval Postgraduate School, graduating in 1966 with a bachelor's in aeronautical engineering.

In September 1967 Phillips was appointed Deputy – later Head – of the Naval Air Systems Command's Program Office (PMA-236) in Washington, D.C., for overall life-cycle management of the North American Rockwell OV-10 Bronco twin-engine turboprop visual reconnaissance light attack and observation aircraft, its associated ground support and test equipment. Now with the rank of major, he directed an office staff of four civilian engineers in the management and coordination of contractors and Navy field activities in the OV-10 design, development, testing, procurement and provisioning, and he was also accountable for the $30 million-plus annual budget to produce the aircraft, each of which cost around $600,000.

In June 1969, an OV-10A was due to be displayed at the Paris Air Show prior to being delivered to Germany, so Maj. Phillips teamed up with a fellow Marine, Lt. Col. Robert L. Lewis, added an extra fuel tank to the aircraft, and established a new point-to-point distance record for light turboprop aircraft by flying non-stop from Stephensville, Newfoundland to RAF Mildenhall in England, covering a distance of 2,522 statute miles in 11 hours and 49 minutes. Apart from experiencing some icing

"Chet" Phillips (as he was known) around the time of his retirement from the USMC. (Photo: Society of Experimental Test Pilots)

and bucking 20-knot headwinds, the flight was fairly uneventful. From England, they flew over to Paris. Later, Phillips personally delivered the OV-10A to its new owner in Germany.

From 1970 to 1972 he commanded Headquarters and Maintenance Squadron 24 in Hawaii, which comprised 45 officers and 400 enlisted men. Next was a year-long assignment as Aircraft Maintenance Officer for the 1st Marine Air Wing in Japan. During the next three years he first headed the Attack Weapons Branch overseeing chase and laser missile control aircraft, then was appointed Director of Programs for the Navy's Pacific Missile Test Center in charge of all missile tests and evaluation. In December 1976, at this end of this assignment, he retired from active service as a lieutenant colonel in the Marine Corps.[32]

Over the next twelve years he served as a pilot and test pilot for the Falcon Jet Corporation operating customer demonstration and delivery flights, and in 1983 was awarded his master's in business administration from Webster University, Missouri. In March 1987 he captained the crew that set a new world point-to-point distance record in a Dassault Falcon 900 aircraft.

In December 1988, Phillips was appointed Director of Flight Operations for the Falcon Jet Corporation based in Little Rock, Arkansas, and he was still working in that position when he died on 19 January 1991, aged 58. Survived by his wife Martha and five children, he was buried in the Little Rock National Cemetery.

REFERENCES

1. NASA *Space News Roundup*, issue June 12, 1963, *Recruiting Opens for 10-15 New Astronauts*, MSC, Houston, TX, pg. 1
2. NASA *Space News Roundup*, issue June 26, 1963, *MSC Extends New Astronaut Recruiting Effort Two Ways*, MSC, Houston, TX, pg. 1
3. Phelps, J. Alfred, *They Had a Dream: The Story of African-American Astronauts*; Chapter 1: *Capt. Edward J. Dwight, Jr.*, Presidio Press, Novato, CA, 1994
4. Aldrin, Edwin E., Jr. with Wayne Warga, *Return to Earth*, Random House, New York, NY, 1973, pp. 147 & 151
5. Schweickart, Russell L, interviewed by Rebecca Wright for NASA JSC Oral History program, Houston, TX, 19 October 1999, pg. 8
6. Scott, David and Alexei Leonov with Christine Toomey, *Two Sides of the Moon: Our Story of the Cold War Space Race*, Simon & Schuster, London, U.K., pg. 82
7. E-mail correspondence with Robert Shumaker, 23 February 2012–28 March 2012
8. Aldrin, Edwin E., Jr. with Wayne Warga, *Return to Earth*, Random House, New York, NY, 1973, pg. 151
9. Cunningham, Walter, *The All-American Boys*, ibooks, New York, NY, 2003, pp. 23-24
10. *Ibid*, pg. 22
11. E-mail correspondence with Robert Shumaker, 23 February 2012–28 March 2012
12. E-mail correspondence with M/Gen. Ken Weir, 10 November 2011–28 May 2012
13. Cernan, Eugene with Don Davis, *The Last Man on the Moon*, St. Martin's Press, New York, NY, 1999, pg. 57
14. Collins, Michael, *Carrying the Fire: An Astronaut's Journeys*, Farrar, Straus and Giroux, New York, NY, 1974, pp. 43–44
15. Telephone interview with William Anders, 17 June 2012
16. Aldrin, Edwin E., Jr. with Wayne Warga, *Return to Earth*, Random House, New York, NY, 1973, pg. 152
17. Slayton, Donald K. and Michael Cassutt, *Deke! U.S. Manned Space from Mercury to the Shuttle*, Forge Books, New York, NY, 1994, pg. 133
18. Cernan, Eugene with Don Davis, *The Last Man on the Moon*, St. Martin's Press, New York, NY, 1999, pp. 59–60
19. Evans, Michelle, *The X-15 Rocket Plane: Flying the First Wings Into Space*, University of Nebraska Press, Lincoln, NE (scheduled release Spring 2013)
20. E-mail correspondence with Carolyn Bell Phillips, 27 March 2012–4 April 2012
21. E-mail correspondence with Cochran family (Ken, Kathleen and Marie), 7 November 2011–3 April 2012

22. The Society of Experimental Test Pilots, Lancaster, CA, *Application for Membership: Darrell E. Cornell*, 9 May 1983
23. Unidentified, undated newspaper clipping supplied to author by the Society of Experimental Test Pilots, *Cornell: Friends Remember a Pro*, by A.L. Randolph (newspaper) staff writer
24. *Ibid*
25. *Los Angeles Times* (newspaper), article, *Northrop Blames Crash of F-20 Fighter on Pilot Error*, by staff writer Ralph Vartabedian, 22 January 1985
26. *Encyclopedia Astronautica*, article, *F-20 Hornet, GI1002*. Website: http://www.f20a.com/gi1002.htm
27. E-mail correspondence with R.R. (Rod) Davis, 16 February 2012–19 April 2012
28. E-mail correspondence with Greg Ebbert, 29 February 2012–16 April 2012
29. E-mail correspondence with RADM George M. Furlong, 9 December 2011–14 July 2012
30. E-mail correspondence with Samuel M. (Murt) Guild, 23 February 2012–13 April 2012
31. E-mail correspondence with Terry Kirkpatrick Loewen, 15 December 2011–15 June 2012
32. The Society of Experimental Test Pilots, Lancaster, CA, *Application for Membership: Charles L. Phillips*, 31 January 1984

8

A few exceptionally good men

There were five more unsuccessful finalists for Group 3, and although three are still with us after many amazing adventures in the service of their nation, Alex Rupp and John Yamnicky had tragically sad ends to their otherwise magnificent lives.

TRADITIONS AND TACTICS IN ORBIT

In one of the most audacious but triumphant space missions to that time, and after the high drama of an earlier launch pad shutdown, the two-man Gemini VI-A spacecraft finally lifted off from Cape Kennedy's Launch Pad 19 atop a Titan II booster on 15 December 1965. The original Gemini VI mission plan had envisaged four proposed dockings with an unmanned Agena target vehicle but its engine exploded in flight. An orbital rendezvous between two orbiting Gemini spacecraft became the hurriedly conceived alternative mission.

The first launch attempt was scrubbed when the Titan booster shut down at the moment of ignition. Sensing that the rocket was still on the pad, Capt. Wally Schirra, USN, accompanied by Maj. Tom Stafford, USAF, opted not to eject. Their courage under extraordinary pressure kept the program on track, and on the third attempt the redesignated Gemini VI-A roared into the skies to chase down Gemini VII, launched eleven days earlier. They rapidly caught up with their sister ship and performed the first non-docking rendezvous in orbit by two manned vehicles. On Gemini VII were Lt. Col. Frank Borman, USAF, and Lt. Cmdr. Jim Lovell, USN.

In late November, during the crews' final training for their joint manned mission, the annual Army-Navy football game had ended in a tense 7-7 draw before 100,000 vocal spectators. As the two Gemini spacecraft carried out their historic *pas de deux*, there came a time when they were nose-to-nose and a rueful smile crossed Borman's face when a handmade sign appeared, pressed up against Stafford's potato-wedge-shaped window. The sign, one of Schirra's renowned "gotcha" pranks, contained just two intimidating words: "Beat Army".

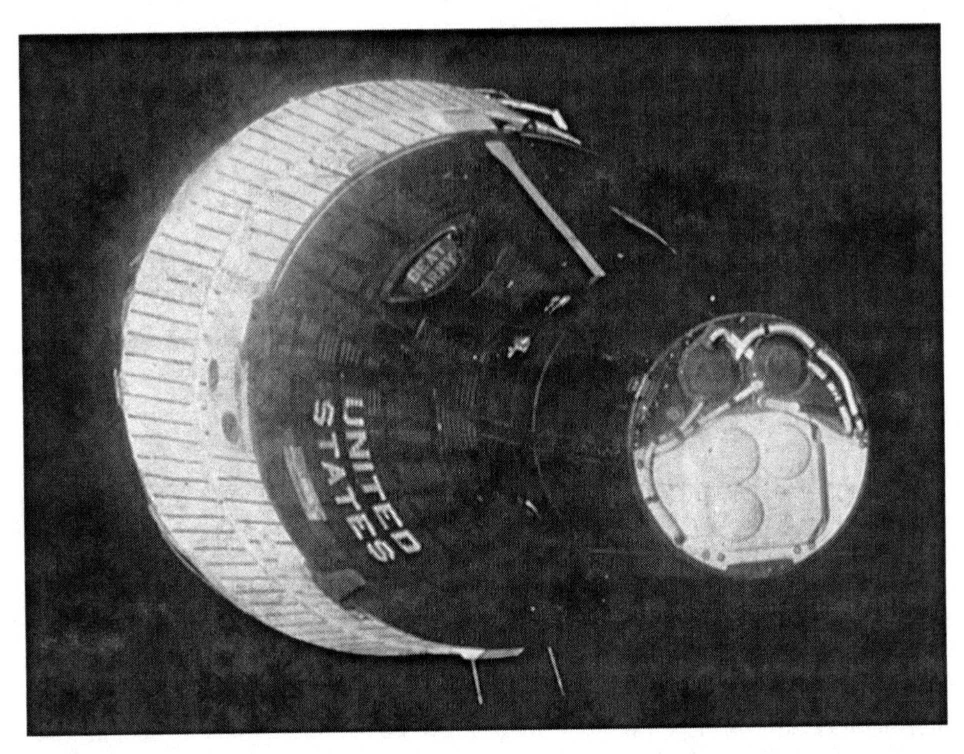

The Gemini VI-A spacecraft with Schirra's salutation to West Point graduate Borman visible in Stafford's window. (Photo: NASA)

THE TALE OF A GOAT

In terms of American sporting tradition, combined with an enduring rivalry, there is little to beat the Army-Navy game. Although considered by some to be just another football match, in reality it almost transcends being called a game. Certainly no other sporting contest boasts deeper foundations or pageantry, and there is no greater prestige and sense of honor than that shared by the victorious team and their parent service. From the outset, the challenging credo "Beat Navy" is vigorously imbued into each new class of West Point cadets, while midshipmen at Annapolis take every opportunity to chant the opposite allegiance. This intense rivalry began in November 1890 when teams from the Military Academy at West Point and the Naval Academy at Annapolis squared off against each other for the first time on the field, with a resounding 24-0 victory going the way of the Navy team against a newly established Army squad.

Symbolizing the two rival academies are their respective four-legged mascots, both being adopted in the late 19th century. The Army has a mule serving as its long-standing mascot, a rather peculiar choice that is said to rest in the animal's long and storied history in military service, whilst the naval academy is represented by a goat.

Legend says that a domesticated goat named Bill was commandeered by enthusiastic team members and their supporters as they made their way up a steep hill to attend the very first Army-Navy game at West Point. In view of the resounding victory, Bill became the academy's mascot, beginning a tradition and introducing a succession of goats bearing that name.

However, on 22 November 1953, just a week before that year's Army-Navy game in Philadelphia, the unthinkable happened when Billy XII was brazenly kidnapped during a commando-style raid on Annapolis.

Future Apollo astronaut Al Worden spoke with an insider's knowledge of the raid in his 2011 autobiography *Falling to Earth*. He was involved as head cheerleader for West Point's Class of '55, who were the ringleaders of this remarkable escapade. "I was head cheerleader for two years," he wrote. "Morale building is a cheerleader's job. Embarrassing the Navy team before big football games was, therefore, a primary objective."[1]

In a meticulously planned covert operation, and having scouted the goat's location that summer, two West Point cadets daringly broke into the Naval Academy at night, cutting a large hole in the chain link fence. They silently made their way to the rear of the football stadium, where Billy XII was housed in a pen beneath the grandstand. After placing a leash on the startled animal they hurriedly made their way back to the fence, beyond which a nervous corporal, recruited for the task from West Point's band, was waiting. He was the owner of the getaway vehicle – a convertible – which was parked near a seawall by the Severn River. Just as they reached the fence the two raiders were startled when challenged by a stern voice. After several heart-stopping moments they realized that they had been accosted by two follow West Pointers who had independently come up with their own plan for kidnapping the goat by boat that same night.

After the moment of mutual astonishment had passed, the cadets joined forces and led Billy XII right around to the far end of the fence, where the other cadets had parked a "getaway" boat. They all piled in with their gold-horned trophy, rowed around a bend and pulled up to the seawall where the convertible was waiting. They pulled the animal ashore and carefully bundled it into the car. The other two cadets then climbed into their boat, gleefully wished them good luck and rowed back up the Severn.

The choice of a convertible proved a wise decision, as the goat gave off a horrible warm stench. As they raced back to West Point, however, the bone-chilling, windy conditions soon convinced them to put the top up, despite their malodorous back-seat passenger. Enjoying his warmer surroundings Billy XII fell asleep. However, in his contentment he kept breaking wind, which quickly filled the car with a pungent, eye-burning stench. Realizing they were running low on fuel, the cadets made a hurried stop at a gas station, whereupon the goat woke and stood up, shaking his head. His horns punched through the soft roof of the convertible and he proceeded to tear it to shreds. The three men and their car finally arrived back at West Point in a sorry state.

That evening, a delighted pandemonium erupted when the three cadets (with the sanction of their commanding officers) presented their hard-won trophy, Billy XII,

Al Rupp, the getaway driver (center) and Ben Schemmer pose with Billy XII and Mr. Jackson, Senior Army Mascot, at the mule pen. (Photo: U.S. Army Signal Corps, courtesy "Tiny" Tomsen)

to the entire Corps in a noisy pep rally in Washington Hall. As prearranged, the cadet rally band played a victory tune as a large cardboard box was broken open to reveal the startled goat.[2] A spontaneous uproar broke out among the astonished cadets and officers. "They went wild, taking off their dress coats, stacking tables and cheering till they were hoarse," recalls former cadet Jay Gould.[3]

The intention had been to return the goat the following Saturday at the game, but when news of the goat snatching eventually reached the press the animal's return was ordered by West Point's complicit but concerned higher command.

As Al Worden recalls, "word got back to the Naval Academy that West Point cadets were to blame. Soon, the phone lines were buzzing between officials from both academies, and I hear that even President Eisenhower got involved." The president – a former West Pointer – had indeed personally intervened and demanded the immediate return of the hapless goat to Annapolis. "We only complied when our commanding officer directly ordered us," Worden notes. "Privately, despite all the ruckus we had caused, I think he was a little proud of us."[4]

It had been a well-orchestrated plan, and skillfully executed. Making the occasion even more memorable was that in the game later that week the Army squad defeated the Navy 20-7.

The first of the two principal goat snatchers was Benjamin Franklin Schemmer of the Class of '54, who went on to become a publisher and the author of four books on military affairs. His co-conspirator was 23-year-old Alexander Rupp, Class of '55,

Al Rupp and Ben Schemmer on stage with Billy XII and an "unnamed helper". (Photo: U.S. Army Signal Corps, courtesy "Tiny" Tomsen)

who dreamt and often talked of one day flying to the Moon. Sadly he would not live long enough to realize this dream, but in a strange twist of fate the head cheerleader of the Class of '55 would follow through when in 1971 Col. Alfred Worden flew to the Moon as the command module pilot on the Apollo 15 mission.

ALEXANDER K. RUPP, USAF

Alexander Kratz Rupp came into the world on 2 July 1930 in the Centre County Hospital in historic Bellefonte, a Victorian-style town nestled in the Nittany Valley of Centre County, Pennsylvania. He and his twin, Charles, were born to Luella Kratz and Dr. Charles A. Rupp, who taught mathematics at Penn State University. He also had two sisters named Susan and Melicent. The first ten years of his life were spent in nearby State College. As his father recalled, "Al had a paper route as a small boy and served it very faithfully, earning and saving his money so he could buy clothes and toys of his own choosing. One day he amazed his parents by coming home with a tuba that was almost as big as he was. It was his school band instrument. He had to practice in the cellar." He would later play the cornet in the local drum and bugle corps, whose uniform bore a striking similarity to the dress uniform of a West Point cadet.

With the outbreak of World War II, Charles Rupp was commissioned a captain in the Army Signal Corps, and was stationed in Washington, D.C., where his family joined him several months later. Young Al now took his education at Central High School in Washington, and not only shone academically but rose to become captain of the cadet corps in his senior year. When he graduated, the Navy awarded him a Holloway scholarship for study at Harvard University. "After two years he had some doubts that the system would ever lead him any higher than lieutenant commander," his father wrote in a heartfelt eulogy to his son in 1965, shortly before his own death, "so he left college and took a job running a bulldozer and later, a gasoline-powered shovel."[5]

It would only ever be an interim measure as Rupp began attending National Guard drills at summer camp, understanding all too well the competitiveness of the exams for admittance to the service academies. His perseverance paid off, his name was high on the 1951 list for appointments to West Point from the National Guard and he joined A-2 Company of the Class of 1955 on 3 July 1951. Fortunately, Rupp found the academics relatively easy. As well, his experiences with the Navy, at college, and on the construction job, would continue to serve him well. He was a Cadet Sergeant and Company Supply Sergeant in his first class year, ran cross-country for two years and became a cheerleader for three years.

His roommate Dick Stemle recalls with fondness a young man with a solid future ahead of him. "To begin with, [Al] was one of the most intelligent persons I've ever met. I really cannot remember seeing him study. He would scan the material and it would be locked away for future reference. He was also a fun-loving individual and seemed to always have a smile on his face. In my opinion, Al could have achieved any goal he established for himself. I believe he would have made it to the Moon had it not been for his accident."[6] However there was also a spirited rebellion inherent in young Rupp that led him and two fellow cheerleaders to kidnap Billy XII and display him at West Point in triumph.

Rupp graduated 63rd in his West Point class on 7 June 1955 and opted to serve in the Air Force. During his flight training in Texas and at Greenville AFB, Mississippi, he also worked on airplane and automobile engines in his spare time at the local base shops.

In one of his regular letters home, Rupp revealed that a base operations officer had been impressed by his flying skills while under instruction, and even more so by his tolerance to g loading. The average limit for a pilot without a suit was 4 to 5 g, but Rupp remained fully conscious at 7 g. "I do find I have a high g resistance and have blacked out every instructor with whom I've flown," he wrote. "My instructor says they've never had another student like me; that the whole base is shook; that even instructors can't get away with what I've done already." Even though he was proving to be a talented, even exemplary pilot Rupp would never knowingly break the rules. "I never buzz, rat race, fly solo formation, or anything else common that gets anyone kicked out," he added. "I fly with flair, imagination, spirit and enthusiasm, and that is admired by all. I've got my check ride with the hottest pilot on the base," he wrote in conclusion to his parents. "That should prove interesting."

Rupp was one of the first in his class to be assigned to a fighter squadron (36th Day-Fighter Wing) in Germany, and was stationed near Bitburg over the next four years. Here he not only became fluent in German but met and married a German girl

Capt. Rupp outside his old school, the former Frazier Street School in State College. (Photo courtesy Karen Rupp Deming)

named Ruth Michaels. While there they had a son named Alexander Michael, born 6 June 1958. Back in the United States they would complete their family with daughter Karen.

As Charles Rupp related, "The Air Force Academy proposed to send Al to school at the University of [Mainz] with a view to his teaching German at the Academy, but Al aspired to a space career. He did succeed in drawing a stateside assignment at Wright-Patterson AFB and there earned a master's in astronautics [at the Air Force Institute of Technology] in 1962. From there he went to Edwards AFB in California, for duty in the Aerospace Research Pilot School program."[7]

On 30 March 1963, Rupp was one of fourteen Air Force captains chosen for the Aerospace Research Pilot School at Edwards. The sixteen-member ARPS IV class was rounded out by one officer from each of the Navy and the Marine Corps. Shortly after studies began on 17 June, some of them, including Rupp, sent in applications to NASA for the third astronaut group. At the end of the intensive seven-month course, the graduates knew that they would be available for assignment

On a return visit to Bellefont following his selection to ARPS, Capt. Rupp was given a 'key to the city' by Mayor George Boughter. (Photo courtesy Karen Rupp Dering)

as a pilot, manager or consultant in any future U.S. manned space flight programs. They were:

Capt. Michael Adams, USAF
Capt. Tommy Bell, USAF
Capt. William Campbell, USAF
Capt. Edward Dwight, USAF
Capt. Frank Frazier, USAF
Capt. Theodore Freeman, USAF
Capt. James Irwin, USAF
Capt. Frank Liethen, USAF
Capt. Lachlan Macleay, USAF
Capt. James McIntyre, USAF
Capt. Robert Parsons, USAF
Capt. Alexander Rupp, USAF
Capt. David Scott, USAF
Capt. Russell Scott, USAF
Lt. Walter Smith, USN
Maj. Kenneth Weir, USMC

Capt. Alex Rupp in an F-104 Starfighter at ARPS, Edwards AFB. (Photo: USAF courtesy Karen Rupp Deming)

By the time the class graduated on 20 December 1963 Rupp knew that he had not been selected for the third astronaut group. Throughout his service career, he was a hard-working and conscientious officer. His promotions to captain and to major were both the result of his endeavors and ambitions. To him, having the stars in his eyes was no idle dream, and there can be little doubt that he would have reapplied the next time NASA sought applications.

Charles Rupp, who had advanced to the rank of lieutenant colonel in the U.S. Army Reserve, noted with pride that his son was always unusually thoughtful of his relatives and friends. "He wrote to his parents faithfully every week; he helped his younger sister through the University of Chicago with a modest allotment, and he gave his mother a miniature of his West Point ring and his insignia as a senior pilot. Friends sometimes remarked that Al lived two lives – his own, and another for his twin brother who drowned in early childhood."[8]

From Edwards, Rupp returned to Wright-Patterson AFB, where he worked with the test program of the Aeronautical Systems Division. He was scheduled to attend the Staff School in Norfolk in August 1965, after which he intended to undertake a leisurely camping trip with his family in Nova Scotia and north-eastern Canada prior to leaving for Vietnam. Sadly, none of this would come to pass.[9]

On Friday, 11 June 1965, Maj. Rupp was to fly an F-102A jet interceptor out of Wright-Patterson on a routine test of a new adaptive flight control manufactured by the Bendix Aircraft Corporation. Having completed his walk-around of the aircraft, he climbed aboard and was airborne at 11:04 a.m. Once over the test area he engaged the adaptive autopilot and made the first of his planned instrument check maneuvers. He then rolled into the first dive-and-pull-out test, reporting his reactions to Flight Test back at base. His second dive was conducted at a lower altitude, and once again he pulled back up hard without incident. At 11:27 a.m., he started into his third dive. On the ground, John Winters, Jr. was working in a building at Camp Canaan when he heard the F-102A make its third dive and pull up maneuver. Then, immediately after the aircraft had passed over his building it exploded, tearing itself apart in the skies over Highland County. The wreckage was scattered over two cornfields near state highway Ohio 134, fifty miles south of Wright-Patterson and about half a mile from the small township of Buford.

Following an extensive investigation, the primary cause of the loss of the pilot and aircraft given in the 29 July 1965 Accident/Incident Report was "material failure due to fatigue in the Number 5 spar resulting in structural breaking of the left wing. Fire and explosion contributed to complete disintegration." It was noted that the wing in question had been reclaimed from a "previously wrecked aircraft". Maj. Alex Rupp, aged 34, was buried in Arlington National Cemetery.[10]

Tragedy would continue to strike the family. On 24 July 1987, Capt. Alexander Michael Rupp (313th TFS, 50th TFW), who, like his father, had joined the Air Force, died when two F-16A aircraft collided over Indian Springs when undergoing training at the Fighter Weapons School in Nevada. Both pilots were killed instantly.[11]

ROBERT H. SHUMAKER, USN

In February 1965, U.S. Navy Lt. Cmdr. Bob Shumaker was shot down by enemy fire on a mission over North Vietnam, breaking his back on impact when his parachute opened just 35 feet above the ground. After spending 2,942 days in captivity, three years of which took place in solitary confinement, he was finally released during Operation Homecoming on 12 February 1973. The Distinguished Service Medal he was subsequently awarded states that during those eight years of deprivation, torture and inhumane treatment, "he never wavered in his devotion and loyalty to the United States and his fellow prisoners".

Robert Harper Shumaker was born on 11 May 1933 in New Castle, Pennsylvania where his parents, Alvah and Eleanor, were a lawyer and a writer, respectively. After graduating from public schools he attended Northwestern University for a year and then the U.S. Naval Academy, where he was a boxer, a cross-country runner and a scholar. On graduating in 1956 he attended flight training and F8 training, eventually joining VF-32 ("Fighting Swordsmen"), an F8U-1 Crusader squadron based at NAS Cecil Field in Jacksonville, Florida.

Completing his service with VF-32 in June 1961, Lt. Shumaker then attended the

Naval Postgraduate School at Monterey, California, leaving three years later with a master's in aeronautical engineering. During this time he was a finalist in the Group 3 astronaut selection but was ruled out by a temporary physical ailment. He was then ordered to F8 Replacement Air Group training with VF-124 ("Gunfighters") at NAS Miramar, California from June to September 1964. In December 1963 he married Lorraine Shaw of Montreal, Canada, and twelve months later, having completed his tour with VF-124, was transferred to VF-154 ("Black Knights"), which was in San Diego, California awaiting deployment orders. On 7 December 1964 – shortly after the birth of his son Grant – the squadron deployed onboard USS *Coral Sea* (CV-43). Beginning on 7 February, aircraft from the carriers *Coral Sea*, *Ranger* (CV-61) and *Hancock* (CV-19) serving with the Seventh Fleet began attacking military barracks and staging areas near Dong Hoi in the southern sector of North Vietnam.

On 11 February 1965, while flying an F8 Crusader, Lt. Cmdr. Shumaker's aircraft was hit by 57-mm cannon fire and went out of control. He ejected, and his parachute only snapped open moments before he struck the ground. The impact broke his back and he was captured soon after, put into a jeep and driven over rough roads to Hanoi, 300 miles away. He had already gained a foretaste of the cruelty of his captors when they placed him in front of a disorderly firing squad; however, the propaganda value associated with being shot down and taken alive meant that the fatal order was never given.

Once they arrived in Hanoi, a white-smocked North Vietnamese ostensibly gave Shumaker a medical examination as a bevy of cameramen snapped their propaganda

Lt. Cmdr. Shumaker is led away by local militia after his F8 Crusader was shot down. He would be imprisoned in the notorious "Hanoi Hilton" for almost eight years. (Photo: Van Bao)

photos, but he received no treatment for his injuries. Eventually his back healed, but it was six months before he could bend. Two years later, his back was reinjured in a session of torture after he refused to participate in a propaganda film for his bayonet-armed captors. As a result of his refusal to comply in any way with his captors he was placed in solitary confinement for three years in a tiny, filthy cell, much of this time clamped in leg irons. "I was lonely, hungry, scared and sick – but I was proud," he recalled. "Proud to be an American, proud of my Navy, proud of my family."

Cmdr. Shumaker is greeted by son Grant following his release from captivity in 1973. (Photo courtesy R. Shumaker)

It was actually Shumaker who first applied the sarcastic nickname "Hanoi Hilton" to the notorious prison. While there, enduring torture, starvation and deprivation, his fellow POW's regarded him as a resister, a leader, and a patriot. He became known as one of those leaders who dubbed themselves the "Alcatraz Eleven". Their notable members included future Congressman Sam Johnson and Senator Jerry Denton, and Vice Presidential candidate, the late VADM James Stockdale.

Their captors had imposed a no-communication policy on them, but as Shumaker later explained, "We thwarted them by developing a 'tap code' which allowed us to clandestinely communicate with our POW neighbor a foot away through a concrete wall using coded knocks that spelled out words." In order to retain their sanity and pass the endless hours in the four-by-nine-foot cells, Shumaker says they did mental exercises without the benefit of pencil or paper, including building houses, studying foreign languages, and solving mathematical problems – all without being detected by their guards. "We also relived pleasant past relationships and even had elaborate fantasy breakfasts each Sunday. We were focused on supporting each other, trying to make life a bit more bearable, and dreaming."

Eventually those dreams became a reality. In February 1973 the emaciated POWs

Capt. Robert H. Shumaker, USN. (Photo: USN courtesy R. Shumaker)

were repatriated. "When we were released," he stated with pride, "we marched to the buses with military precision."

Promoted to the rank of commander during his captivity, Shumaker underwent a period of recuperation at the Naval Hospital in San Diego. He resumed his studies at the Naval Postgraduate School in Monterey, California, which led to him completing his master's degree and then going on to gain a doctorate in electrical engineering in June 1977.

Bearing the rank of captain, Shumaker became a government project manager for smart missiles at the Naval Air Systems Command in Washington, D.C. At the end of his involvement in the project in June 1983 he was promoted to commodore and then served as the Superintendent of the Naval Postgraduate School in Monterey, California for three years. His final assignment in the Navy was as a rear admiral at the Pentagon, where he was responsible for coordinating the research efforts of the Navy's air, surface, electronics and space activities.

RADM Shumaker retired on 1 February 1988 and was made an Assistant Dean at George Washington University and later Associate Dean of the Center for Aerospace Sciences at the University of North Dakota, from which position he retired in 1991. He was subsequently elected President of the Quantico Flying Club and worked as a flight instructor. He also constructed a Glasair experimental airplane, which he flew to Alaska and the Bahamas. In April 2011 he was presented with the Distinguished Graduate Award from the U.S. Naval Academy.

"Paradoxically, I learned a lot about life from my experience as a prisoner of war

A recent photograph of Bob Shumaker holding a picture of himself as a young fighter pilot. (Photo courtesy *This Emotional Life*)

in Vietnam," he recently mused. "Those tough lessons learned within a jail cell have application to all those who will never have to undergo that particular trauma. At some point in life everybody will be hungry, cold, lonely, extorted, sick, humiliated, or fearful in varying degrees of intensity. It is the manner in which you react to these challenges that will distinguish you."

Admiral Shumaker's military decorations include the Distinguished Service Medal, two Silver Stars, four Legions of Merit, the Distinguished Flying Cross, a Bronze Star, and two Purple Hearts. He and his wife now live in Fairfax Station, Virginia where his hobbies are golfing and flying. Their son Grant is a neurosurgeon in Sioux City, Iowa.[12]

LESTER R. SMITH, USN/USNR

Lester Robert Smith, who prefers to be known as "Bob", was born in San Francisco, California on 11 October 1929. He was the only son of Lester Blake Smith, a Navy signalman, radioman and electrician's mate, and the former Olive Lucille Thompson, a ladies' beautician, and a brother to Betty Jean Smith, who was born in December 1930.

Growing up in Salt Lake City, Utah, where his parents were born and married, he attended Lowell, Ensign and McKean elementary schools before the family relocated to California. There he studied at Mount Vernon Junior High School and completed grades 10 to 12 at Los Angeles High School, where he lettered in varsity swimming. In the latter years of high school he and several of his friends "came under the spell" of a recruiter who talked them into the V-5 (Naval Aviation Cadet) Program. Among other commitments, this cadetship would entail enrollment at college, taking on two years of flight school, and then completing the last two years of college before being commissioned.

In 1947, at the age of seventeen, Smith was sworn into the U.S. Navy as a Seaman Apprentice. He completed a two-year course at the University of Los Angeles in just seventeen months and was ordered to active duty in February 1949 to commence his flight training as a midshipman. He gained his Wings of Gold in September 1950 and joined VF-52 ("Knight Riders") as a flying midshipman. Having transitioned earlier in the year from the F4U Corsair to the F9F-2 Panther, VF-52 was the first Navy jet squadron in combat. On being commissioned an ensign, Smith joined the squadron aboard USS *Valley Forge* (CV-45) on 15 October 1951 on a delayed nine-month deployment to the Korean War zone. His second combat deployment with VF-52 began on 30 March 1953, this time on the straight-decked USS *Boxer* (CVA-21). For his service in the Korean conflict, during which he flew 132 combat missions, Smith was awarded the Distinguished Flying Cross, six Air Medals and the Navy/Marine Commendation Medal with "V" (Valor) and the Legion of Merit (two awards), along with many unit and service medals.

In November 1953, following the armistice, *Boxer* returned to the United States and VF-52 settled into NAS Miramar, California. Early the following year there was a pleasant surprise in store for the squadron when they were selected by

Commander, Naval Air Forces, Pacific Fleet to assist in making the film *The Bridges of Toko-Ri* for Paramount Pictures, based on a novel by James A. Michener. As one of eight pilots assigned to the movie, which accurately reflected the actual combat over North Korea, Lt. Smith did much of the flying for the role of Lt. Harry Brubaker, played by William Holden. An added bonus was Paramount inviting the participating pilots to Hollywood, where the film was being produced, to meet the stars and attend a formal dinner. In the summer of 1954, Smith was recommended by his superior officer to attend the Navy Test Pilot School at Patuxent River, Maryland. He joined Class 13, which included Lt. Scott Carpenter, who would go on to become one of the Mercury astronauts. In 1955, Smith was introduced to his future wife Carmella ("Carmen") Marie DeLuca on a blind date. After graduating later in the year, he remained at the Naval Air Test Center for two years in the Carrier Suitability Branch of Flight Test Division.

He was then sent to the NROTC unit at Georgia Tech in Atlanta, Georgia, where he got his bachelor's and master's degrees in aeronautical engineering with highest honors in December 1957 and September 1958 respectively. While studying for his master's he resigned from active duty and became a lieutenant in the Naval Reserve. He joined Chance Vought Aircraft (later Ling-Temco-Vought), where he worked for 29 years, holding many positions including Chief Project Engineer F8U Crusader, Director of A7 Corsair programs, and Director for the B-2 Spirit stealth bomber.

In 1963 Smith was one of 34 finalists for NASA's third group of astronauts, but was unsuccessful. He reapplied in 1965 but was turned down for being two months over the age limit.

During his time with Chance Vought, Smith rose steadily through the ranks of the Naval Reserve and in December 1979 was promoted to rear admiral (lower half). His promotion to rear admiral (upper half) three years later was the highest rank held by a reservist at that time.

After retiring from Vought, he became Vice President, Program Development for the Turbomeca Engine Corporation, Vice President, Business Development for EV Power, involved with electrically powered vehicles, and Executive Director of the Central Electric Vehicle Coalition, a non-profit association which promotes electric vehicles. He is currently CEO of EV TECH (Electric Vehicle Technology) in Dallas, Texas.

In the Naval Reserve, Admiral Smith served six tours as a commanding officer: VF-701, which was named the best fighter squadron in 1966; two Naval Air Station units; a maintenance unit; an aeronautical engineering unit; and Volunteer Training Unit senior staff. From 1981 to 1984 he was Commander, Naval Reserve Readiness Command, Region 11. This comprised 14 area commands with over 125 subordinate commands and 7,000 sailors in three states. He was awarded the Legion of Merit twice, and, for his combat service in Korea, the Distinguished Flying Cross, six Air Medals and the Navy/Marine Commendation Medal with "V" (Valor), together with many other unit and service medals. He retired from the Naval Reserve in September 1986 having completed 39 years of service.

Since then, Smith has been involved in numerous community activities. Among these, he has coached junior swimming teams, been an active member of the Dallas

RADM Lester R. Smith, USNR. (Photo courtesy RADM L.R. Smith)

Police Department's Volunteers in Patrol unit, and an officer or board member for non-profit organizations that raised money for youth scholarships and awards. He has served as Association of Naval Aviation Regional Vice President; Naval Reserve Association (and past National President); Naval Order of the United States (past National Commander); Navy League of the United States (past National President, Sea Cadets); Naval War College, Regional Vice President; and Frontiers of Flight Aviation Museum (board member). He was the founding Chairman of the Greater Dallas Veterans Foundation, and has served three times as President of the Dallas Military Ball – a fundraiser for military personnel and families (and board member for nine years), and a past Grand Marshal for the Dallas Veteran's Day Parade. He served as a judge for the Winston School Solar Car competitions. He also helped to raise money for the Hella Shriner's Hospitals for Children and the Texas Scottish Rite Hospital for Children for more than ten years.

Admiral Smith is currently a member of the Schreiber Memorial Church. He and

Carmen have a daughter Sherri, and three grandchildren, Christopher, Stephen and Ashley.[13]

ROBERT J. VANDEN-HEUVEL, USAF

Robert Jan Vanden-Heuvel came into the world on 16 March 1930, one of four sons of Renzo and Rose Vanden-Heuvel of Staten Island, New York. His older brother, George, was a P-51 Mustang "ace" with the 8th Air Force during World War II and was credited with nine kills. Another brother, Theodore ("Teddy"), was killed during a high-speed P-51 bailout while undergoing pilot training at Punta Gorda, Florida in 1945. His other brother, John, served in the National Guard.

In his youth, Robert attended the PS 29 school on Staten Island prior to graduating from Curtis High School. In 1938 he saw the Clark Gable/Spencer Tracy movie *Test Pilot*, and the next year wrote an 8th grade essay saying he wanted to be a test pilot. Because both George and Teddy were graduates of Georgia Institute of Technology (Georgia Tech), there was never any doubt about which college Robert was going to attend.

In June 1951 he graduated Georgia Tech with a degree in aeronautical engineering and was commissioned a 2nd lieutenant in the Air Force Reserve. He entered the Air Force at Langley AFB, Virginia on 22 July 1951, undergoing basic pilot training in the T-6 Texan at Spence Air Base in Georgia from 19 November to July 1952 and Perrin AFB, Texas through to 21 September. He then went on to advanced training in the Lockheed T-33 Shooting Star at Craig AFB, Alabama, where he was awarded his wings on 19 December 1952. He was promoted to 1st lieutenant on 22 April 1953. The following month he was transferred to Nellis AFB, Nevada for combat training in the F-86 Sabre jet. In June he was given supplemental F-86 combat training at George AFB, California. "Triple jet ace Joe McConnell was my flight commander, and taught me to do things with the F-86 that allowed me to outmaneuver anyone I encountered," he reflected. He finished his training for the F-86 at Clovis AFB, New Mexico in June 1954, then went on leave to Staten Island for what he calls "his most important mission". On 23 January 1954 he married Shirley Pedersen at St. Teresa's on Staten Island. She was the "girl next door", but her door was one street over from his.

In October 1954 he was deployed to Hahn Air Base, Germany, attached to the 561st Fighter Squadron of the 388th Fighter-Bomber Wing, which became the first nuclear weapons organization on the continent of Europe. On 26 February 1955 he was leading a two-plane training flight. His wingman was not looking at the lead aircraft whilst crossing from right to left and collided with Vanden-Heuvel's F-86. The wingman lost the greater part of one wing and had to bail out, while Vanden-Heuvel unknowingly had his right elevator ripped off in the collision. "Because the pitot tube was damaged I had no airspeed indicator," he says. "I was able to maintain control and land in formation with an escort from another F-86. The escort pilot did not tell me about the elevator loss and when I asked him why, after landing, he said, 'I didn't want to scare you!'"

Vanden-Heuvel's damaged F-86 following the collision. (Photo: USAF courtesy Vanden-Heuvel family)

At Eglin AFB, Maj. William Garvey briefs experimental test pilots. Capt. Vanden-Heuvel is second from the right. (Photo: USAF)

Capt. Robert J. Vanden-Heuvel, USAF, in April 1964. (Photo: USAF courtesy Vanden-Heuvel family)

It was during this deployment that Robert and Shirley's first son, Doug, was born on 9 August 1955. After transitioning to the F-100 Super Sabre, in October of that year he was transferred along with the 388th FBW to Étain Air Base, France, which was one of several bases the U.S. Air Force had set up and maintained in France as part of NATO. On 19 April 1957 he was promoted to reserve captain. In December of that year the 388th was redesignated the 49th FBW. In 1958 he was attached to the 435th Fighter Squadron at George AFB, California, where he upgraded to the F-104 Starfighter. He was promoted to regular captain on 9 February 1959, and on 26 May daughter Theresa was born.

Vanden-Heuvel's next assignment on 18 November 1959 was to Edwards AFB, California as a member of the fourteen-strong Class 60-A of the Air Force Test Pilot School. After graduating on 19 August 1960 he was assigned to Eglin AFB, Florida, where he served as a test pilot for the next five years flying the T-33, F-100, F-104 and the F-105. During this time, Michael and Mary were welcomed into the family on 4 October 1961 and 9 April 1963 respectively. That June, soon after Mary's birth, Vanden-Heuvel heard that NASA was seeking a third astronaut group and he decided to apply. As he explains, his rationale for wanting to join the astronaut program was purely a domestic one. "I had heard that the first astronauts were given homes, so that made me want to join the program, even though Apollo had three seats; I hated airplanes with more than one seat."

He took the astronaut physical in August 1963 and it happened that his roommate during the testing was Lt. Cmdr. Roger Chaffee, who was later killed in the Apollo spacecraft fire. But a common illness ended his chances. "While taking the astronaut physical I was bedridden for two days with the flu. Because of that, I didn't do well on the later physical endurance tests. I'd like to believe that's why I wasn't chosen. After failing to be selected I returned to Eglin for my best Air Force assignment ever – quadruple currency in the F-100, F-104, F-105, and the T-33." On 22 December 1964 he was further promoted to the rank of major.

In February 1965, Vanden-Heuvel was transferred to Wright-Patterson AFB, Ohio where he took part in Project Rough Rider, which entailed thunderstorm penetrations in an instrumented F-100. He needed to go where the storms were, so that spring he was based at Tinker AFB, Oklahoma, and in the summer at Patrick AFB, Florida. Before he left, he was able to welcome his daughter Cindy into the world on 12 April 1965. For his participation in Project Rough Rider he was awarded the Distinguished Flying Cross in 1966 (the only non-combat award permitted for the DFC). Over his career, he was also awarded four Air Medals.

From 1968 to 1969 he underwent combat training in the F4 Phantom at George AFB, being promoted to lieutenant colonel on 1 July 1968. He was then transferred to the Philippines for jungle survival training prior to going to Vietnam, where he spent half of his tour based at Da Nang. All his missions were to Laos, none to North Vietnam. The remainder of his tour was at Headquarters, 7th Air Force and proved to be the only non-flying assignment of his career. Following his return from Vietnam Vanden-Heuvel was reassigned to Kirtland AFB, New Mexico. From 1970 he spent four years with the 4900th Test Group, flying the F-100 and F4. His final assignment was at Eglin AFB, Florida from January 1975 to September 1979, flying the F4.

On 1 September 1979 Vanden-Heuvel retired from active service, having logged a total of 6,064 hours' flying time. Aged 49, he had no desire to fly for a commercial airline because he preferred to fly alone and wanted the controls in his own hands. In 1982 he asked his friend Ron Walker, who owned a banner flying company, if he had need of another pilot. The response was an enthusiastic affirmative, and for the next sixteen years Vanden-Heuvel was a banner plane pilot. In this often-hazardous job he logged 9,400 hours, 6,400 of them in his favorite aircraft, the single-engine two-seat Citabria.

As he says, "I flew banner airplanes along the beach and at football games. It was the most dangerous flying of my career. I was flying planes that were maintained by incompetent mechanics and recycled parts. The only flameout landing of my life was on a golf course along the beach. My guardian angel had the engine quit as I was over a long unoccupied course, with a 15-knot headwind blowing. I dropped the banner at one end and stopped at the other. I folded the banner up and put it in the plane while the course manager went to the Destin airport for five gallons of aviation fuel. I was out of there before the authorities showed up. My picture made the paper the next day."

After that, Vanden-Heuvel wisely decided to retire at the end of the season and has never flown again. When asked if he had enjoyed a good life in aviation, despite not being selected as an astronaut, the answer came quickly. "I may not have become an astronaut, but I had one hell of a career."[14]

THE DAY THE WORLD CHANGED FOREVER

Through the large plate-glass windows at Washington Dulles International Airport's Gate D26, John Yamnicky would have observed a pre-departure flurry of activity at the American Airlines 757-223 aircraft as it was checked, catered, and in all other ways prepared for the scheduled daily service to Los Angeles International Airport. There were very few passengers waiting at the gate that Tuesday morning, and he would have been anticipating a comfortable flight with every likelihood of having a couple of seats to himself on the transcontinental flight. It had been an early morning start; he had left his Waldorf home at 4:30 a.m. and, customary for him, was wearing his Annapolis ring and eye patch bearing a 'dragon' motif.

At around 7:30 the crew began passing through the gate to board and prepare their flight. Captain Charles Burlingame and First Officer David Charlebois were soon followed by four smartly dressed flight attendants; three females and one male.

Twenty minutes before the scheduled departure time of 8:10 a.m., the boarding call was aired in the lobby for American Airlines Flight 77 to Los Angeles, and it became obvious that the aircraft, with a seating capacity of 188 passengers, was only about one-third full. John Yamnicky boarded and made his way down to his seat, and may have begun reading a newspaper as he waited for the flight to depart. The date on the front page of the newspapers that fateful morning was 11 September 2001.

Right on schedule the front left passenger door was closed, the engines started up prior to push-back from the gate, and the standard pre-flight safety announcements made. As the attendants checked that the cabin was secure for take-off, the 757 rolled out to active runway 30 where it paused before continuing, the pilots waiting for a take-off clearance from the tower. Once this was given, the aircraft turned onto the runway and the pilots applied full thrust to the engines. Slowly at first, the 757 began trundling down the 10,500-foot concrete runway, rapidly picking up speed. At 8:20 a.m., with a gentle pull on the pilot's control column, the aircraft's nose lifted and Flight 77 roared into the skies in a normal, smooth take-off.

THE TRAGEDY OF FLIGHT AA77

On board were six crewmembers and fifty-eight passengers, comprising twenty-five women, five children ranging in ages from five to eleven, and thirty-four men. At 71 years of age, John Yamnicky was the oldest male passenger. Most people were flying from Dulles, Virginia to California for the usual variety of reasons, mostly business. That day John was on his way to Los Angeles for Veridian Engineering, a Virginia-based military contractor, where he worked on fighter aircraft and air-to-air missile programs. But there were five males on board Flight 77 that day with just one deadly purpose – to hijack the aircraft and fly it directly into the Pentagon building as part of a massive and coordinated al-Qaeda terrorist attack against the United States.

The last normal radio call made by the American Airlines flight crew to air traffic control was registered at 8:51 a.m. Three minutes later it was noted with concern that the 757 had deviated from its planned flight path and was unexpectedly heading in a southerly direction. In reconstructing the events on board, it is known that the five terrorists left their seats brandishing knives and box cutters that they had managed to carry through airport security checks. While a couple stormed the cockpit and seized control of the aircraft, the others forced the startled crew members and passengers to move to the rear of the aircraft.

One of the al-Qaeda operatives had trained as a pilot in the United States and held a commercial pilot's license. He quickly reset the 757's autopilot for a direct route to Washington D.C., then disabled the transponder that automatically transmitted the aircraft's position and heading. Neither American Airlines nor the Indianapolis Air Traffic Control was able to establish contact with the aircraft.

Possibly unknown to the hijackers, two people managed to make calls on their cell phones. At 9:12 a.m. Renee May, one of the flight attendants, called her mother and gave vital information concerning the hijacking and the hijackers; details her mother passed straight to American Airlines. In addition, Barbara Olsen, a passenger, was able to speak to her husband twice and provide information on the hijackers. During their second conversation, her husband asked where they were and Barbara told him they were flying over a residential area.

By this time, air traffic controllers and the military knew that American Airlines Flight 11 had hit the North Tower of the World Trade Center in New York, and that United Airlines Flight 175 had been hijacked after departing Boston.

Five miles west-southeast of the Pentagon, Flight 77 began a 330-degree turn and descended rapidly. On the flight deck, the terrorist pilot gave the aircraft full throttle and aimed it like a missile directly at the Pentagon. At 9:37:46 a.m. the jet slammed into the western side of the building at the first-floor level, disintegrating and sending a massive ball of fire into the air. Everyone on the aircraft died, as did 125 people in the building.[15]

In due course, John Yamnicky was buried as a hero in Section 64 at Arlington National Cemetery.

A NATION MOURNS

At the time of writing, Master Sergeant Jennifer Yamnicky is serving as an egress technician with the District of Columbia Air National Guard's 113th Maintenance Squadron, currently based in Bagram, Afghanistan. A major involvement in her work focuses on F-16 ejection seats and canopies, and saving the lives of military pilots when the loss of their aircraft becomes inevitable. As the younger daughter of John Yamnicky, she told the author she feels good about usefully contributing

Capt. John D. Yamnicky, USN (Photo: USN)

through her efforts "in the fight against the people that brought this horrible, senseless act to our home. Dad loved the military and loved that I joined. The ironic thing is that he was a test pilot in the Navy and had several close calls in aircraft, but always survived them."[16]

On the morning of 11 September 2001, Jennifer Yamnicky, who was at that time a Technical Sergeant, was working at Andrews AFB, Maryland when the unbelievable news of an aircraft hitting the North Tower of the World Trade Center reached her office. Then the horrors of that day began to rapidly escalate.

"We'd already heard one airplane had crashed into the World Trade Center, but we figured it was a Cessna," she recalled. "When the second plane hit, things went haywire around here." Now on high military alert, her unit quickly covered all the windows, locked all the doors and secured the area. Even as updates of the shocking events reached Jennifer and her unit, there was worse news yet to come.

"We'd all been so busy securing the building that I hadn't had time to think about Dad," she reflected. Then someone told her reports were coming in of a commercial airliner being deliberately flown into the Pentagon. Jennifer made her way outside through the back door of a hangar and looked in the direction of the Pentagon, some twelve miles distant. She could see black billows of smoke rising ominously into the sky, and the first fingers of fear for her father began to creep up her spine. She knew he was flying out of Dulles Airport bound for Los Angeles that day, but despite her growing apprehension tried to reassure herself in the knowledge that vast numbers of commercial flights were operating the air routes that day. Then she learned that the aircraft flown into the Pentagon had originally been bound for California. Terrified of the answer that she might get, Jennifer reached for her office phone. "I decided to call Mom to ask her what flight he was on."

Jennifer knew that it was her father's regular habit to place a copy of his itinerary on the refrigerator. However, her mother, also in a deep panic of uncertainty, said he must have forgotten to do it this time. "I kept telling myself it wasn't his plane, but I guess I kind of knew," Jennifer said. "I knew it wasn't good, because there's no way he wouldn't have gotten in touch with us. But I kept imagining long lines at the phones with people calling family to let them know they were okay."

By this time, Jennifer's unit had gone onto a full alert, with 24-hour operations ordered. As she tried to get more information, F-16 fighter jets were being scrambled into the skies, the pilots now on high homeland defense mode. Her supervisor said to go home, get a change of clothes, and prepare herself for a long and arduous day. It was a journey she will never forget. "When I drove home, I was so nervous I was shaking."

Ten minutes after pulling up at her home, the telephone rang. "I knew I didn't want to answer it. I picked up the phone and it was my Mom." All she could hear for the next several moments was the sound of deep, chilling sobs, and tears flooded into her eyes. Then her mother managed to say a single, strangled word – "Jennifer". This served to confirm the worst. "She couldn't say anything else, but she didn't have to. I told her I'd be right over. Then I called my boss and told him I wouldn't be coming back to work." Only now did the full impact hit Jennifer in waves of savage emotion. "Basically, I lost control," she said. "I screamed and beat on the walls. No way could they have taken him from us."

After ten minutes, much of Jennifer's deep anguish had been spent and she knew that she had to be strong for her mother. "I put myself back together and then I drove to Mom's house. I did a pretty good job of keeping it together. I called people that I needed to call. I went through the motions. Later that night when I was alone, I broke down again. We're a close-knit family, so it was a very emotional time."

For a while Jennifer could not accept that her father would never be coming home. Not until two weeks later, when the family – who had supplied DNA samples for the purpose of victim identification – was told that his remains had been found. "After that I said, 'Okay, I give up. He's definitely not coming home.'"

To ease the grief, Jennifer immersed herself in her work, and the support of her whole unit carried her through the most difficult time she had ever known. She also began going to church, which she realized would have pleased her father, and found herself developing stronger bonds within the military.

"Dad always said life is for the living," Jennifer added. "He wouldn't want anyone hanging out, carrying on, grieving and not getting things done. At work I was busy. It was good for me. I needed the distraction.

"I really try to stay strong because that's the way he was. He was very strong in his faith and very strong for his family.

"People say I'm a lot like my Dad, and that makes me feel good. I do my best to stay busy, but every day I think about what I can do to make him proud of me."[17]

His widow, Jann, explained, "I never did worry about him when he was flying. I had such confidence in his ability. He always said he would die in a plane crash, so when he quit flying, it was a huge relief." She admits that recalling the tragic events of 9/11 is always difficult and painful. "But I don't want it forgotten," she insists. "It changed my whole life. It changed life here in America. I don't know if at my age I'll ever get over it. There is so much I thought I could take care of that I can't take care of without John."

Jann says that her husband's Annapolis gold ring was not among the items found and catalogued at the crash site, but she treasures the miniature Annapolis ring – "my engagement ring" – inscribed inside "My love always: John."

Likewise, her love for him endures. "John was the most honorable and honest man I have ever met. And a devout Catholic. He was truly a man to be admired," she told the author. "One thing that he was always very proud of was the fact that he was a member of the Society of Experimental test Pilots, an honor only given to those who have done something exemplary in test pilot work, and he always wore his SETP pin.

"He loved the farm work and driving the tractor after we moved to the farm. He said it was because you could see the results of the hard work, not like the work he was in at the time. He was very compassionate, and always available to help anyone in need. He was especially aware of the problems of his enlisted men, and was there for them. He was able to be on the same level with whoever he happened to be with, whether admirals or enlisted men, women or children. They all loved him. He always brought back presents for his 'girls' when we traveled, and they loved him. No male chauvinism there; he truly respected women and admired their abilities.

"We all still miss him, all his stories, his gentleness and ability to learn how to do what had to be done."[18]

Their son John David is in full agreement. "This guy was the head of the family; he made everyone feel safe. If he ever talked about accomplishing something, it was as a group or a team. He was a modest man."[19]

His brother Mark said that while his father ran a tight ship at home, it was always with customary fairness to all, and this included allocating chores to the children. Mark did not mind the discipline of doing these at all, and in fact when his friends would call around it was not uncommon for him to say he would be with them "when my chores are complete". He also recalls that their father would always have them up early for Mass on Sundays. "Seven o'clock when it was available, so that we could get back and complete our chores."[20]

A COLLEAGUE RECALLS

Joe Sutliff is now a professional cartoonist based in Washington, D.C., and his work appears in various publications. Over many years with defense contractor Veridian he came to know John Yamnicky well. "I was a co-worker, but he had a lot of friends closer than me," Joe told the author. "He was the kind of person everyone *wanted* as a friend.

"John was the guy you would pick out of a crowd if you were looking for a hero. A big man, but never imposing, he projected strength and confidence. I never saw him angry, but when things were screwing up all around him, it was John who would quietly and calmly set things on the right track. For many years after 9/11, I could not bear the thought of flying, because I would think of John and wonder about what he went through on that flight. Eventually I was reminded by John's daughter that, as a combat pilot, John would be focused on getting control of the situation and saving it. That's how he was, and that's how I like to remember him."

In recognition of his lost friend and colleague, Joe once drew the cartoon tribute reproduced here. After it was published he discovered that the dragon motif he had painted on an eye patch (mentioned in one panel) was actually completed for another friend at work who was attending a costume party disguised as John Yamnicky. "So my dragon was a duplicate," he added, "not the original."[21]

ADDED PAIN

As if losing a husband and father were not enough, there was more pain in store for the Yamnicky family. On 11 September 2003 a remembrance party – "a celebration of life" – was held on the outside patio of a waterside restaurant with many of John's friends and relatives in attendance. Two of his closest friends were there that night; Veridian work colleague Dennis ("Doc") Plautz and his wife Priscilla. They had been of great help and comfort to the family after John's loss, making sure that everything was taken care of. Priscilla had been of particular assistance to Jann. "She arranged the reception after the funeral, took care of any detail that needed looking after; was just the best of everything," Jann reflected. "I loved her dearly, and we were always in close contact." They had arrived at the restaurant in separate vehicles, as Dennis had a reunion function to attend that night, although Priscilla would be staying on at the remembrance party before driving home.

The following morning John David answered the telephone at his mother's house, where he had stayed overnight before returning to Virginia. When he suddenly went pale and silent, Jann instantly knew something had happened. "I could tell something was wrong. He didn't want to tell me, but he finally said that Priscilla had been in a single-car accident. I asked how she was and he said, 'She's dead.' I cry even now thinking about it. Such a lovely and loving person."[22]

The tribute to John Yamnicky drawn by Joe Sutliff. (Courtesy Joe Sutliff)

A LIFE OF HONOR

Today, Lorraine Dixon, the older Yamnicky daughter, works for the government in the same Air-to-Air Missiles Program Office that her father was supporting when he was killed. "That happened four years ago," she revealed. "Not by my design, but it happened anyway."

For Lorraine and her family, the deep pain of tragic loss still exists. "I relive that morning and that day over and over and over again, and I get sick with the memories. They never go away. The closer the anniversary date comes, the more frightened I become. I look at my young children and cry yet again, knowing that my father will not see them grow up. I cry knowing how much he loved them, and how very proud he was of his grandchildren. My son and daughters used to climb all over him – he was so big, and they were so tiny. And he was so proud.

"I go crazy sometimes thinking that my beloved father, who spent his entire life fighting terrorism and evil, could have come to his end at the hands of people he spent his life trying to protect the rest of us against.

"My father was a man who commanded love and respect from all avenues. He could talk to anyone, making them feel as if they were the only person in the world. The best part about him was that he really meant it. He exuded love and confidence, and everyone loved him for it."

John Yamnicky applied this attitude in all areas of life, including his community contributions. Typically, he served on the board of directors at St. Mary's Academy, his daughter Lorraine's all-girl high school. "Can you imagine what it was like to have your dad on the board of directors?" she recalled with a smile. "I couldn't get away with anything!" In 1981 St. Mary's merged with Ryken boys' school, and he played a major role in that. The school now awards an annual Memorial Scholarship in his name to assist an academically talented but financially challenged student.

He joined the Knights of Columbus in 1981 and was appointed to the position of Grand Knight in July 1987. He was also a member of the Thomas Manor Assembly, 4th Degree, the highest Degree of the Order. The Supreme Council bestowed on him the highest award a council could receive, that of Star Council, in recognition of his leadership and guidance. He also chaired campaigns for the mentally handicapped through the Knights of Columbus Tootsie Roll Drives, raising more funds than any previous chairman.

His magnanimity and selflessness also extended to his youth program work as a member of the Elks Lodge in California, Maryland, where he was instrumental in the installation of a swimming pool at the Lodge in order to provide swimming lessons and organized meets for the children of southern Maryland. He was equally proud of his volunteer contributions to the De La Brooke Foxhounds Hunt Club, where he and Jann were members for 25 years. Jann and Jennifer often rode with the hunt club, but he was a non-participant, preferring instead to work voluntarily on the club's support staff – usually as the bartender.

"The one absolutely wonderful thing about my father – he was so proud of all of us, and it showed in everything he did or said," Lorraine said with heartfelt emphasis. "I feel for those who were never close, for whatever reason, to their

John D. Yamnicky, patriot and devoted family man. (Photo courtesy Yamnicky family)

parents, their families. I know I should be thankful for what I had, and I truly am, but it doesn't take away the grief and pain that I feel, today and every day, when I remember, yet again."[23]

American patriot and legislator Henry Clay once said: "Of all the properties which belong to honorable men, not one is so highly prized as that of character." Character is what John David Yamnicky had in abundance. A gentle, loving and compassionate man on one hand, and a determined protector of his nation and its ideals on the other, he had humble origins. But through passion, rigorous dedication and his renowned perseverance he rapidly transformed himself from a farmer's son into a talented and highly resourceful fighter pilot and test pilot. In his desire to serve his nation as an astronaut, his character and reputation stood tall, but in the end it was really only his physical height that prevented him from possibly realizing this particular dream.

Had he been selected as an astronaut, John Yamnicky's determination may well have carried him to the Moon and other glories. We shall never know. But those who do remember him speak highly, and with passion, of his nobleness of character and Christian values, his love of family, his affinity with the land, and his unquenchable desire always to serve his nation proudly, and to the best of his considerable talents and ability.

His close friend Dennis Plautz said it all. "If you could have an example of what you would want to be and how you'd like your life to progress, from work to family and everything, there just isn't anybody else you could have as an example."[24]

Vale, and God Bless: John D. Yamnicky, 6/8/1930–9/11/2001

REFERENCES

1. Worden, Al with Francis French, *Falling to Earth: An Apollo 15 Astronaut's Journey to the Moon*, Smithsonian Books, Washington, D.C. 2011, pg. 21
2. McWilliams, Bill, *A Return to Glory: The Untold Story of Honor, Dishonor & Triumph at the United States Military Academy, 1951-1953*, Warwick House Publishing, Lynchburg, VA, 2000
3. Gould, Jay, online article in West Point football site: *For What They Gave on Saturday Afternoon*. Website: http://forwhattheygaveonsaturdayafternoon.com/wp-1954/jay-gould
4. Worden, Al with Francis French, *Falling to Earth: An Apollo 15 Astronaut's Journey to the Moon*, Smithsonian Books, Washington, D.C. 2011, pp. 21-22
5. Rupp, Dr. Charles, undated eulogy for Alexander Kratz Rupp, USMA '55 in *The West Point Connection*. Website: http://defender.west-point.org/service/display.mhtml?u=20226&i=40424
6. Stemle, Dick, letter posted at West Point *Al Rupp Eulogies*, pages 7-8. Website: http://www.west-point.org/class/usma1955/R/RuppA.htm
7. Rupp, Dr. Charles, undated eulogy for Alexander Kratz Rupp, USMA '55 in *The West Point Connection*. Website: http://defender.west-point.org/service/display.mhtml?u=20226&i=40424
8. *Ibid*
9. E-mail correspondence with Karen Rupp Deming, 13 November 2011–25 April 2012
10. USAF Accident/Incident Report (AF711), *Major Alexander K. Rupp, USAF*, 29 July 1965
11. *F-16 Losses and Ejections, 1978-1989*, entry for 24 July 1987. Website: http://www.ejection-history.org.uk/Aircraft-by-Type/F-16/USAF/f_16_USAF_80s.htm
12. E-mail correspondence with Robert H. Shumaker, 23 February 2012–28 March 2012
13. E-mail correspondence with RADM Lester R. (Bob) Smith, 8 February 2012–9 March 2012
14. E-mail correspondence with Terry Jones, 22 February 2012–19 March 2012
15. Wikipedia online encyclopedia, *American Airlines Flight 77*. Website: http://en.wikipedia.org/wiki/American_Airlines_Flight_77
16. E-mail correspondence with Jennifer Yamnicky, 30 October 2011–2 February 2012
17. Barela, Tim, *Fateful Flight*, USAF News Agency article, courtesy of Jennifer Yamnicky
18. E-mail correspondence with Jann Yamnicky, 5 November 2011–19 April 2012
19. E-mail correspondence with John D. Yamnicky, Jr., 4 February 2012
20. E-mail correspondence with Mark Yamnicky, 1 February 2012
21. E-mail correspondence with Joe Sutliff, 1-2 May 2012
22. E-mail correspondence with Jann Yamnicky, 5 November 2011–19 April 2012
23. E-mail correspondence with Lorraine Yamnicky Dixon, 30 October 2011–7 July 2012
24. E-mail correspondence with Dennis Plautz, 23 January 2012

9

The Fourteen

At a press conference held at the Manned Spacecraft Center, Houston on 18 October 1963, Deke Slayton introduced America's newest astronauts, taking to thirty the total assigned to NASA's astronaut training center. After silence was called for, Slayton ran through some selection details:

"Okay, I guess you all know why you are here today and why we are here. We would like to introduce the new group of astronauts that we've been in the process of selecting for about the last four months.

"As a preliminary I might just give you a brief rundown on how we ended up with these fourteen individuals. I think you're all familiar with what our basic criteria was this time, versus the last. The only – what you might consider major – change, was that we went to operational pilots with over a thousand hours of flight time. We did not require, or make mandatory, the requirement for experimental test pilots. Based on these criteria we ended up with approximately 500 military and 225 civilian applications. Out of these we ended up with 71 military recommended to us, and 65 civilians who were eligible. We pre-screened this group and selected 34 out of it for interviews. In the process of the medical exams at Brooks, six of these were dropped as having medical problems which were unacceptable, which left us with 28, and out of that 28 we ended up with this group of 14 we have here today."

One by one the new astronauts took the microphone and offered a brief account of their birthplace, and current places of residence and occupation. Then Slayton asked for questions from the assemblage, most of which were entirely predictable. The men were asked for their religious affiliations, whether they had been Boy Scouts, and if their wives were happy with the news of their selection. Sole bachelor C.C. Williams caused the biggest laugh of the session by responding to that question with, "None of my lady friends have voiced any objection." The introduction and questions from the floor lasted 35 minutes. With the formal proceedings concluded, the press was free to move close and take photographs.[1] In fact the way that Mike Collins remembered the press conference was that "it was mainly a picture-taking and get-acquainted session, a very pleasant day that had been a long time coming".[2]

Summaries provided to the press pointed out that the recruitment effort had been announced by NASA on 5 June, with a deadline of 1 July for applications. After that date the following actions were taken prior to making the selection:

- 15 July: Deadline for the receipt of all papers requited of the applicants.
- 17-20 July: The selection committee considered the applications and selected 34 candidates for detailed evaluation.
- 25 July: Medical examinations of those men were started by the School of Aerospace Medicine, Brooks AFB, Texas, under contract from the Manned Spacecraft Center. Working with the resulting data, MSC medical authorities selected 28 candidates as medically qualified.
- 2-7 September: These men were called to Houston for examinations relating to engineering and space sciences, technical interviews and final evaluation.[3]

SERVICE DETAILS

The new group of astronauts was composed of seven volunteers from the Air Force, four from the Navy, one from the Marine Corps, and two civilians.

The Air Force officers were Maj. Edwin E. Aldrin, Jr., 33, assigned at Houston, Texas; Capt. William A. Anders, 30, and Capt. Donn F. Eisele, 33, both assigned to Kirtland AFB, New Mexico; and Capt. Charles A. Bassett II, 31, Capt. Theodore C. Freeman, 33, Capt. David R. Scott, 31, and Capt. Michael Collins, 32, all assigned at Edwards AFB, California.

The Navy officers were Lt. Cmdr. Richard F. Gordon, Jr., 34, and Lt. Eugene A. Cernan, 29, both assigned at Monterey, California; Lt. Alan L. Bean, 31, Cecil Field, Florida; and Lt. Roger B. Chaffee, 28, Wright-Patterson AFB, Ohio.

The sole Marine was Capt. Clifton C. Williams, Jr., 31, of Quantico, Virginia. He was also NASA's first bachelor astronaut.

NASA's Group 3 astronauts. Back row, from left: Mike Collins, Walt Cunningham, Donn Eisele, Ted Freeman, Dick Gordon, Rusty Schweickart, Dave Scott and C.C. Williams. Front row: Buzz Aldrin, Bill Anders, Charlie Bassett, Alan Bean, Gene Cernan and Roger Chaffee. (Photo: NASA)

The two civilians were R. Walter Cunningham, 31, a research scientist for Rand Corporation at Van Nuys, California, then a captain in the Marine Corps Reserve; and Russell L. Schweickart, 27, a research scientist from Lexington, Massachusetts, working at the Massachusetts Institute of Technology in Cambridge, Massachusetts.[4]

There now follows a biographical study of each of the fourteen new astronauts, as compiled by NASA and handed out to members of the press at the 18 October 1963 news conference.[5]

EDWIN E. ALDRIN, JR., USAF

Maj. Edwin E. Aldrin, Jr., 206 Confederate Way, El Lago, Texas, was born in Glen Ridge, New Jersey, January 20, 1930. He is the son of Col. and Mrs. Edwin E. Aldrin (USAF retired), 180 Walnut, Montclair, New Jersey, currently living at their summer home, 38 First Avenue, Manasquan, New Jersey. He completed his secondary education at Montclair High School.

Aldrin graduated third in a class of 475 from the United States Military Academy at West Point, New York, in 1951 with a bachelor of science degree, transferred to the Air Force, and in 1963 received a doctor of science degree in astronautics from the Massachusetts Institute of Technology at Cambridge, Massachusetts. His thesis concerned manned orbital rendezvous.

He is 5 feet 10 inches tall, weighs 165 pounds, and has blond hair and blue eyes. His wife is the former Joan Ann Archer, daughter of Mr. and Mrs. Michael Archer, 50 Edgewood, Ho-Ho-Kus, New Jersey. The Aldrins have three children: James M., 8; Janice Ross, 6; and Andrew John, 5.

Prior to his appointment as an astronaut, Aldrin's last assignment with the Air Force was with the Space Systems Division's Field Office at Manned Spacecraft Center in Houston where he was doing work concerning integrating Department of Defense experiments in the Gemini-Titan II flights. Before that, he served as an engineer in the Gemini Target Division of the Space System Division with work centered on a study effort performed by Lockheed Aircraft Corporation concerning the maneuver capabilities of the Agena target vehicle. He has amassed more than 2,500 hours flying time, including more than 2,200 hours in jet aircraft. On duty in Korea, he was credited with two enemy fighter kills.

Aldrin is a member of the American Institute of Aeronautics and Astronautics; Sigma Gamma Tau, Aeronautical Engineering Society; Tau Beta Pi, National Engineering Society; and Sigma Xi, National Science Research Society.

WILLIAM A. ANDERS, USAF

Capt. William A. Anders, who observed his 30th birthday yesterday, was born in Hong Kong, where his father was based on military duty, October 17, 1933. He lives at 10420 Princess Jeanne NE, Albuquerque, New Mexico. He is the son of

Cmdr. and Mrs. Arthur F. Anders (USN retired), 4602 Resmar Drive, La Mesa California.

Anders graduated from the U.S. Naval Academy at Annapolis, Maryland, in 1955 with a bachelor of science degree. On graduation, he was commissioned in the Air Force. He received a master of science degree from the Air Force Institute of Technology at Wright-Patterson AFB, Ohio. He has done additional graduate work at Ohio State University.

He is 5 feet 8 inches tall, weighs 150 pounds, and has brown hair and blue eyes. Anders is married to the former Valerie Elizabeth Hoard, daughter of Mr. and Mrs. Henry G. Hoard, 2481 Bonita, Lemon Grove, California. They have four children: Alan Frank, 6; Glen Thomas, 5; Gayle Alison, 3; and Gregory Michael, 1.

His last assignment was as a nuclear engineer-instructor pilot at the Air Force Weapons Laboratory, Kirtland AFB, New Mexico. He had technical management responsibility for space and space reactor radiator shielding and radiation effects programs.

Anders has logged more than 1,800 hours flying time, including almost 1,600 hours in jet aircraft. He is a member of Tau Beta Pi, National Engineering Society; and the American Nuclear Society.

CHARLES A. BASSETT II, USAF

Capt. Charles A. Bassett II, who lives at 6848 Lindbergh, Edwards, California, was born in Dayton, Ohio, December 30, 1931. His mother, Mrs. Belle James Bassett, lives at 4419 Groveland Royal Oaks, Michigan. He received his secondary education at Berea, Ohio.

Bassett attended Ohio State University from 1950 to 1952 and Texas Technological College, Lubbock, Texas from 1958 to 1960. He received a bachelor of science degree in electrical engineering with high honors from Texas Tech. Since then he has done graduate work at the University of Southern California. He entered the Air Force in October 1952.

He is 5 feet 9½ inches tall, weighs 160 pounds, and has brown hair and brown eyes. He is married to the former Jean Marion Martin, daughter of Mr. and Mrs. Wiley O. Martin of Hesperia, California. They have two children; Karen Elizabeth, 6; and Peter Martin, 2.

His last Air Force assignment was as experimental test pilot and engineering test pilot in the Fighter Projects Office at Edwards AFB, California. Bassett is a graduate of the Aerospace Research Pilot School and the Air Force Experimental Test Pilot Course.

He has logged almost 2,800 hours flying time, with almost 2,100 hours in jet aircraft included. He is a member of the American Institute of Aeronautics and Astronautics; and Phi Kappa Tau.

ALAN L. BEAN, USN

Lt. Alan L. Bean, 4371 Water Oak Lane, Jacksonville, Florida, was born in Wheeler, Texas, March 15, 1932. His parents, Mr. and Mrs. Arnold H. Bean, live at 3100 Bellaire Drive West, Fort Worth, Texas.

Bean received his high school diploma from Paschal High School in Fort Worth in 1950 and a bachelor's degree in aeronautical engineering from the University of Texas in 1955 and was commissioned in the Navy.

He is 5 feet 9½ inches tall, weighs 150 pounds, and has brown hair and hazel eyes. His wife is the former Sue Ragsdale, daughter of Mr. and Mrs. Edward B. Ragsdale, 6914 Hyde Park Drive, Dallas, Texas. The Beans have two children: Clay Arnold, 7; and Amy Sue, born this year.

Bean's last Navy assignment was with Attack Squadron 44 at Cecil Field, Florida, as an A4 attack replacement pilot. He graduated from the Navy Test Pilot School at Patuxent and served as project officer on various aircraft for Navy preliminary evaluation, initial trials, and final board of inspection and survey trials at Patuxent from 1960 to 1963. He also attended the School of Aviation Safety at the University of Southern California.

He has more than 2,000 hours flying time, including about 1,800 in jet aircraft.

EUGENE A. CERNAN, USN

Lt. Eugene A. Cernan, 1410 Via Marettimo Way, Monterey, California, was born in Chicago, Illinois, March 14, 1934. His parents, Mr. and Mrs. Andrew G. Cernan, reside at 939 Marshall, Bellwood, Illinois. He received his high school diploma from Proviso High School at Maywood, Illinois.

Cernan attended Purdue University at West Lafayette, Indiana, and graduated in 1956 with a bachelor of science degree in electrical engineering. He entered the Navy the same year. Since 1961 he has been a student at the U.S. Naval Postgraduate School at Monterey and is currently a candidate for a master's degree in aeronautical engineering. Prior to his last assignment he was a member of Attack Squadrons 126 and 113 at Miramar, California, Naval Air Station.

He is 6 feet tall, weighs 175 pounds, and has brown hair and blue eyes. He is married to the former Barbara Jean Atchley of Corpus Christi, Texas, whose mother, Mrs. Jackie Mae Atchley, lives at 112 John A. Street, in Baytown, Texas. They have a daughter, Teresa Dawn, born this year.

Cernan has logged more than 1,400 hours flying time, including more than 1,200 hours in jet aircraft. He is a member of Tau Beta Pi, National Engineering Society.

ROGER B. CHAFFEE, USN

Lt. Roger B. Chaffee, 1960 Redstone Drive, Fairborn, Ohio, was born in Grand Rapids, Michigan, February 15, 1935. His parents, Mr. and Mrs. Donald L. Chaffee, live at 3710 Hazelwood SW, Grand Rapids, and he attended Central High School in that city.

Chaffee attended the Illinois Institute of Technology in Chicago, Illinois, for one year, then transferred to Purdue University. He graduated with a bachelor of science degree in aeronautical engineering in 1957 and entered the Navy in August that year. His last Navy assignment started in January 1963 as a student at the Air Force Institute of Technology at Wright-Patterson AFB, Ohio, where he is working toward a master of science degree in reliability engineering. Prior to entering AFIT, he was safety officer and quality control officer for heavy Photographic Squadron 62 at the Jacksonville, Florida, Naval Air Station.

He is 5 feet 9½ inches tall, weighs 157 pounds, and has brown hair and brown eyes. Chaffee is married to the former Martha Louise Horn, whose parents, Mr. and Mrs. Henry W, Horn, live at 1801 Dorchester Place, Oklahoma City. They have two children: Sheryl Lyn, 5; and Stephen Bruce, 2.

Chaffee has logged nearly 1,700 hours flying time, including more than 1,400 hours in jet aircraft. He is a member of Tau Beta Pi, National Engineering Society; Sigma Gamma Tau; and Phi Kappa Sigma.

MICHAEL COLLINS, USAF

Capt. Michael Collins, 6766 Rickenbacker Drive, Edwards, California, was born in Rome, Italy, October 31, 1930, where his father, M/Gen. James L. Collins (USA deceased), served as military attaché. His mother, Mrs. James L. Collins, now resides at 2126 Connecticut Avenue NW, Washington, DC. He attended Albans School in Washington in 1948.

Collins attended the United States Military Academy, graduated in 1952 with a bachelor of science degree, and chose an Air Force career. His last assignment was as an experimental flight test officer at the Air Force Flight Test Center, Edwards AFB, California. In that capacity he tested performance and stability and control characteristics of Air Force aircraft, primarily jet fighters.

He is 5 feet 10½ inches tall, weighs 168 pounds, and has brown hair and brown eyes. Collins married the former Patricia Mary Finnegan of Boston, Massachusetts, in Chambley, France, in 1957. They have three children: Kathleen, 4; Ann Stewart, 2; and Michael Lawton, born this year.

Collins has flown more than 3,000 hours, including more than 2,700 hours in jet aircraft. He is a member of the Society of Experimental Test Pilots.

R. WALTER CUNNINGHAM

R. Walter Cunningham of 6640 Rubio Avenue, Van Nuys, California, was born in Creston, Iowa, March 16, 1932. His parents, Mr. and Mrs. Walter W. Cunningham, reside at 1022 Nowita Place, Venice, California, and he completed his secondary education at Venice High School.

He joined the Navy in January 1951 and went into flight training in July 1952. He joined a Marine squadron in 1953 and remains a Marine air reservist with the rank of captain, flying with VMA-134 at the Los Alamitos, California, Naval Air Station.

Cunningham, one of the two civilians selected, has been a research scientist for Rand Corporation. He received from the University of California at Los Angeles a bachelor of arts degree in physics with honors in 1960 and a master of arts degree in physics in 1961. He is currently completing requirements for a doctorate in physics at UCLA.

While working for Rand Corporation, he performed error analysis and feasibility studies of defense against submarine-launched ballistic missiles and problems of the Earth's magnetosphere. His latest work at UCLA has concerned development, testing and analysis of results of a tri-axial search coil magnetometer which will be flown aboard the first NASA Orbiting Geophysical Observatory satellite.

He is 5 feet 10 inches tall, weighs 165 pounds, and has blond hair and blue eyes. Cunningham is married to the former Lo Ella Irby of Anaheim, California, whose mother, Mrs. Nellie Marie Maynard, lives at 2371 Ventura Boulevard, Oxnard, California.

Cunningham has logged almost 2,000 hours of flying time, including more than 1,350 hours in jet aircraft. He is a member of Sigma Pi Sigma, the American Geophysical Union; and Sigma Xi, National Science Research Society.

DONN F. EISELE, USAF

Capt. Donn F. Eisele, 2059A Crossroads Place, Kirtland AFB, New Mexico, was born in Columbus, Ohio, June 23, 1930. Mr. and Mrs. Herman E. Eisele, his parents, live at 248 N. Murray Hill Road, in Columbus, and Eisele attended West Senior High School there.

He attended the United States Naval Academy, and received a bachelor of science degree in 1952. He opted for an Air Force career. In 1960 he received a master of science degree in astronautics from the Air Force Institute of Technology at Wright-Patterson AFB, Ohio.

Eisele is 5 feet 9 inches tall, weighs 150 pounds, and has brown hair and blue eyes. He is married to the former Harriet Elaine Hamilton of Gnadenhutten, Ohio. Her parents, Mr. and Mrs. Harry D. Hamilton, live at 156 Moravian Avenue SW in that community. The Eiseles have three children: Melinda Sue, 9; Donn Hamilton, 7; and Matthew Reed, 2.

His last Air Force assignment was as flight commander and experimental test pilot at the Air Force Special Weapons Center at Kirtland AFB. In this capacity he flew experimental and developmental test flights in jet aircraft in support of special weapons development programs.

He has amassed more than 2,500 hours flying time, with more than 2,100 hours in jet aircraft. Eisele is a member of Tau Beta Pi, National Engineering Society.

THEODORE C. FREEMAN, USAF

Capt. Theodore C. Freeman, 6757 Rickenbacker Drive, Edwards, California, was born in Haverford, Pennsylvania, February 18, 1930. His parents, Mr. and Mrs. John Freeman, live near Lewes, Delaware, where Freeman completed his secondary education in 1948.

He attended the University of Delaware at Newark for one year, then entered the United States Naval Academy and graduated in 1953 with a bachelor of science degree. He then elected to serve with the Air Force. He received a master of science degree in aeronautical engineering from the University of Michigan in 1960.

Freeman is 5 feet 10½ inches tall, weighs 139 pounds, and has brown hair and brown eyes. He is married to the former Faith Dudley Clark, whose parents, Mr. and Mrs. Walter E. Clark, Jr., live on Grassy Hill Road, Orange, Connecticut. They have a daughter, Faith Huntington, 9.

His last assignment was as a flight test aeronautical engineer and experimental flight test instructor at the Air Force's Aerospace Research Pilot School at Edwards AFB, California. He served primarily in performance flight testing and stability testing areas.

He has logged more than 3,000 hours flying time, including more than 2,000 hours in jet aircraft. Freeman is a member of the American Institute of Aeronautics and Astronautics.

RICHARD F. GORDON, JR., USN

Lt. Cmdr. Richard F. Gordon, Jr., 1106 Spruance Road, Monterey, California, was born in Seattle, Washington, October 5, 1929. Mrs. Richard F. Gordon, his mother, lives at 7336 17th Street NE in that city. He completed his secondary education at North Kitsap High School, Poulsbo, Washington.

Gordon received a bachelor of science degree in chemistry from the University of Washington in 1951 and entered the Navy in August that year. At the time of his selection as an astronaut he was a student at the U.S. Naval Postgraduate School at Monterey. Gordon is a graduate of the All-Weather Flight School and the Navy's Test Pilot School. Prior to entering the Monterey school he was assigned to Fighter Squadron 96 at the Miramar, California, Naval Air Station, where he had served as flight safety officer, assistant operations officer and ground training officer.

He is 5 feet 7 inches tall, weighs 150 pounds, and has brown hair and hazel eyes. He is married to the former Barbara Jean Field of Seattle, Washington. Her parents, Mr. and Mrs. Chester W. Field, live near Freeland, Washington. The Gordons have six children: Carlee Elizabeth, 9; Richard F. III, 8; Lawrence Joseph, 6; Thomas Alan, 4; James Edward, 3; and Diane Marie, 2.

Gordon has logged nearly 2,800 hours flying time, with almost 2,000 hours in jet aircraft. He won the Bendix Trophy Race from Los Angeles to New York in 1961.

RUSSELL L. SCHWEICKART

Russell L. Schweickart, who will observe his 28th birthday next Friday, was born in Neptune, New Jersey, October 25, 1935. He now lives at 4 Third Street, Lexington, Massachusetts. Mr. and Mrs. George L. Schweickart, his parents, live at 6 Eighth Avenue, Seagirt, New Jersey.

After receiving his secondary education at Manasquan (NJ) High School, he attended Massachusetts Institute of Technology where he received a bachelor of science degree in aeronautical engineering in 1956 and a master of science degree in aeronautics and astronautics in 1963. His thesis was on stratospheric radiance.

Schweickart entered the Air Force in 1956 and became a pilot. He went on inactive duty in 1960 to return to MIT, but was called up for another year of active duty in 1961. He holds the rank of captain in the Massachusetts Air National Guard.

He is 6 feet tall, weighs 158 pounds, and has red hair and blue eyes. He is married to the former Clare Grantham Whitfield, whose parents, Mr. and Mrs. Randolph Whitfield, live at 2540 Dellwood Drive NW, Atlanta, Georgia. They have two daughters: Vicki Louise, 4; and Elin Ashley, 2; and twin sons: Russell Brown and Randolph Barton, 3.

Prior to his selection as an astronaut Schweickart was a research scientist at the Experimental Astronomy Laboratory at MIT. His duties there included research in upper atmosphere physics and applied astronomy as well as research in star tracking and stabilization of stellar images. He has logged more than 1,250 hours flying time, including almost 1,100 hours in jet aircraft.

DAVID R. SCOTT, USAF

Capt. David R. Scott, 107 16th Street, Edwards, California, was born in San Antonio, Texas, June 6, 1932. His parents, B/Gen. and Mrs. Tom W. Scott (USAF retired), now live at 8438 Paseo Del Ocaso, La Jolla, California.

He attended the University of Michigan for one year, then entered the United States Military Academy and received a bachelor of science degree in 1954. At West Point he finished fifth in a class of 633, and chose an Air Force career.

He attended Massachusetts Institute of Technology from 1960 to 1962 and earned both a master of science degree in aeronautics and astronautics, and an engineer of aeronautics and astronautics degree. His thesis concerned interplanetary navigation. At the time of his selection for the astronaut program he was a student at the Air Force Aerospace Research Pilot School at Edwards AFB, California.

Scott is 6 feet tall, weighs 190 pounds, and has blond hair and blue eyes. He is married to the former Ann Lurton Ott, daughter of B/Gen. and Mrs. Isaac W. Ott (USAF retired), who live at 115 Lagos Avenue, San Antonio, Texas. The Scotts have two children: a daughter, Tracy Lee, 2; and a son, William Douglas, born this year.

He has logged more than 2,300 hours flying time, including nearly 2,100 hours in jet aircraft. Scott is a member of Tau Beta Pi, National Engineering Society; Sigma Xi, National Science Research Society; Sigma Gamma Tau; and Sigma Chi.

CLIFTON C. WILLIAMS, JR., USMC

Capt. Clifton C. Williams, Jr., stationed at Quantico, Virginia, was born in Mobile, Alabama, September 26, 1932. He is the son of Mr. and Mrs. Clifton C. Williams who reside at 115 Mohawk Street, Mobile, and he attended Murphy High School there.

Williams attended Spring Hill College from 1949 to 1951, then transferred to Auburn University where he completed his college work and graduated in 1954 with a bachelor of mechanical engineering degree. He entered the Marines in August 1954. Williams is a graduate of the Navy Test Pilot School at Patuxent, Maryland, and is currently a student at the Marine Corps Intermediate Staff and Command School at Quantico.

He is 6 feet tall, weighs 187 pounds, and has brown hair and brown eyes. Williams is the only single astronaut selected for the NASA program to date.

Prior to his assignment to the Marine School, he served at Patuxent as the F8 project officer, A4 project officer, and short airfield tactical support officer.

Williams has logged more than 1,800 hours flying time, including more than 1,300 hours in jet aircraft. He is a member of Sigma Chi, Pi Tau Sigma, National Honorary Mechanical Society; Tau Beta Pi, National Engineering Society; and an associate member of the Society of Experimental Test Pilots.

MIXED FORTUNES

As Dave Scott told the author, he was delighted to have been chosen as one of the new astronauts, with one qualifier. "I was very fortunate to have been selected with a great group of aviators. However, I was disappointed that Mike Adams was not selected; he was a good friend and an exceptional pilot. It would be interesting to learn why he was passed over."[7]

Without being given access to the medical records clarifying Adams's exclusion, it can only be surmised that this resulted from the F-104 Starfighter engine failure incident, when both men had been slightly injured and hospitalized following the loss of the aircraft – mostly for observation. Adams certainly had not suffered any serious injury after ejecting from the crippled aircraft, and in fact was snapped up for the Air Force's MOL program not long after. He passed all the official medical tests with no problems recorded. But with the known caution of the program physicians back then, it appears the only thing to make a specific difference in NASA's astronaut selection was the fact of his recent and superficial ejection injuries. Despite never finding or understanding the answer to this question, Scott did derive some satisfaction in knowing that Adams stayed on at Edwards AFB and had a stellar – if tragically short – career, including flying the X-15 into space and achieving his Air Force astronaut wings, albeit posthumously.

MOON BOUND COMPARITIVE FACTS TABLE: GROUPS 1, 2 AND 3

Place of Birth:

The seven 1959 Group 1 (Mercury) astronauts were born in Colorado, Oklahoma, Ohio, Indiana, New Jersey, New Hampshire and Wisconsin.

The nine 1962 Group 2 astronauts came from Ohio (2), Texas (2), California, Pennsylvania, Indiana, Illinois and Oklahoma.

The fourteen 1963 Group 3 astronauts were born in Italy and Hong Kong, Texas (2) Ohio (2), New Jersey (2), Illinois, Michigan, Iowa, Washington, Alabama and Pennsylvania.

Altogether, the 30 astronauts selected in the first three groups came from 15 U.S. states and two were born overseas (Anders in Hong Kong and Collins in Italy). Those born in the United States came from Ohio (5), Texas (4) New Jersey (3), Illinois (2), Pennsylvania (2), Oklahoma (2), Indiana (2), and one each from Colorado, New Hampshire, Wisconsin, California, Michigan, Iowa, Washington and Alabama.

Physical:

The average age of the Group 1 astronauts at the time of selection was 34.5 years; Group 2, 32.5 years; and Group 3, 31 years.

The average weight of the Group 1 astronauts was 159 pounds; Group 2, 161.5 pounds; and Group 3, 162 pounds.

The average height of the Group 1 astronauts was 69.79 inches; Group 2, 69.94 inches, and Group 3, 70.1 inches.

Flight experience:

There was a natural decline in the total number of flight hours logged by the respective groups due to the fact that they were progressively younger. The 1959 pilots had an average of more than 3,500 hours flying time, with 1,700 hours in jet aircraft. For Group 2 the average was 2,800 hours, including 1,000 hours in jets, and in Group 3 the average was more than 2,300 hours, with 1,800 in jet aircraft.

Month of Birth:

Although it bears little relevance to the selection process, the most common birth month within the first three astronaut groups – a total of 8 out of 30 – was March; three from Group 1, two from Group 2, and three from Group 3.[6]

THE "FOURTEENTH MAN"

In June 2012, while being interviewed for this book, Walt Cunningham revealed an amusing tale concerning his own selection in the third astronaut group. With a wide grin he confided, "You know, I'm probably the only one who knows exactly where he was ranked out of the fourteen people selected in that group."

Cunningham went on to relate that one of his good friends from those times was Paul Haney, then NASA's Chief of Public Affairs at MSC. Haney, who died in 2009, was a former Washington journalist who went on to become the 'voice' of manned space flight for NASA's Gemini and Apollo missions. "About ten years ago, Paul sent me an email saying he had *finally* gotten around to reading my book [*The All-American Boys*] which was published back in 1977. He said how much he'd enjoyed it, and then intrigued me by saying, 'You do know of course that you were Number 14 to be selected?' Now Paul was a real insider back then, so I pursued the story further with him. All I knew was I had no idea where I'd been ranked.

"Paul then told me that about a week before the announcement of the 1963 group, MSC Director Dr. Gilruth had called a meeting so Deke Slayton could brief his staff on those who had been selected, including brief biographies. At the time the meeting began, Paul was on the phone elsewhere, and when he finally walked in Deke was just finishing up. As Deke completed his briefing on the newly selected astronauts, Max Faget wanted to say something." Maxime Faget, Director of Engineering and Development at MSC, was not only chiefly responsible for the design of the Mercury spacecraft but also contributed to the Gemini and Apollo spacecraft, and later to the space shuttle. "Max stuck his hand up," Cunningham continued. "Now, as well as being a major influence in just about everything, Max was a totally numerically oriented guy, so everyone valued his opinion. He said, 'Deke, you can't do that. That's thirteen names!' Deke was puzzled and replied, 'Why not? What's the big deal about thirteen names?' Max then said, 'It's just not a good number, Deke; you've got to have either twelve or fourteen – not thirteen.'

"So Deke thought about it, and after a while he said, 'Well, I don't want to leave out this last guy, whoever he was. If we have to have fourteen ...' He pulled out another of his folders and said, 'Here he is, then – this is the next in line.' And the name on that folder was Walter Cunningham. So that's how Paul Haney let me know that I was the fourteenth guy chosen!"[8]

REFERENCES

1. Internet Archive, NASA Audio Collection, *14 New Astronauts Press Conference*. Website: http://archive.org/details/14NewAstronautsPressConference
2. Collins, Michael, *Carrying the Fire: An Astronaut's Journeys*, Farrar, Straus and Giroux, New York, NY, 1974, pg. 45
3. NASA New Release MSC-180-63, headed *For Release After 5:00 PM (EDT), October 18, 1963*
4. *Ibid*

5. NASA *Space News Roundup*, issue October 30, 1963, *14 New Astronauts Introduced at Press Conference*, MCS, Houston, TX, pp. 1, 4, 5, 7
6. NASA *Space News Roundup*, issue October 30, 1963, *Old and New, This is How They Compare*, MSC, Houston, TX, pp. 5,7
7. Interview with David Scott by author, Tucson, AZ, 1 June 2012
8. Interview with Walt Cunningham by author, Tucson, AZ, 2 June 2012

10

Patience and Persistence

Some would make it, many would not. Only 14 of the 34 Group 3 finalists would be selected, so 20 candidates had to shake off their disappointment and get on with their lives. There was encouragement for some, and they would try again later. However, Group 4 selected by NASA in June 1965 comprised solely of six men designated as scientist-astronauts. It would not be until 10 September 1965 that the agency initiated the process to select another 10 to 15 astronauts as Group 5.

APPLYING CRITERIA

For the Group 5 selection the criteria were similar to those applied to the third group in 1963, but the maximum age was raised by two years to 36. This kept a few former applicants such as Jim Irwin in the race. As with the second group, NASA would give preference to test pilots but made an allowance for others with only 1,000 hours of high-performance jet flight. And the space agency was again seeking candidates who not only had the ability and attitude to operate in high-stress environments but also had engineering expertise in extremely technical development programs.

Having eliminated the requirement for a test pilot school certificate, NASA began requesting recommendations from a large number of organizations such as industrial aerospace firms, the Air National Guard, various reserve organizations, the military services, the Society of Experimental Test Pilots, the Airline Pilots' Association, the Federal Aviation Administration, and the three NASA field centers which used pilots in their work – the Langley Research Center in Hampton, Virginia, Ames Research Center in Moffett Field, California, and Flight Research Center, Edwards, California. The other organizations included Boeing in Seattle, Washington, Douglas Aircraft in Santa Monica, California, General Dynamics in New York, General Electric in New York, Lockheed Aircraft in Burbank, California, Republic Aviation in Farmingdale, New York, Westinghouse in Pittsburgh, Pennsylvania, McDonnell Aircraft in St. Louis, Missouri, Ling-Temco-Vought in Dallas, Texas, Northrop Space Laboratories in Hawthorne, California, and Grumman Aircraft Engineering in Bethpage, New York.

ASTRONAUT ASPIRANTS

By the final date of 1 December 1965 a total of 351 applications had been received. Interestingly, six of the applicants were women pilots, whilst another was from a determined Lt. Frank K. Ellis, USN, a 32-year-old amputee who had lost both his legs in the crash of a Grumman F-9 Cougar on 11 July 1962. After recovering and being granted flight status in Service Group III – flying dual control aircraft only and accompanied by a qualified co-pilot – Lt. Ellis attended the U.S. Naval Postgraduate School in Monterey, California and it was during this time he applied to NASA for consideration as a Group 5 astronaut, arguing that the loss of his legs did not impair his flying ability, and there was no need for legs in the weightlessness of space. "I learned that I was one of the fifty pilots selected by the Navy," he wrote in an article for the Spring 1996 edition of *Foundation* magazine. "However, when the list was sent to the Chief of Naval Operations my name was no longer on it. A letter sent with the list to NASA said I was not technically qualified for the astronaut program; but from the standpoint of motivation, background, training and experience, would have ranked number five on the list."[1]

Another applicant was physicist Don Leslie Lind. He had applied for the third astronaut group in 1963 armed with a bachelor's in physics and the information that he was working on his doctorate. Unfortunately when Lind applied he did not meet the minimum requirement of 1,000 hours of jet flying time, falling 150 hours short, and his application was set aside. In 1965, already a qualified naval aviator and now with his doctorate in high-energy nuclear physics from the University of California at Berkeley, he applied for the scientist-astronaut group, but missed out again because he was 79 days over the mandatory age limit.

Eventually the Group 5 list was narrowed down to 159 finalists, and they were asked to attend Brooks Air Force Base in San Antonio, Texas for exhaustive physical and psychological examinations. Those who passed had then to report to Houston for further tests and interviews conducted by Deke Slayton, John Young and Michael Collins. Most of the questions concerned the applicant's background and motivation, but they were also asked why they wished to become an astronaut and a variety of technical questions.

After their interview the finalists were given further written tests, and that day, as Charlie Duke recalls, was filled with added drama and misfortune for the established astronaut corps. "I will never forget that on the day of the tests we were told that two astronauts – Charlie Bassett, whom I had met at MIT, and Elliot See – had just been killed in a plane crash at Lambert Field in St. Louis .... This tragic news put a damper on everyone's enthusiasm."[2]

THE ANNOUNCEMENT

For those selected it was a time of great joy and anticipation, but for others the news came as a real blow. RADM George Furlong says missing out on selection a second time was difficult to accept. Years earlier, while serving as a Landing Signals Officer

Lt. Frank Ellis, a student at the Naval Postgraduate School in Monterey, California. He lost his legs in an accident, but still applied for astronaut selection with NASA. (Photo: UPI)

on the aircraft carrier *Shangri-La*, he and squadron buddy Fred Baldwin befriended fellow aviator Gene Cernan and as time passed Cernan and his wife Barbara became close friends with George and Ry Furlong – a friendship that exists to this day. In his book, *The Last Man on the Moon,* Cernan speaks often and fondly of his former squadron colleague, and says he had high hopes that his friend's disappointment in missing out on Group 3 would be tempered by being chosen in Group 5 in 1966. But as Cernan looked through the list of the nineteen who had made it, the one name he'd hoped to see on there was missing. "For several months, I had campaigned on behalf of Skip Furlong, but he didn't make the final cut, and the reason was never known. We had drunk many a beer, talking about the day we would be the first men to walk on the Moon together. It didn't work out that way. He wouldn't go at all, and I would be the last. My disappointment was as great as his."[3]

Although Furlong could accept missing out the first time, falling short once again was a tough thing to swallow.

"Having been through the process once before for the third group I had high expectations of being selected the second time around. Geno and I had been selected for the 1963 finals coming out of postgraduate school and had decided to become the first two astronauts to land on the Moon. Obviously I wasn't selected then or during the 1966 selection, as they selected the 'wrong half' each time! As a result, Geno was the last man on the Moon and I never made it at all. Our track record was not very good. Non-selection was the biggest disappointment of my life. By then I was a lieutenant commander and figured two things kept me from being selected. First, at six foot one inch I was probably the tallest candidate to make the finals. And I had an aunt who prayed every night that I *not* be selected! Also, by the time I was notified of non-selection I was in the first of five deployments to Vietnam. Geno used to say that, vicariously, guys like me fought the wars for guys like him and he flew in space for us. A nice arrangement."[4]

On 4 April 1966, NASA finally announced the names of nineteen new astronauts to join in the space agency's thrust to the Moon and beyond. One of those who had persisted through three selection processes and finally been successful was physicist-pilot Don Lind. Nine of those selected would travel to the Moon on Apollo missions, and three – Charlie Duke, Jim Irwin and Ed Mitchell – would walk on its surface.

Those selected were:

Vance D. Brand
Lt. John S. Bull, USN
Maj. Gerald P. Carr, USMC
Capt. Charles M. Duke, Jr., USAF
Capt. Joe H. Engle, USAF
Lt. Cmdr. Ronald E. Evans, Jr., USN
Maj. Edward G. Givens, Jr., USAF
Fred W. Haise, Jr.
Maj. James B. Irwin, USAF
Don L. Lind
Capt. Jack R. Lousma, USMC

Lt. Thomas K. Mattingly, II, USN
Lt. Bruce McCandless II, USN
Lt. Cmdr. Edgar D. Mitchell, USN
Maj. William R. Pogue, USAF
Capt. Stuart A. Roosa, USAF
John L. Swigert, Jr.
Lt. Cmdr. Paul J. Weitz, USN
Capt. Alfred M. Worden, USAF

Four of those selected this time had also been unsuccessful finalists for Group 3 back in 1963 – civilian test pilot Vance D. Brand, Lt. Cmdr. Ronald E. Evans, Jr., USN, Maj. James B. Irwin, USAF, and a second civilian who had been a finalist for Group 2 selection, John L. ("Jack") Swigert, Jr.

VANCE D. BRAND

Born in Longmont, Colorado on 9 May 1931, Vance DeVoe Brand was the son of Rudolph William Brand and the former Donna DeVoe. Following graduation from Longmont High School in 1949 he undertook studies at the University of Colorado, gaining his bachelor of business degree on 6 June 1953. Over the next four years he served as a Marine Corps officer, completing his flight training in 1955 as a naval aviator and then being assigned a 15-month tour of duty in Japan as a jet fighter pilot.

After his release from active duty, Brand returned to the University of Colorado, graduating on 2 June 1960 with a bachelor's in aeronautical engineering. He then joined the Lockheed Aircraft Corporation, initially as a flight test engineer on the Navy's P-3A Orion aircraft. Throughout the period 1957 to 1964 he also served as a pilot in the USMC Reserve and as a fighter squadron pilot in the Air National Guard. Then in 1962 Lockheed sent him to the Navy Test Pilot School at Patuxent River, Maryland as a member of Class 33, from which he graduated in February 1963. He was encouraged by Lockheed to apply for NASA's third astronaut group later in the year and made the final 34, but was not selected. He remained with Lockheed at Palmdale, California as an experimental test pilot on F-104 Starfighter programs for Canada and West Germany. In 1964 he got his master's in business administration from the University of California, Los Angeles. When finally accepted for NASA's Group 5 astronauts as a civilian pilot on 4 April 1966 he was working for Lockheed at the company's F-104G flight test center in Istres, France as an experimental test pilot and leader of a company flight-test advisory group. Additionally, he still held a commission in the USAF Reserve.

Following his initial training as an astronaut, Brand trained as a specialist on the command and service module (CSM) and in 1968 completed thermal vacuum tests of this spacecraft along with fellow astronauts Joe Kerwin and Joe Engle. He replaced the medically disqualified John S. Bull on the support crew for Apollo 8 and then served as backup command module pilot on Apollo 15, which would have projected

Vance D. Brand. (Photo: NASA)

him into that role on Apollo 18 had this flight not been canceled due to budget cuts imposed by Congress. After specific training he was the backup commander for the final two Skylab missions and in 1972 was awarded a place as a crew member on the Apollo-Soyuz Test Project flight in July 1975.

Brand was then assigned work on the development of the space shuttle, and on 17 March 1978 was assigned to command STS-5, the first operational flight of a space shuttle, launched on 11 November 1982. He also commanded STS 41-B, launched on 3 February 1984, and STS-35, launched on 2 December 1990.

After training for two canceled missions, Brand departed the Astronaut Office in 1992 to work on the National Aerospace Plane (NASP) project at Wright-Patterson AFB, Ohio, later transferring to NASA's Dryden Research Center at Edwards AFB, California as Assistant Chief of Flight Operations. He retired from NASA in January 2008.[5]

RONALD E. EVANS, USN

Although he was born in rural St. Francis in the northwest corner of Kansas, Ronald Ellwin Evans always considered the state's capital Topeka his home town. Born to Clarence E. ("Jim") Evans and the former Marie Priebe on 10 November 1933, he attended the St. Francis Grade School until completing the eighth grade. In 1947 the family moved to Topeka, and while at Highland Park High School he became a member of the National Honor Society, a track star, and an all-Conference football guard. He received a bachelor's degree in electrical engineering from the University of Kansas in 1956. That same year he entered the U.S. Navy through ROTC. He was commissioned an ensign at the end of the course, and in June 1957 he completed his flight training as a naval aviator.

As a fighter pilot in VF-142 based at NAS Miramar, California, he participated in two WestPac deployments on USS *Ranger* (CV-61) and USS *Oriskany* (CV-44) in 1959 and 1960 respectively, flying the F8U-1 Crusader. From January 1961 to June 1962 he was a combat flight instructor for the F8 aircraft with VF-124. He applied for NASA's third astronaut group in 1963 but was not chosen. The following year he

Ronald E. Evans. (Photo: NASA)

received his master's degree from the U.S. Naval Postgraduate School and returned to the WestPac with VF-51, commanded by Cmdr. James B. Stockdale. He flew over 100 combat missions in F8U-2NE Crusaders during seven months operating off USS *Ticonderoga* (CV-14) – the same carrier as would pick up his Apollo 17 crew after their mission to the Moon in 1972. He was the Electronics Officer for VF-51 when notified of his selection by NASA as an astronaut in April 1966.

On 20 November of that year he received his first assignment as a support crew member for Apollo 1, and, owing to events, subsequently for Apollo 7. He was also a member of the support crew for Apollo 11. In 1971 he served as backup command module pilot for Apollo 14, and under the normal process then in place was rotated onto the prime crew for Apollo 17. It began on 6 December 1972 with a spectacular night-time launch from the Kennedy Space Center. As Evans orbited the Moon solo in CSM *America*, mission commander Gene Cernan and geologist Harrison Schmitt, a Group 4 scientist-astronaut, conducted the final moonwalks of the 20th century. On the way back to Earth, Evans made a televised hour-long spacewalk.

Apollo 17 was the first and only space mission for Ron Evans. He retired from the Navy in April 1976 with the rank of captain, and from NASA in March the next year. After working for the Western American Energy Corporation in Scottsdale, Arizona until 1978 he joined Sperry Flight Systems in Phoenix, becoming their Director for Space Systems Marketing. He later formed his own consulting company.

On 7 April 1990, Evans' wife Janet discovered that he had died in his sleep of an apparent heart attack in their Scottsdale home. A memorial service was conducted four days later at Scottsdale's Valley Presbyterian Church.[6]

JAMES B. IRWIN, USAF

Growing up with a fascination for aviation, young James Benson Irwin could never have guessed that one day he would become the eighth person to walk on the Moon. He was born in Pittsburgh, Pennsylvania on 17 March 1930 to James Irwin and the former Elsie Strebel. The fact that he was born on St. Patrick's Day delighted his father, a First World War infantryman who hailed from Pomeroy in County Tyrone, Northern Ireland. A plumber by trade, his father was employed as a steamfitter at the Carnegie Museums of Pittsburgh, running the huge turbines that generated power for the immense building. Young Jim's interest in flying began while he was in second grade, when a neighbor gave him a superb model aircraft, which he treasured. His interest intensified when his father took him on outings to a nearby county airfield to watch airplanes take off and land.

When he was eleven years old, the family moved to New Port Richey and then to Orlando, Florida. Later moves took him to Roseburg, Oregon and to Salt Lake City, Utah where he graduated from East High School in 1947. He won an appointment to the U.S. Naval Academy, from which he graduated on 1 June 1951 with a bachelor's degree in naval science. Irwin then opted to join the Air Force, was commissioned a 2nd lieutenant, and took his flight training in T-6 Texan propeller aircraft at Hondo

AFB and later Reece AFB, both in Texas. He was then assigned as a P-51 Mustang pilot to a squadron in Yuma, Arizona, which subsequently transitioned to the T-33 Shooting Star jet fighter. After obtaining master of science degrees in aeronautical engineering and instrumentation engineering from the University of Michigan, Irwin spent three years as a project officer in the AIM-47 Falcon air-to-air missile program at Wright-Patterson AFB, Ohio.

Capt. Irwin was a member of Class 60-C at the Test Pilot School at Edwards AFB, California, and after graduation in April 1961 stayed at Edwards as a Test Director in the armament section. Soon thereafter, he was told that he was in line to become the first test pilot on the new top-secret Mach 3 Lockheed YF-12A "which we called the A-11" and was a precursor to the SR-71 Blackbird. Prior to taking up this project, he decided to get some extra flying in light airplanes and also qualify for his instructor's license. His pupil one weekend was M/Sgt. Sam Wyman, who had undergone flight training during the Second World War without flying solo. As Irwin recalled in his 1973 memoir, *To Rule the Night*, his student was "nervous and tended to overreact".

As the time approached for Wyman to attempt his first solo landing, he became increasingly worse in his technique, and his landings were not good enough to permit him to fly by himself. His first attempt was clearly not good, landing tail wheel first and bouncing. On the fourth or fifth attempt at landing, disaster struck. "I was in the back seat," Irwin recalled. "He pulled the plane up too abruptly, turned it too tightly onto the crosswind, and we went into an uncontrollable flat spin." The light aircraft crashed into the desert. Wyman's head was smashed into the front panel. Irwin hit the back of the front seat with his head turned sideways. "I was wearing tennis shoes, and when we hit and the front seat collapsed on my feet, particularly my right foot. The seat came down right above my ankle, giving me a compound fracture, with the bones sticking out through the flesh. I had two broken legs, a broken jaw, and a head injury. I was pinned in the wreckage. Fortunately the plane did not catch on fire."

Sam Wyman was critically injured, but survived. Irwin came perilously close to having his leg amputated, but after fourteen months of physiotherapy he was restored to flying status. In the meantime, he had had to relinquish his role as the first test pilot of the YF-12A. He was accepted to ARPS, graduated, and applied to NASA for the third intake of astronauts, but missed the final selection cut. "They didn't tell you why," he pointed out, "but I am sure it was because of that recent accident on my record." At the time of his selection into NASA's Group 5 astronauts in 1966, Irwin was serving as the Chief of the Advanced Requirements Branch at Headquarters, Air Defense Command in Colorado Springs.

After his astronaut training, Irwin became involved in the systems testing program for the lunar module. In 1969 he served on the support crew for Apollo 10 and was then named as backup lunar module pilot on Apollo 12. He rotated onto the prime crew for Apollo 15, with mission commander Dave Scott and command module pilot Al Worden. His first and only space flight began with the launch of Apollo 15 on 26 July 1971. He and Scott spent almost three days on the lunar surface and conducted three excursions using the lunar roving vehicle. Post-flight, the crew was assigned to back up Apollo 17, but administrative concerns about the sale of

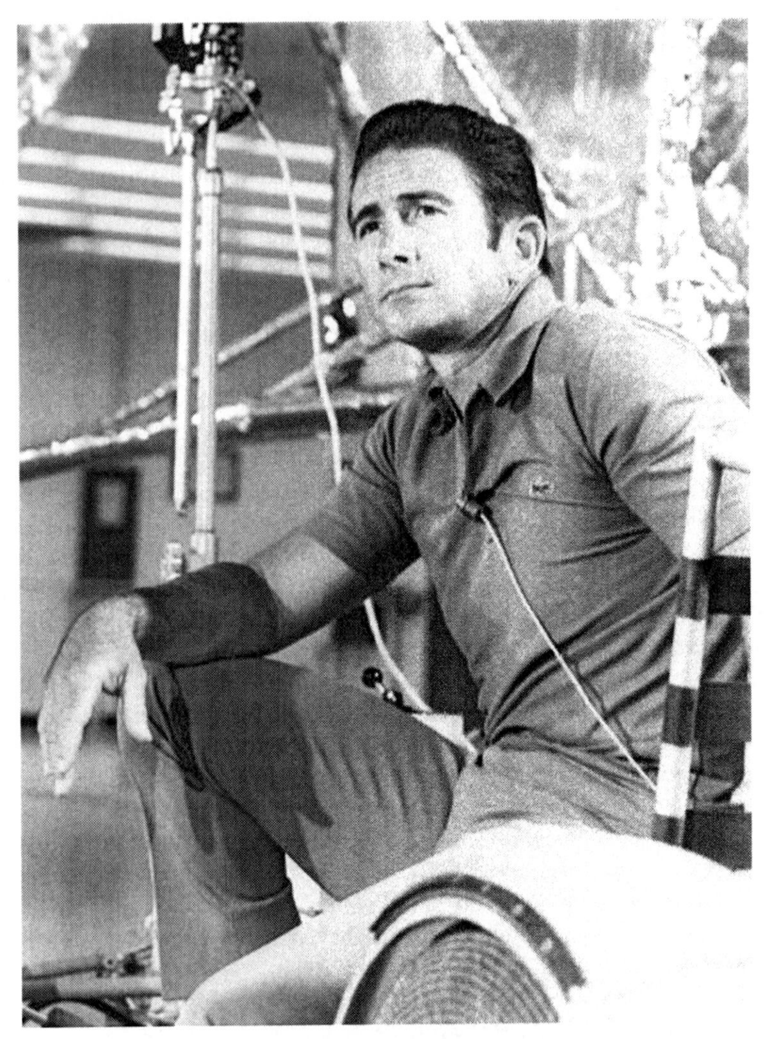

James B. Irwin. (Photo: NASA)

signed envelopes which they had carried in space as part of their personal kit resulted in all three men being reprimanded by NASA.

Irwin retired from the Air Force and then from NASA on 1 July 1972. He founded the non-profit High Flight Foundation in Colorado in order to discuss and debate his religious convictions at speaking engagements and seminars, explaining how these had been enhanced by his voyage to the Moon. He also led five annual expeditions to Mount Ararat in Turkey in an unfulfilled quest to find the remains of the legendary Noah's Ark.[7]

By this time, he had a history of heart problems and succumbed to a massive heart attack on 8 August 1991. He passed away in the Valley View Hospital in Glenwood Springs, Colorado, aged just 62. He was buried with full military honors at Arlington

National Cemetery a week later. He was the first loss from the twelve men who had set foot on the Moon.

JOHN L. SWIGERT, JR.

John Leonard ("Jack") Swigert, Jr. was born in Denver, Colorado on 30 August 1931. He attended the Blessed Sacrament School and Regis Jesuit East High Schools and then enrolled at the University of Colorado, where he played varsity football. On graduating with a bachelor's in mechanical engineering, he joined the Air Force in 1953. After the pilot training program at Nellis AFB, Nevada he was assigned a tour as a fighter pilot in Japan and Korea. He left active service in 1956.

From 1957 to 1964 he was a research engineering test pilot for Pratt & Whitney Aircraft. He served as a jet fighter pilot with the Massachusetts Air National Guard from September 1957 to March 1960 and the Connecticut Air National Guard from April 1960 to October 1965. He applied to NASA for astronaut selection in 1962 and 1963, being rejected both times due to his lack of test pilot experience or advanced schooling. From 1964 to 1966 he was an engineering test pilot for North American Rockwell. One of his involvements was demonstrating the Rogallo wing as a feasible ground landing system for returning space vehicles and astronauts, for which he was named co-recipient of the AIAA Octave Chanute Award for 1966.

In 1965 Swigert received a master's degree in aerospace science from Rensselaer Polytechnic Institute in New York, and the following year was accepted by NASA as one of nineteen new astronauts. In 1967 he got a master's in business administration from the University of Hartford, Connecticut.

As an astronaut, Swigert worked as a specialist in the Apollo command module and served on the support crew for Apollo 7 and later for Apollo 11. In August 1969 he was selected as command module pilot for the Apollo 13 backup crew, but when a suspected illness grounded Ken Mattingly from the prime crew three days before the mission's launch, he was replaced by Swigert. An explosion in the service module on the way to the Moon forced the mission to be aborted. He then joined Jim Lovell and Fred Haise in converting lunar module *Aquarius* into a "lifeboat". By sheer good fortune, Swigert had helped to write the procedure for this. After three days of high drama, command module *Odyssey* splashed down safely in the Pacific on 17 April 1970.

Although assigned to the Apollo-Soyuz Test Project along with Tom Stafford and Deke Slayton, Swigert was removed from the crew following a failure to disclose his participation in a scheme to sign and sell collectable postal items. He subsequently resigned from NASA and served as Executive Director of the Committee on Science and Technology of the U.S. House of Representatives until 31 August 1977. He was offered a chance to return to NASA as a space shuttle pilot but declined. Following forays into the private sector he became a candidate for the House of Representatives in February 1982 and was elected in November. Unfortunately he found during this time that he had bone cancer and died on 27 December, eight days before he was to have taken office.[8]

John L. ("Jack") Swigert. (Photo: NASA)

Among his numerous honors, space-suited statues of Jack Swigert were placed in the U.S. Capitol's Visitor Center, and the Concourse B train platform at Colorado's Denver International Airport.

REFERENCES

1. Ellis, Frank K., article, *No Man Walks Alone*, Foundation magazine, Spring 1996
2. Duke, Charlie and Dotty Duke, *Moonwalker*, Oliver-Nelson Books, Nashville, TN, 1990, pg. 76
3. Cernan, Eugene with Don Davis, *The Last Man on the Moon*, St. Martin's Press, New York, NY, 1999, pg. 92
4. E-mail correspondence with RADM George Furlong, 9 December 2011–18 July 2012
5. Hawthorne, Douglas B., *Men and Women of Space*, entry: *Vance (DeVoe) Brand*, Univelt, Inc., San Diego, CA, 1992

6. Hawthorne, Douglas B., *Men and Women of Space*, entry: *Ronald (Ellwin) Evans*, Univelt, Inc., San Diego, CA, 1992
7. Irwin, James B., with William A. Emerson, Jr., *To Rule the Night: The Discovery Voyage of Astronaut Jim Irwin*, A.J. Holman Company, New York, NY, 1973
8. Hawthorne, Douglas B., *Men and Women of Space*, entry: *John (Leonard) Swigert, Jr.*, Univelt, Inc., San Diego, CA, 1992

11

"Before this decade is out"

Following their formal introduction as NASA's third astronaut group on 18 October 1963 the fourteen new men were told they would not be required to report to MSC until the beginning of January, and that their formal training would commence on 2 February. This would give them time to complete any current assignments and search out or build a new family home convenient to the space center.

They departed Houston, proud and determined to be a part of what President John F. Kennedy had described in a speech at Rice University the previous year as setting sail on "this new ocean" of space exploration. They were to be leading players in the young president's national commitment of achieving a manned lunar landing before the decade was out, and they looked forward to the challenge. Just four weeks after their press conference they joined their fellow Americans in mourning the tragic loss of their beloved president, assassinated in Dallas on Friday, 22 November.

TRAINING BEGINS

When the new recruits began their training on 2 February 1964, NASA had twenty-nine astronauts on its books – down one due to John Glenn's resignation the previous month in order to enter private business. He attended some of the classes but was not formally included due to his other obligations and the certainty that he would not fly in the Apollo program. Scott Carpenter would soon be granted leave to participate in the Navy's Project Sealab. Deke Slayton was prohibited from solo flying due to his minor heart ailment. In November 1963 he had resigned his commission as a major in the Air Force in order to focus on his role as Director of Flight Crew Operations at the Manned Spacecraft Center. Early in 1964 Alan Shepard was also grounded with a troublesome inner ear ailment, diagnosed as Ménière's Disease, causing him to lose the first manned Gemini mission that would have seen him fly with rookie astronaut Tom Stafford. To keep Shepard active in the astronaut corps he was given the role of Chief Astronaut, and would work together with Slayton in this capacity through the 1960s as he fought to regain flight status. Some radical surgery in 1969 corrected his ailment, and two years later he walked on the Moon as the commander of the Apollo 14 mission.

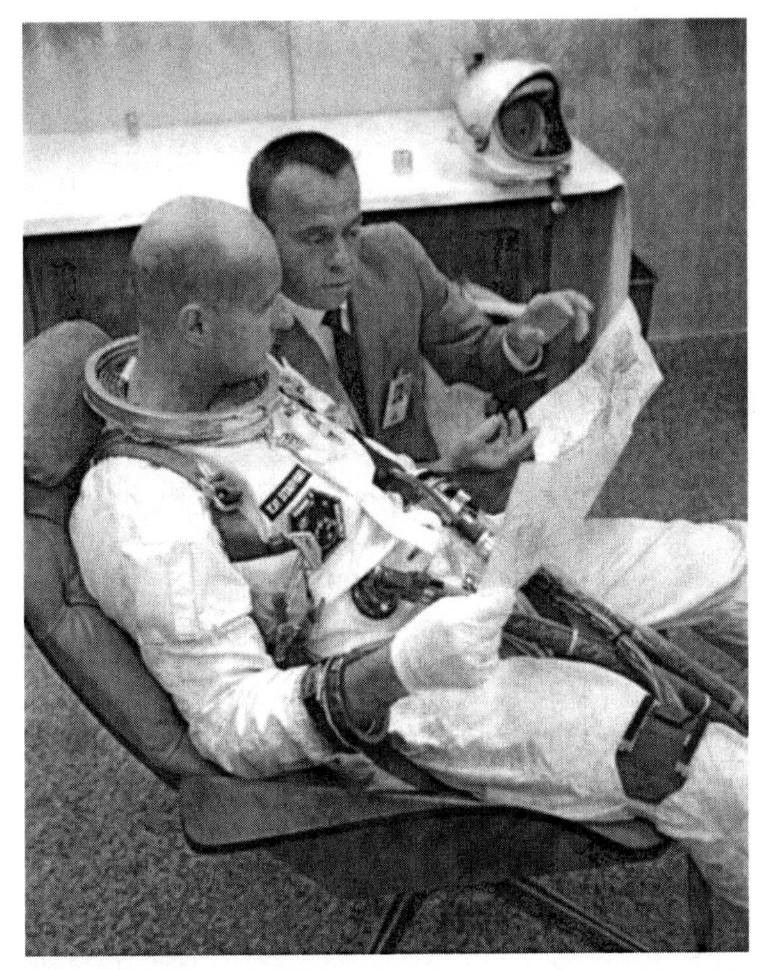

Had Alan Shepard not contracted Ménière's Disease, he and Tom Stafford would have flown the first Gemini mission. (Photo: NASA)

Greeting the astronaut trainees was a new initial training phase, similar to that of the previous groups, with a 20-week series of briefings, lectures, academic studies, computer sessions and field trips that would end in June. After this they were to start concentrated activities in relation to the Gemini program and be assigned to specific areas of responsibility.

While much of their training pertained to Project Gemini, they would also receive expert tuition in the science of geology specifically aimed at Project Apollo. Having been through much of this training, the resident astronauts would not participate in these activities, they would instead use the Gemini simulators to hone their skills for the upcoming flights. As well, because the overall training program would not permit all of the astronauts to be involved in field trips at the same time, they were often split into smaller groups. One discipline that all available astronauts were required to participate in was known as Series I of the geology program, involving a number of field trips across the United States. Altogether, the astronauts would undergo about 58 hours of classroom instruction in addition to their field trips. The purpose of this training was to imbue them with sufficient knowledge to sensibly collect samples of the lunar surface. The course emphasized the study of impact features and volcanism, because of the probable importance of those phenomena on the structure of the lunar surface.

One portion of the course, covering Principles of Terrestrial and Lunar Geology, was supervised by Dr. E. Dale Jackson of the U.S. Geological Survey (USGS), and covered geologic processes, stratigraphy, Earth and Moon structures and land forms, geologic mapping and geophysical properties of the Earth and Moon. The other part, Elements of Mineralogy and Petrology, was taught by three NASA geologists of the Lunar Surface Technology Branch at the Manned Spacecraft Center; Drs. Ted Foss, Uel Clanton and Bert King, Jr. It dealt with the study of minerals and rocks expected to be present on the lunar surface, and trained the astronauts to recognize the most significant samples for return to Earth for laboratory analysis.[1]

Group 3 Science Training Schedule (As compiled by David J. Shayler)[2]

Topic	*Attendance*	*Sessions*	*Total hours*
Geology Series I	All astronauts	25	50
Mineralogy and petrology	All astronauts	13	26
Flight mechanics	Group 3	18	36
Astronomy (three-hour sessions)	Group 3	5	15
Digital computers	Group 3	6	12
Gemini onboard computers	All astronauts	11	22
Rocket propulsion systems	Group 3	6	12
Upper atmosphere and space physics	Group 3	6	12
Aerodynamics	Group 3	2	4
Guidance and navigation	Group 3	15	30
Communications	Group 3	4	8
Medical aspects of space flight	Group 3	6	12
Meteorology	Group 3	2	4

ALONG THE GEOLOGY TRAIL

Topics covered in the geology training lectures began with introductory physical geology, mineralogy, petrology, and the Moon itself, mostly presented by MSC staff and geologists from the USGS. The lecturers knew that the trainees were extremely busy people and they must not waste their time with ill prepared or poorly presented material, so they had to cut to the quick of their presentations and laboratory work to focus on what the future lunar astronauts would need to know. It was the equivalent of a one-semester college course on land forms and geologic processes, minerals and their origins, and how to undertake topographic and geologic mapping. As geologist Bert King observed, "My first lecture was delivered in February 1964. I had to cover everything they needed to know about mineralogy in a one-hour lecture. This was a tall order. I got through the material, leaving a lot of gaps out of necessity."

Although the geologists continued to lecture from time to time, they found that intensive instruction on field trips was far more productive. As King explains, "Each instructor was assigned two or three astronauts as his students for each field trip or field exercise. That way the astronauts were not as shy about asking questions, and the instructor got to know the strengths and weaknesses of each one. We commonly made three field trips out of one. That is, all the geologist instructors, including an expert on the local area, took one trip to outline the work and compile the program. Then, usually two trips were taken with different groups of astronauts because all of them could not get together simultaneously .... We chose a set of field trips where there would normally be two or three days of instruction. We started with basic sites and progressed to more complicated areas and exercises. For our first field trip we selected the Grand Canyon."[3]

USGS instructor Dale Jackson (wearing hat) overlooking the Grand Canyon with astronauts Rusty Schweickart, Elliot See and Ted Freeman. (Photo: NASA)

This two-day field trip began on 5 March and was led by Ed McKee of the USGS, supported by seven other geologists from the USGS and NASA. Eighteen astronauts from all three groups took part: Aldrin, Anders, Armstrong, Bassett, Bean, Carpenter, Cernan, Chaffee, Collins, Cunningham, Eisele, Freeman, Gordon, Schweickart, Scott, See, Shepard and Williams. They were accompanied by Paul Haney, who was a Public Affairs Officer at the Manned Space Center, and two NASA photographers. The astronauts were instructed on fundamental geologic stratigraphic concepts such as layering, superposition, ages and structures primarily by show-and-tell discussions and demonstrations.[4]

Charlie Bassett riding a mule out of the Grand Canyon. (Photo: USGS)

On the first day everyone walked down the Kaibab trail from the canyon's south rim in groups comprising three or four astronauts with their own geologist pointing out the features and answering questions. As they literally traveled back through the history of the canyon the geologists described the distinctive features, rock types and processes. The next day they all made their way up the Bright Angel trail in the same groups. This time the onus was on the astronauts to identify and discuss different features based on the previous day's observations. Having arrived at Indian Gardens on the path out of the canyon, many of the group elected to ride mules the rest of the way while the more determined, led by Alan Shepard, hiked out. A week later, 12-13 March, the remaining eleven astronauts similarly trekked the Grand Canyon trails. Even though the area bore no resemblance to what they might see, find, or describe on the Moon, it provided an interesting and obviously scenic introduction to geology and field work.

Two weeks later a second geologic expedition took place from 2-3 April, this time led by Bill Muehlburger of the University of Texas and much the same support team

Gene Cernan inspects an exposed incline of sedimentary rock during a field trip to the Texas Big Bend Country. (Photo: NASA)

as before. This time, in typically hot weather, the field trip took a slightly different group of eighteen astronauts to the Marathon Basin and Big Bend of Texas, with an introduction to volcanic rocks along the Rio Grande River west of Big Bend National Park. As before, a second group of nine astronauts would make the same expedition several days later, on 15-16 April.

Bert King said many important lessons were learned on this particular trip. One occurred when Neil Armstrong hit a basalt outcrop with his rock hammer, slightly cutting his arm with a flying rock chip. "This was not the correct way to collect rock samples," King pointed out. "Even a small cut in a space suit could have serious consequences." In the cool evening of the second day, the geologists and astronauts enjoyed cold beers over a Tex-Mex meal and later played a few hands of poker. "On our way back to Marfa to board the NASA aircraft, a number of us crossed the Rio Grande into Mexico to make a 'booze run'," King added, "taking advantage of duty-free, low Mexican prices."[5]

Other field trips expanded the geologic training for the astronauts. They were sent to landscapes near Flagstaff and to the Kitt Peak Observatory, both in Arizona, where they were given a close look at the Moon using a telescope. Their final Series I field trip was to the Philmont Boy Scout Ranch near Cimarron, New Mexico for four days of geology training in early June. This trip consisted of mapping, studying geological formations of the area and seismic studies using geophysics instruments.

Series II of geology studies began in September 1964 and ran through to May the

The astronaut group at the Philmont Boy Scout Ranch, 3-6 June 1964. From left: Conrad, Aldrin, Gordon, Freeman, Bassett, Cunningham, Armstrong, Eisele, Schweickart, Lovell, Collins, Cernan, See, White, Chaffee, Cooper, Williams, Anders, Scott and Bean. (Photo: NASA)

At the Philmont Boy Scout Ranch Joel Watkins of the USGS discusses gravity meter readings with Dave Scott, Neil Armstrong and Roger Chaffee. (Photo: NASA)

following year. During this second phase the astronauts visited the Newbury Crater, which is a large shield volcano, and Valles Grande near Los Alamos in New Mexico, where Ted Freeman suffered a bad sprain to his ankle. Less than a week later, on 31 October 1964, he was killed when his T-38 ran into a flock of Canadian snow geese near Ellington AFB, Texas. Beginning again in January 1965 the astronauts visited various sites of volcanic interest in Hawaii, artificial craters at the Nevada Test Site, and the impact structure at Meteor Crater in Arizona.

The final phase of their training, Series III, took the astronauts to Katmai, Alaska, Iceland, Medicine Lake, California, the Zuni Salt Lake in New Mexico and finally, in November 1965, the volcanic craters at Pinacates in Mexico.

No one expected the astronauts to become fully qualified field geologists, but the key thing was that they learned to identify, interpret, and evaluate rocks and features in terms of their geological context, to collect both representative and exotic samples, and to describe to their tutors what they had found using correct terminology.

Bill Anders and Charlie Bassett during geology training in Iceland in 1965 with an unidentified geologist. (Photo: NASA)

OTHER TRAINING

In addition to intensive geology training, the fourteen Group 3 astronauts completed two-day celestial recognition and navigational training at the Morehead Planetarium on the campus of the University of North Carolina, which began on 26 March 1964. Interestingly, technicians at the planetarium developed simplified replicas of Gemini flight modules and tools for astronaut training at their facility, generally employing plywood or cardboard. They even constructed a wooden mockup which represented aspects of the Gemini spacecraft and mounted it on a barber's chair to enable trainees to perform elementary changes in pitch and yaw. This rudimentary training may have assisted and even helped to save astronauts' lives by reconfiguring balky

Ed White and Jim McDivitt check out celestial navigation equipment at the Morehead Planetarium in North Carolina. (Photo: NASA)

navigation systems using celestial fixes taken manually. For instance, after suffering a fault with the automated navigational controls of his Mercury spacecraft, Gordon Cooper used the stars to guide his re-entry, and manual star sightings were also carried out by the Apollo 12 and 13 missions.[6]

Like their predecessors, the new astronauts received jungle survival training in Panama in June 1964 and desert survival training near Stead AFB in Nevada two months later. The jungle training took place at the U.S. Caribbean Air Command's Tropic Survival School, Albrook AFB in the Panama Canal Zone, with the fourteen new astronauts accompanied by Pete Conrad from the second group as the astronaut supervisor.

As before, the astronauts lived for three days in the Panama rain forest using only items that would be in the survival kit of the Gemini spacecraft. This included a life raft, a canteen with four pints of water, a combination flashlight and strobe light, a signal mirror, fishing line and hooks, a sewing kit, a flint and steel fire starter with cotton balls, a whistle, and a compass. There was also a desalination kit, sun glasses, compresses, medication kit, zinc oxide ointment, sea dye marker, and a one-foot-long machete. Additionally, the astronauts had a 25-foot personal parachute for use if they were forced down in the jungle. The only difference between this and the previous survival course was that emergency food rations and a pocketknife were eliminated from the survival kit in order to reduce weight and volume in the spacecraft.

Gene Cernan seems to be pointing the way out of the Panamanian jungle to teammate Dick Gordon. (Photo: NASA)

Ray Zedekar, a pre-flight training assistant, said food rations were taken from the kit after the jungle exercise with sixteen astronauts the previous year. "There was enough food in the jungle," he said. The astronauts' favorite jungle dishes, according to Zedekar, were the iguana lizard, small jungle pigs and deer, and fish. A delicacy of the plant life was palm heart.[7]

Jungle campsites, set far apart so the astronauts could not be seen or overheard by others, were assigned to pairs of men to represent Gemini flight crews. An Air Force instructor visited two teams during the day to ensure that they were doing alright, but slept elsewhere. On their final day the men had to leave the jungle, and it was up to each pair to find their own way out. The best way, as taught by the instructors, was to inflate their raft and float down the many tributaries to a main stream and then to the base camp.

ASTRONAUT ASSIGNMENTS

Prior to embarking on the desert survival phase of their training, all of the astronauts were handed specific assignments by Deke Slayton. It was also announced that Alan Shepard had been appointed chief of the Astronaut Office, a role which Slayton had been temporarily filling in addition to his other duties. The Gemini group was headed by Gus Grissom and included John Young, Wally Schirra and Tom Stafford. The Apollo group was headed by Gordon Cooper and included Jim McDivitt (command and service module), Pete Conrad (lunar module and cockpit layout), Frank Borman (all booster vehicles), and Ed White (control systems, communications systems and instrumentation). In the operations and training area, Neil Armstrong was selected to head the group composed of Elliot See (mission planning, guidance and navigation) and Jim Lovell (recovery and crew systems). Scott Carpenter, who was on duty leave with the Navy, was not given a specific assignment.

The new astronauts were issued individual assignments:

Aldrin: mission planning, trajectory analysis and flight plans
Anders: environmental control systems, radiation and thermal protection
Bassett: training and simulators as main duties
Bean: recovery systems
Cernan: spacecraft propulsion and the Agena target vehicle
Chaffee: communications and the Deep Space Network
Collins: pressure suits and extravehicular experiments
Cunningham: electrical and sequential systems; non-flight experiments in other programs
Eisele: attitude and translation control systems
Freeman: boosters
Gordon: cockpit integrations
Schweickart: future manned space programs and inflight experiments
Scott: guidance and navigation
Williams: range operations and crew safety.[8]

TURNING UP THE HEAT

The Group 3 astronauts reported for desert survival training on Monday, 10 August and spent that day in classroom instruction from the 3637th Combat Crew Training Squadron (Survival and Special Training). The following morning they had another three hours of instruction and then at midday the astronauts and their instructors were taken by helicopter to the main desert site at Carson Sink, situated 40 miles east of Reno, Nevada. They spent that afternoon and Wednesday morning in demonstrations of the erection of shelters, the improvisation of clothing, and signaling equipment. Wednesday afternoon they moved to two-man sites, spending the next day-and-a-half practicing survival in conditions such as might be encountered if a Gemini capsule were to parachute into a desert region.

Roger Chaffee, Dick Gordon and C.C. Williams practice erecting a shelter tent using a Gemini parachute. (Photo: NASA)

As with earlier groups, the astronauts used their parachutes to fashion tents and loose clothing to protect them from the fierce desert sun. They practiced firing flares and using hand mirrors and other signaling devices. As for food, they could eat only what they could catch in the desert.

There was compass work on the final day, 14 August. This prepared the astronauts to locate their capsule containing food, water and signaling equipment, in the event of ejecting from it and landing using their personal parachutes. The intensive course ended later that day and the astronauts returned to Stead AFB for a course review and critique. After a press conference for local news media representatives, the astronauts flew back to MSC.[9]

Astronauts Charlie Bassett, Buzz Aldrin and Ted Freeman practice eating and drinking during brief periods of weightlessness aboard the KC-135. (Photo: NASA)

Three days later, six men were the first of their group to participate in weightless training aboard an Air Force KC-135 that flew a dizzying series of parabolas over Dayton, Ohio. This training lasted two days. Involved in these first tests were Aldrin, Bean, Bassett, Cunningham, Freeman and Schweickart. In periods of weightlessness that lasted between 20 and 30 seconds, the astronauts were given some idea within the padded fuselage of the modified Boeing jet aircraft of how they might experience the environment of space without gravity.

During their brief periods of weightlessness, the men played catch using a large medicine ball, ate and drank from special packs, and ran from one end of the long cabin to the other in 16-foot strides. They also tried to make their way around whilst blindfolded, and discovered that it was almost impossible to discern up from down without any visual aids. Finally, they donned a backpack fitted with air jets to propel them from place to place. This worked but was somewhat clumsy to operate. Due to mechanical problems with the KC-135, the other eight astronauts of this group had to wait for their chance to fly at a later date.[10]

The astronauts also received parachute training at Ellington AFB in September, in towed and free descent flights. As with their predecessors, the ground landing phase of the parachute training was followed by a water landing phase. Later that month twelve of the group – Anders and Eisele were unable to attend – traveled to Naval Air Station, Pensacola, Florida for a one-day extension of their water survival training. The course was conducted by Jack Martin of the Naval School Pre-flight Swimming and Water Survival Division. Here, like the Group 2 astronauts, they would become acquainted with the device known as the Dilbert Dunker, a cockpit-like steel box fashioned from an SNJ trainer, mounted two stories high on a steeply inclined track. As they were informed, it was used to train naval aviators in techniques for escaping from aircraft after crashing in water – or in this case, astronauts from a capsule. Each astronaut would ride the Dilbert Dunker into the facility's swimming pool twice; once in flight coveralls and parachute, and again while wearing a Gemini space suit. After a period of instruction they were strapped

NAS Pensacola, 22 September 1964. Astronaut Walt Cunningham in the Dilbert Dunker immediately after it was released. (Photo: NASA)

in, a signal bell rang, and then the instructor's thumb jerked upward. Once released, the mockup cockpit clattered down the track into 12 feet of water where, after impact, it was mechanically overturned. Navy instructor divers closely monitored the situation as each inverted man released himself from his shoulder harness and made his way to the surface, making sure that he didn't get into difficulties or panic.

While at the facility, they were given further instruction in the correct techniques for releasing a parachute harnesses after splashing down, boarding a life raft, and mounting helicopter rescue slings and seats.[11]

RUSTY'S "BIRTHDAY SUIT"

On Sunday, 25 October 1964, astronaut Rusty Schweickart spent his 29th birthday in a Gemini space suit during an extended evaluation of biomedical equipment intended for use by astronauts. Around the clock for eight days, unable to shower or shave, he wore the suit for a series of tests and simulations to determine the

Russell ("Rusty") Schweickart. (Photo: NASA)

compatibility of a man and the biomedical recording equipment over an extended period.

Doctors and engineers had attached biosensors to Schweickart's body and he had suited up at the NASA Flight Research Center in Edwards, California at 8:00 a.m. on 19 October for a series of one-hour flights in an F-104B Starfighter jet aircraft, flying zero g and positive g maneuvers while biomedical data was recorded. The holding van in which he was suited up was used for flight preparation, sleeping, rest, eating and waste management. He consumed special Gemini food during the test. On the evening of 21 October, clad in his suit and carrying a self-contained cooling unit, he flew from Edwards to Ellington AFB, Texas and slept in a hospital bed at MSC. Then, following his third night in the suit, he began a 96-hour simulated flight in the MSC Gemini Mission Simulator, pursuing as near as possible the duties that would be undertaken during a Gemini space mission.

On his birthday, members of the biomedical staff at MSC jokingly presented the astronaut with a packet of dehydrated pea soup. But then they brought out a birthday cake delivered the previous night by his wife Clare, together with a gift toy razor and a bar of deodorant soap. Unfortunately for him, the exercise meant that he could only consume freeze dehydrated space foods and he looked on in envy as members of the test team devoured his cake.

On Monday, 26 October, Schweickart was flown to the Ames Research Center in California for three modified re-entry g profiles in a centrifuge. Following these tests he finally shed the Gemini suit, showered, shaved and had a good night's sleep ahead of returning to MSC the following day.[12] It was all part of preparing for a mission in space.

REFERENCES

1. NASA *Space News Roundup*, issue March 18, 1964, *Arizona's Mile Deep Grand Canyon Scene of Astronaut Geology Training,* MSC, Houston, TX, pg. 3
2. Shayler, David J., *Gemini: Steps to the Moon*, Praxis Publishing Co., Chichester, U.K., 2001, pg. 99
3. King, Dr. Elbert A., *Moon Trip: A Personal Account of the Apollo Program and its Science*, University of Houston, TX, 1989, pg. 27
4. Jones, Eric M. and William Phinney, *Apollo Lunar Surface Journal: Apollo Geology Field Exercises, Appendix E, Early Training*, 2002
5. King, Dr. Elbert A., *Moon Trip: A Personal Account of the Apollo Program and its Science*, University of Houston, TX, 1989, pg. 29
6. Morehead Planetarium and Science Center, *History Notes: Astronaut Training (Parts I&II),* University of North Carolina, 2002
7. NASA *Space News Roundup*, issue June 24, 1964, *15 Astronauts Taking Survival Course in Jungles of Panama this Week*, MSC Houston, TX, pg. 8
8. NASA *Space News Roundup*, issue July 8, 1964, *Astronauts are Given Specific Assignments*, MSC Houston, TX, pp. 1&3
9. NASA *Space News Roundup*, issue August 5, 1964, *Desert Survival Training Sessions to Start Monday for 14 Astronauts*, MSC Houston, TX, pg. 8

10. NASA *Space News Roundup*, issue September 2, 1964, *Six Astronauts Get 'Feel' of Space in Jet Aircraft Weightless Flights*, MSC Houston, TX, pg. 8
11. NASA *Space News Roundup*, issue September 30, 1964, *'Dilbert Dunker' Gives Astronauts a Ride in Florida Water Survival Training Course*, MSC Houston, TX, pg. 8
12. NASA *Space News Roundup*, issue October 28, 1964, *Schweickart has Birthday While Confined in Space Suit*, MSC Houston, TX, pg. 8

12

For some, the glory

As Gemini training intensified, the training aids became increasingly complex and mission-oriented. Deke Slayton also began formulating two-man crews, both prime and backup, for the first flights in the Gemini series. With Glenn departing NASA, Carpenter on leave with the Navy, and medical problems ruling himself and Shepard (belatedly) out of the flight roster he only had Mercury astronauts Grissom, Schirra and Cooper ready to act as mission commanders. All nine of the second group were available to him, and the new group of fourteen would be fully trained and ready for selection in the latter part of the program. The only real problem for Slayton was the questionable training ethic of Gordon Cooper, which had brought the man into an uneasy conflict with the unforgiving NASA administration. For that reason, Slayton decided for the time being to leave Cooper out of consideration for the initial Gemini flights and adopt a wait-and-see attitude.

In considering those who would be paired on Project Gemini, Slayton introduced a crewing system which would work – and work well – right through that program as well as Apollo. The prime crew for each mission would be announced, along with a backup crew. Assuming they were not elevated to the prime crew for some medical, operational, or other reason, the fully trained backup crew would then skip the next two missions and become the prime crew on the following flight. With this in mind, and after meetings with his fellow Mercury astronauts on crewing and compatibility, in late 1963 Slayton put together the first Gemini crew manifest.[1]

Mission	*Prime crew*	*Backup cres*
Gemini 3	Shepard & Stafford	Grissom & Borman
Gemini IV	McDivitt & White	Conrad & Lovell
Gemini V	Schirra & Young	Armstrong & See
Gemini VI	Grissom & Borman	Stafford & Group 3 pilot
Gemini VII	Conrad & Lovell	White & Group 3 pilot
Gemini VIII	Armstrong & See	Young & Group 3 pilot

Note: The first three Gemini missions had Arabic numbers, and thereafter Roman numbers.

When Shepard became unavailable for the first Gemini mission owing to his inner ear ailment, Slayton advanced Grissom as the flight commander. But Slayton decided he would rather utilize Borman's services on a later, longer-duration mission that Slayton thought would be a better role for him. He reckoned that John Young was a compatible pilot for Grissom, and they now became the first Gemini crew. A further reshuffling of crews saw Cooper assigned with Pete Conrad for the long-duration Gemini V mission. Schirra, known for his intense dislike of lengthy space missions, was reassigned to Gemini VI, along with Tom Stafford. However, it was decided that NASA would name crews incrementally during the program. Hence MSC Director Robert Gilruth announced only the prime and backup crews for the first Gemini mission – a three-orbit test of the spacecraft and its systems – as Gus Grissom and John Young, along with backups Wally Schirra and Tom Stafford.

RENDEZVOUS AND DOCKING SIMULATIONS

One of the prime objectives of Project Gemini was rendezvous and docking with an unmanned Agena target vehicle previously placed into orbit. To this end, the Gemini

Jim Lovell acquaints himself with the Gemini spacecraft element of the Rendezvous Docking Simulator at the Langley Research Center. (Photo: NASA Langley)

Langley's full-scale Gemini spacecraft simulator. (Photo: NASA)

crews would work to perfect the docking maneuver of two vehicles by training on a facility known as the Translation and Docking Simulator (TDS) located in MSC's long, high-ceilinged Building 260. It recreated all of the motions required to perform a docking in space, having been designed to simulate responses to the firing of the Gemini Orbit Attitude and Maneuver System (OAMS) thrusters. All the instruments and controls used during maneuvering responded the same as for the operational spacecraft. The motion was smooth and the "feel" was the same as would occur in orbit, apart from the factor of weightlessness. Visually, the view out of the window was identical to that of the actual spacecraft, while the harsh simulated solar illumination gave a realistic visual sense of the light angles and shadows in space.

Training was also carried out on the newly completed, and far more sophisticated, Rendezvous Docking Simulator in Hangar 1244 at the Langley Research Center in Hampton, Virginia. This full-scale simulator was a massive, intricate structure using an overhead carriage and cable-suspended gimbal system which could accommodate full-scale models of the Gemini (and later Apollo) spacecraft. It enabled the pilots to simulate closing in on and docking with an Agena. The target vehicle was suspended at one end of a track attached to the hangar roof, while a movable vehicle was driven along the track to move toward the target during a docking maneuver.

A little later, Langley added the Lunar Landing Research Facility (LLRF) which allowed Apollo astronauts to fly in a simulated lunar environment, practicing piloting difficulties they might encounter in the final 150 feet of descent to the surface of the Moon. Built in 1965, it was basically an A-frame with a gantry used to

The Langley facility allowed Gemini astronauts to practice rendezvous and docking techniques. (Photo: NASA Langley)

Langley's Rendezvous Docking Simulator would later be used to train Apollo astronauts in procedures to dock the lunar module with the command module. (Photo: NASA Langley)

manipulate a full-scale Lunar Excursion Module Simulator (LEMS). This suspension system also enabled astronauts to practice walking in lunar gravity, as its base was modeled with fill material that it was thought would resemble the surface of the Moon.

A HIGH-RISK OCCUPATION

Inevitably the challenge of space flight, just as in testing high-performance aircraft, involves immense risk and unforeseen hazards. As Mercury astronaut Gus Grissom prophetically observed while in training for the Apollo 1 mission, "There's always a possibility that you can have a catastrophic failure, of course. This can happen on any flight. It can happen on the first one as well as the last one. You just plan as best you can to take care of all these eventualities and get a well-trained crew, and [then] you go fly." Within weeks, Grissom and his two crewmembers were dead; their lives lost under the most horrifying of circumstances.[2]

While tragedies often clouded America's many triumphant footsteps to the Moon, the list of accomplishments of the Group 2 and 3 astronauts is impressive. Between them they flew seventeen Gemini missions and twenty Apollo missions (discounting the Apollo-Soyuz Test Project in 1975); they flew sixteen missions to the Moon, and seven walked upon its surface.

In November 1965, Roger Chaffee (in helmet) receives Lunar Landing Simulator instruction from Maxwell Goode, a scientist at the Langley Research Center. (Photo: NASA Langley)

The Lunar Landing Research Facility at Langley included a device to enable Apollo astronauts to simulate walking in the one-sixth gravity of the Moon. (Photo: NASA Langley)

Neil Armstrong descends the ladder of the Lunar Module Training Vehicle in training for Apollo 11. (Photo: NASA)

Here is what fate had in store for the nine Group 2 members:

- *Neil Armstrong:* Flew on Gemini VIII with Dave Scott. Mission cut short after serious thruster problems in orbit. On Apollo 11, Armstrong became the first person to walk on the Moon on 20 July 1969.
- *Frank Borman:* Long-duration Gemini VII mission which served as a target for the rendezvous by Gemini VI. He commanded Apollo 8 in 1968, the first manned spacecraft to fly around the Moon.
- *Pete Conrad:* Followed his long-duration Gemini V flight with another successful mission on Gemini XI. He walked on the Moon during Apollo 12 in 1969 and later commanded the first mission to the Skylab space station.

C.C. Williams, Charlie Bassett and Elliot See discuss the aborted liftoff of Gemini VI in the Mission Operations Control Room, 15 December 1965. All three would later die in accidents involving T-38 jets. (Photo: NASA)

- *Jim Lovell:* A total of four space missions, including Gemini VII and Gemini XII. He flew around the Moon on Apollo 8, and as commander of Apollo 13 safely brought home his crippled spacecraft.
- *Jim McDivitt:* Commanded Gemini IV and later flew the Earth-orbital test of the complete Apollo spacecraft on Apollo 9 in March 1969.
- *Elliot See:* Selected to command Gemini IX, but was killed in a T-38 accident along with Group 3 astronaut Charlie Bassett.
- *Tom Stafford:* Flew Gemini VI and Gemini IX, commanded the Apollo 10 mission in lunar orbit, and in 1975 commanded the Apollo spacecraft that participated in the Apollo-Soyuz Test Project.
- *Ed White:* Made America's first spacewalk on Gemini IV. Died in launch pad fire on 26 January 1967 while training for Apollo 1.
- *John Young:* Became the first astronaut to fly six space missions: two Gemini flights (Gemini 3 and Gemini X), two Apollo flights (Apollo 10 and Apollo 16, in the latter case walking on the Moon), and two flights of space shuttle *Columbia* (STS-1 and STS-9).

Although three of their number would die in airplane accidents and a fourth in a launch pad fire before they could make their first space flight, the Group 3 astronauts nevertheless enjoyed a spectacular series of missions:

The crew of the ill-fated Apollo 1 mission; Gus Grissom, Ed White and Roger Chaffee. (Photo: NASA)

- *Buzz Aldrin:* Flew on Gemini XII and set a record for three spacewalks on a single mission. In July 1969 he was the lunar module pilot on Apollo 11 and walked on the Moon with Neil Armstrong.
- *Bill Anders:* His only space mission was the triumphant lunar-orbiting flight of Apollo 8.

Bill Anders and Jim Lovell in front of the Lunar Module Simulator. (Photo: NASA)

- *Charlie Bassett:* Was to have flown on Gemini IX and make a spacewalk, but was killed in a T-38 accident together with Elliot See. He would likely have gone on to command an Apollo mission.
- *Alan Bean:* Walked on the Moon with Pete Conrad on the Apollo 12 mission. He later commanded the second resident crew of the Skylab space station.
- *Gene Cernan:* On Gemini IX he made with extreme difficulty the spacewalk originally planned for Charles Bassett. Flew in lunar orbit on Apollo 10 and commanded Apollo 17, the final lunar landing mission.

Gene Cernan with Snoopy, the mascot of Apollo 10. (Photo: NASA)

- *Roger Chaffee:* His first space flight was to have been Apollo 1, but he was killed in the fatal launch pad fire on 26 January 1967.
- *Mike Collins:* Flew with John Young on Gemini X, and in July 1969 was the command module pilot on Apollo 11.
- *Walt Cunningham:* Only flew one mission, Apollo 7, which was the first manned test flight of the Apollo spacecraft in Earth orbit.
- *Donn Eisele*: Like Cunningham, he only flew the one mission, Apollo 7.
- *Ted Freeman:* Was in line for a later Gemini mission but was killed when his T-38 hit a flock of snow geese on 31 October 1964 and he remained with the aircraft too late to bail out and save himself.

In what may have been the last photo taken of him, Ted Freeman (right) is seen walking through the training area with Buzz Aldrin the night before Freeman died. (Photo: NASA)

- *Dick Gordon:* Flew with his buddy Pete Conrad on Gemini XI, and again on Apollo 12 along with Alan Bean. Would have commanded Apollo 18 if that mission had not been canceled.
- *Rusty Schweickart:* His only space flight was on the Earth-orbiting Apollo 9 mission along with Jim McDivitt and Dave Scott.
- *Dave Scott:* After the near-fatal Gemini VIII mission with Neil Armstrong he flew as command module pilot on Apollo 9. As commander of Apollo 15 in 1971 he drove the first lunar roving vehicle on the Moon.
- *Clifton Williams:* Known as "C.C.", he was assigned as lunar module pilot for what would become the Apollo 12 mission, but died on 5 October 1967 when his controls of his T-38 aircraft jammed.

ON TO TOMORROW

With astronaut training for the Gemini and Apollo missions in full swing in 1963 and 1964, and an ambitious forecast of numerous lunar landing missions of several days' duration, as well as Earth-orbital flights projected well beyond the end of the decade, NASA's Office of Manned Space Flight had to consider future astronaut recruitment.

One thing that was abundantly clear was the need for true scientific exploration of the Moon. Instead of training test pilots as geologists, it was felt that qualified and experienced scientists could be recruited into the astronaut corps and given necessary piloting skills. And there was an ongoing need for highly respected scientists from a variety of disciplines, including physicians, to serve on long-duration flights on the envisaged space stations. With this in mind, starting in February 1964 MSC officials called a series of meetings with representatives of the National Academy of Sciences to discuss and define criteria, qualifications and training for a recruitment of several scientist-astronauts.[3]

With this process agreed, in October NASA invited applications from scientists wishing to become astronauts. They were required to have a doctorate in medicine, engineering or one of the natural sciences, but did not require to have any piloting experience because those selected would be given a year's flight training with the Air Force. In June 1965 the agency's Group 4 intake selected six scientists, but as events transpired only one would make the journey to the Moon. However, three of their number would serve aboard the Skylab space station.

A fourth group of nineteen pilot-astronauts was selected by NASA in April 1966 and several of them made Apollo flights as command or lunar module pilots, with three walking on the Moon.

The next year, in August 1967, NASA selected another group of eleven scientist-astronauts, but the cancellation of the three final Apollo lunar landing missions and cutbacks in other programs rendered these men superfluous to the future needs of the space agency. Recognizing this, the disappointed men even began to call themselves the XS-11 (Excess Eleven). Eventually four quit the astronaut corps out of frustration in order to resume their studies, while the remainder had to wait over a decade before eventually flying on space shuttle missions.

The public enthusiasm and romance with Apollo and space exploration had waned rapidly after the Apollo 11 lunar landing, and subsequent budget cuts meant that the series concluded with Apollo 17 in December 1972 – a mere four years after Apollo 8 blazed the trail to the Moon.

The only Group 2 astronaut to fly after that was Pete Conrad, who commanded a 28-day mission in 1973 that rescued the damaged Skylab space station. His fellow moonwalker on Apollo 12, Alan Bean, became the final member of Group 3 to fly in space when he commanded the second Skylab crew later that year.

The two Skylab missions would prove to be the final space mission for both men, and the last for two magnificent groups of pioneering NASA astronauts.

REFERENCES

1. Slayton, Donald K. and Michael Cassutt, Deke! *U.S. Manned Space: From Mercury to the Shuttle*, pg. 138, Forge Books, New York, NY, 1994
2. French, Francis and Colin Burgess, *In the Shadow of the Moon: A Challenging Journey to Tranquility*, pg. 145, University of Nebraska Press, Lincoln, NE, 2003
3. Compton, William David, *Where No Man Has Gone Before: A History of Apollo Lunar Exploration Missions*, pg. 65, NASA History Series, Washington, D.C., 1989

Appendix 1

NASA Group 2 Astronaut Finalists

Name	*Service & rank (1962)*		*Born*	*Died*	*Final rank*
ARMSTRONG, Neil Alden	Civilian	–	05.08.1930	–	–
ASLUND, Roland Erhard	USN	LCDR	20.07.1927	20.10.1987	CAPT
BEAN, Alan LaVern	USN	LT	15.03.1932	–	CAPT
BIRDWELL, Carl, Jr.	USN	LCDR	06.01.1928	12.04.1996	CAPT
BORMAN, Frank Frederick II	USAF	MAJ	14.03.1928	–	COL
COLLINS, Michael	USAF	CAPT	31.10.1930	–	B/GEN
CONRAD, Charles, Jr. + +	USN	LT	02.06.1930	08.07.1999	CAPT
DICKEY, Roy Sterling	USAF	CAPT	31.03.1929	–	COL
EDMONDS, Thomas Eugene	Civilian	–	1928	–	–
FITCH, William Harold	USMC	CAPT	06.11.1929	–	L/GEN
FRITZ, John Merritt	Civilian	–	15.08. 1930	–	–
GEIGER, William John	USMC	CAPT	06.01.1931	–	COL USMCR
GLUNT, David Latta, Jr.	USN	LT	09.03.1930	12.09.2001	CAPT
GORDON, Richard Francis, Jr.	USN	LT	05.10.1929	–	CAPT
JOHNSON, Orville Christian	Civilian	–	24.12.1928	–	–
KELLY, William Perry, Jr.	USN	LCDR	15.08.1928	20.10.2002	CDR
LOVELL, James Arthur, Jr. + +	USN	LCDR	25.03.1928	–	CAPT
McCANNA, Marvin George, Jr.	USN	LT	22.03.1929	11.05.1986	CAPT
McDIVITT, James Alton	USAF	CAPT	10.06.1929	–	B/GEN
MITCHELL, John Robert Cummings + +	USN	LT	16.08.1929	31.05. 2011	CAPT
NEUBECK, Francis Gregory	USAF	CAPT	11.04.1932	–	COL
RAMSEY, William Edward	USN	LCDR	07.09.1931	–	VADM
SEE, Elliot McKay, Jr.	Civilian	–	23.07.1927	28.02.1966	–
SMITH, Robert Wilson	USAF	CAPT	11.12.1928	19.08.2010	LTCOL
SOLLIDAY, Robert Edwin + +	USMC	CAPT	04.12.1931	–	LTCOL
STAFFORD. Thomas Patten	USAF	CAPT	17.09.1930	–	L/GEN
SWIGERT, John Leonard, Jr.	Civilian	–	30.08.1931	27.12.1982	–
UHALT, Alfred Hunt, Jr.	USAF	CAPT	03.06.1931	–	COL
WEIR, Kenneth Wynn	USMC	CAPT	20.10.1930	–	M/GEN
WHITE, Edward Higgins II	USAF	CAPT	14.11.1930	27.01.1967	LTCOL
WRIGHT, Richard Lee	USN	LCDR	25.12.1929	–	CDR
YOUNG, John Watts	USN	LCDR	24.09.1930	–	CAPT

(+ + Previously applied for Mercury Group 1 astronaut selection)

Appendix 2

NASA Group 3 Astronaut Finalists

Name	*Sercie & (1963)*	*Rank*	*Born*	*Died*	*Final rank*
ADAMS, Michael James	USAF	CAPT	05.05.1930	15.11.1967	MAJ
ALDRIN, Edwin Eugene, Jr.	USAF	MAJ	30.01.1930	–	COL
ANDERS, William Alison	USAF	CAPT	17.10.1933	–	M/GEN
BASSETT, Charles Arthur II	USAF	CAPT	30.12.1931	28.02. 1966	CAPT
BEAN, Alan LaVern	USN	LT	15.03.1932	–	CAPT
BELL, Tommy Ila, Jr.	USAF	CAPT	04.12.1930	14.02.2011	B/GEN
BRAND, Vance DeVoe	Civilian	–	09.05.1931	–	–
CERNAN, Eugene Andrew	USN	LT	14.03.1934	–	CAPT
CHAFFEE, Roger Bruce	USN	LT	15.02.1935	27.01.1967	LCDR
COCHRAN, John Kenneth	USMC	CAPT	23.07.1931	22.09.2008	LTCOL
COLLINS, Michael	USAF	CAPT	31.10.1930	–	M/GEN
CORNELL, Darrell Eugene	Civilian	–	19.08.1932	10.10.1984	–
CUNNINGHAM, Ronnie Walter	USMCR	CAPT	16.03.1932	–	COL
DAVIS, Reginald Roderick	USAF	CAPT	09.03.1933	–	COL
EBBERT, Donald George	Civilian	–	24.05.1930	10.11.2011	–
EISELE, Donn Fulton	USAF	CAPT	30.06.1930	01.12.1987	COL
EVANS, Ronald Elwin	USN	LT	10.11.1933	06.04.1990	CAPT
FREEMAN, Theodore Cordy	USAF	CAPT	18.02.1930	31.10.1964	CAPT
FURLONG, George Morgan, Jr.	USN	LT	23.11.1931	–	RADM
GORDON, Richard Francis, Jr.	USN	LT	05.10.1929	–	CAPT
GUILD, Samuel Murton, Jr.	USAF	CAPT	03.01.1930	–	LTCOL
IRWIN, James Benson	USAF	CAPT	17.03.1930	08.08.1991	COL
KIRKPATRICK, James Earl, Jr.	Civilian	–	12.11.1929	04.06.2005	–
PHILLIPS, Charles Laurence	USMC	CAPT	13.01.1933	19.01.1991	LTCOL
RAMSEY, William Edward	USN	LT	07.09.1931	–	VADM
RUPP, Alexander Kratz	USAF	CAPT	02.07.1930	11.06.1965	MAJ
SCHWEICKART, Russell Louis	Civilian	–	25.10.1935	–	–
SCOTT, David Randolph	USAF	CAPT	06.06.1932	–	COL
SHUMAKER, Robert Harper	USN	LT	11.05.1933	–	RADM
SMITH, Lester Robert	USNR	LT	11.10.1929	–	RADM
SWIGERT, John Leonard, Jr.	Civilian	–	30.08.1931	27.12.1982	–
VANDEN-HEUVEL, Robert Jay	USAF	CAPT	16.03.1930	–	LTCOL
WILLIAMS, Clifton Curtis, Jr.	USMC	CAPT	26.09.1932	05.10.1967	MAJ
YAMNICKY, John David	USN	LCDR	08.06.1930	11.09.2001	CAPT

Appendix 3

Space flights by Group 2 Astronauts

Name	*Space flights*	*Spacecraft*	*Mission*	*Launch date*	*Landing date*	*Mission Duration (dd:hh:mm:ss)*
Neil A. Armstrong	2	Gemini VIII	GT-8	16.03.1966	16.03.1966	00:10:41:26
		Apollo 11	AS-506	16.07.1969	24.07.1969	08:03:18:35
Frank Borman	2	Gemini VII	GT-7	04.12.1965	18.12.1965	13:18:35:17
		Apollo 8	AS-503	21.12.1968	27.12.1968	06:03:00.42
Charles Conrad, Jr.	4	Gemini V	GT-5	21.08.1965	29.08.1965	07:22:55:14
		Gemini XI	GT-11	12.09.1966	15.09.1966	02:23:17:08
		Apollo 12	AS-507	14.11.1969	24.11.1969	10:04:36:25
		Skylab 2	SL-2	25.04.1973	22.06.1973	59:11:09:04
James A. Lovell, Jr.	4	Gemini VII	GT-7	04.12.1965	18.12.1965	13:18:35:17
		Gemini XII	GT-12	11.11.1966	15.11.1966	03:22:34:31
		Apollo 8	AS-503	21.12.1968	27.12.1968	06:03:00:42
		Apollo 13	AS-508	11.04.1970	17.04.1970	05:22:54:41
James A. McDivitt	2	Gemini IV	GT-4	03.06.1965	07.06.1965	04:01:56:31
		Apollo 9	AS-504	03.03.1969	13.03.1969	10:01:00:53

Name	*Space flights*	*Spacecraft*	*Mission*	*Launch date*	*Landing date*	*Mission Duration (dd:hh:mm:ss)*
Thomas P. Stafford	4	Gemini VI	GT-6A	15.12.1965	16.12.1965	01:01:51:43
		Gemini IX	GT-9A	03.06.1966	06.06.1966	03:00:50:20
		Apollo 10	AS-505	18.05.1969	26.05.1969	08:00:03:23
		Apollo ASTP	AS-210	15.07.1975	24.07.1975	09.01.28:00
Edward H. White II	1	Gemini IV	GT-4	03.06.1965	07.06.1965	04:01:56:31
John W. Young	6	Gemini III	GT-3	23.03.1965	23.03.1965	00:04:53:00
		Gemini X	GT-10	18.07.1966	21.07.1966	02:22:46:39
		Apollo 10	AS-505	18.05.1969	26.05.1969	08:00:03:23
		Apollo 16	AS-511	16.04.1972	27.04.1972	11:01:51:05
		Columbia	STS-1	12.04.1981	14.04.1981	02:06:20:52
		Columbia	STS-9	28.11.1983	08.12.1983	10:07:47:24

Space flights by Group 3 Astronauts

Name	*Space flights*	*Spacecraft*	*Mission*	*Launch date*	*Landing date*	*Mission duration (dd:hh:m:ss)*
Edwin E. Aldrin, Jr.	2	Gemini XII	GT-12	11.11.1966	15.11.1966	03:22:34:31
		Apollo 11	AS-506	16.07.1969	24.07.1969	08:03:18:35
William A. Anders	1	Apollo 8	AS-503	21.12.1968	27.12.1968	06:03:00:42
Alan L. Bean	2	Apollo 12	AS-507	14.11.1969	24.11.1969	10:04:36:25
		Skylab 3	SL-3	28.07.1973	25.09.1973	59:11:09:04
Eugene A. Cernan	3	Gemini IX	GT-9A	03.06.1966	06.06.1966	03.00:50:20
		Apollo 10	AS-505	18.05.1969	26.05.1969	08:00:03:23
		Apollo 17	AS-512	07.12.1972	19.12.1972	12:13:51:59
Michael Collins	2	Gemini X	GT-10	18.07.1966	21.07.1966	02:22:46:39
		Apollo 11	AS-506	16.07.1969	24.07.1969	08:03:18:35
R. Walter Cunningham	1	Apollo 7	AS-205	11.10.1968	22.10.1968	10:20:09:03
Donn F. Eisele	1	Apollo 7	AS-205	11.10,1968	22.10.1968	10:20:09:03
Richard F. Gordon, Jr.	2	Gemini XI	GT-11	12.09.1966	15.09.1966	02:23:17:08
		Apollo 12	AS-507	14.11.1969	24.11.1969	10:04:36:25
Russell L. Schweickart	1	Apollo 9	AS-504	03.03.1969	13.03.1969	10:01:00:54
David R. Scott	3	Gemini VIII	GT-8	16.03.1966	16.03.1966	00:10:41:26
		Apollo 9	AS-504	03.03.1969	13.03.1969	10:01:00:54
		Apollo 15	AS-510	26.07.1971	19.07.1971	12:13:51:59

Non-Selected Group 3 Finalists on Later Space Flights
(All chosen in NASA's Group 5)

Name	*Space flights*	*Spacecraft*	*Mission*	*Launch date*	*Landing date*	*Mission duration (dd:hh:m:ss)*
Vance D. Brand	4	Apollo ASTP	AS-210	15.07.1975	24.07.1975	09:01:28:00
		Columbia	STS-5	11.11.1982	16.11.1982	05:02:14:26
		Challenger	STS-41B	03.02.1984	11.02.1984	07:23:15:55
		Columbia	STS-35	02.12.1990	10.12.1990	08:23:05:08
Ronald E. Evans	1	Apollo 17	AS-512	07.12.1972	19.12.1972	12:13:51:59
James B. Irwin	1	Apollo 15	AS-510	26.07.1971	07.08.1971	12.07.11.53
John L. Swigert, Jr.	1	Apollo 13	AS-508	11.04.1970	17.04.1970	05:22:54:41

Group 2 & 3 Astronauts Lost Before Flying in Space

Name	*Group*	*Assignment*	*Died during mission training*
Elliot M. See, Jr.	2	Gemini IX (GT-9)	Killed in T-38 jet accident with Capt. Charles Bassett, 28.02.1966
Charles A. Bassett II	3	Gemini IX (GT-9)	Killed in T-38 jet accident with Elliot See, 28.02.1966
Roger B. Chaffee	3	Apollo 1 (AS-204)	Launch pad fire during mission training, 27.01.1967
Theodore C. Freeman	3	Not assigned	Killed in T-38 jet accident, 31.10.1964
Clifton C. Williams	3	Apollo 12 (AS-507)	Killed in T-38 accident, 05.10.1967

Appendix 4

Group 2 & 3 Finalists with ARPS or MOL Qualification

Groupe	Name	Service/rank	ARPS/MOL
3	Adams, Michael J.	Capt., USAF	ARPS Class 4 MOL Group 1
3	Bassett, Charles A. II	Capt., USAF	ARPS Class 3
3	Bell, Tommy I.	Capt., USAF	ARPS Class 4
2	Birdwell, Carl, Jr.	LCDR, USN	ARPS Class 2
2	Borman, Frank	Maj., USAF	ARPS Class 1
2/3	Collins, Michael	Capt., USAF	ARPS Class 3
3	Freeman, Theodore C.	Capt., USAF	ARPS Class 4
3	Irwin, James B.	Capt., USAF	ARPS Class 4
2	McDivitt, James A.	Capt., USAF	ARPS Class 1
2	Neubeck, Francis G.	Capt., USAF	ARPS Class 3 MOL Group 1
3	Rupp, Alexander K.	Capt., USAF	ARPS Class 4
3	Scott, David R.	Capt., USAF	ARPS Class 4
2	Smith, Robert W.	Capt., USAF	ARPS Class 2
2	Uhalt, Alfred H., Jr.	Capt., USAF	ARPS Class 3

(ARPS = USAF Aerospace Research Pilot School)
(MOL = USAF Manned Orbiting Laboratory)

Bibliography

Aldrin, Edwin E. With Wayne Warga, *Return to Earth*, Random House, New York, NY, 1973

Atkinson, Joseph D., Jr. and Jay M. Shafritz, *The Real Stuff: A History of NASA's Astronaut Recruitment Program*, Praeger Publishers, New York, NY, 1985

Borman, Frank with Robert J. Serling, *Countdown: An Autobiography*, Silver Arrow Books, New York, NY, 1988

Burgess, Colin, *Selecting the Mercury Seven: The Search for America's First Astronauts*, Praxis Publishing Ltd., Chichester, UK, 2011

Burgess, Colin and Rex Hall, *The First Soviet Cosmonaut Team: Their Lives, Legacy and Historical Impact*, Praxis Publishing Publishing Ltd., Chichester, UK, 2009

Burgess, Colin and Kate Doolan, *Fallen Astronauts: Heroes Who Died Reaching for the Moon*, University of Nebraska Press, Lincoln, NE, 2003

Cassutt, Michael, *Who's Who in Space* (International Space Year Edition), Macmillan Publishing Company, New York, NY, 1987

Cernan, Eugene with Don Davis, *The Last Man on the Moon*, St. Martin's Press, New York, NY, 1999

Collins, Michael, *Carrying the Fire: An Astronaut's Journey*, Farrar, Straus and Giroux, New York, 1974

Compton, William David, *Where No Man Has Gone Before: A History of Apollo Lunar Exploration Missions*, NASA History Series (NASA SP-4214), Washington, D.C., 1989

Conrad, Nancy and Howard A. Klausner, *Rocket Man: Astronaut Pete Conrad's Incredible Ride to the Moon and Beyond*, New American Library, New York, NY, 2005

Cunningham, Walter, *The All-American Boys*, ibooks/Simon & Schuster Inc., New York, NY, 2003

French, Francis and Colin Burgess, *In the Shadow of the Moon: A Challenging Journey to Tranquility, 1965-1969*, University of Nebraska Press, Lincoln, NE, 2007

French, Francis and Colin Burgess, *Into that Silent Sea: Trailblazers of the Space Era, 1961-1965*, University of Nebraska Press, Lincoln, NE, 2007

Hansen, James R., *First Man: The Life of Neil A. Armstrong*, Simon & Schuster Inc., New York, NY, 2005

Hawthorne, Douglas B., *Men and Women of Space*, Univelt Inc., San Diego, CA, 1992

Irwin, James B. With William A. Emerson, Jr., *To Rule the Night: The Discovery Voyage of Astronaut Jim Irwin*, A.J. Holman Company, New York, NY, 1973

Lovell, Jim and Jeffrey Kluger, *Lost Moon: The Perilous Voyage of Apollo 13*, Houghton Mifflin Co., New York, NY, 1994

Phelps, J. Alfred, *They Had a Dream: The Story of African-American Astronauts*, Presidio Press, Navato, CA, 1994

Scott, David and Alexei Leonov with Christine Toomey, *Two Sides of the Moon: Our Story of the Cold War Space Race*, Simon & Schuster Inc., London, UK, 2004

Shayler, David J., *Gemini: Steps to the Moon*, Praxis Publishing Ltd., Chichester, UK, 2001

Shayler, David J., *Space Rescue: Ensuring the Safety of Manned Spacecraft*, Praxis Publishing Ltd., Chichester. UK, 2009

Slayton, Donald K. and Michael Cassutt, *Deke! U.S. Manned Space: From Mercury to the Shuttle*, Forge Books, New York, NY, 1994

Sobel, Lester (ed.), *Space: From Sputnik to Gemini*, Facts on File, Inc., New York, NY, 1965

Stafford, Tom and Michael Cassutt, *We Have Capture*, Smithsonian Institution Press, Washington, D.C., 2002

Thompson, Neal, *Light This Candle: The Life & Times of Alan Shepard, America's First Spaceman*, Crown Publishers, New York, NY, 2004

Wagener, Leon, *One Giant Leap: Neil Armstrong's Stellar American Journey*, Forge Books, New York, NY, 2004

Index

(A number in italics indicates a photograph)

About the Author

Australian author Colin Burgess grew up in Sydney's southern suburbs. Initially working the wages department of a major Sydney afternoon newspaper (where he first picked up his writing bug) and as a sales representative for a precious metals company, he subsequently joined Qantas Airways as a passenger handling agent in 1970 and two years later transferred to the airline's cabin crew. He would retire from Qantas as an onboard Customer Service Manager in 2002, after 32 years' service. During those flying years several of his books on the Australian prisoner-of-war experience and the first of his biographical books on space explorers such as Australian payload special Dr. Paul Scully-Power and teacher-in-space Christa McAuliffe had already been published. He has also written extensively on spaceflight subjects for astronomy and space-related magazines in Australia, the United Kingdom and the United States.

In 2003 the University of Nebraska Press appointed him series editor for the ongoing *Outward Odyssey* series of 12 books detailing the entire social history of space exploration, and was involved in co-writing three of these volumes. His first Springer-Praxis book, *NASA's Scientist-Astronauts*, co-authored with British-based space historian David J. Shayler, was released in 2007. *Moon Bound* will be his fifth title with Springer-Praxis, for whom he is currently researching three further books for future publication. He regularly attends astronaut functions in the United States and is well known to many of the pioneering space explorers, allowing him to conduct personal interviews for these books.

Colin and his wife Patricia still live just south of Sydney. They have two grown sons, two grandsons and a granddaughter.

CPSIA information can be obtained at www.ICGtesting.com
Printed in the USA
LVOW11s0228291013

359042LV00003B/7/P